Annals of Mathematics Studies

Number 137

Modern Methods
in Complex Analysis

Edited by

Thomas Bloom
David Catlin
John P. D'Angelo
Yum-Tong Siu

PRINCETON UNIVERSITY PRESS

PRINCETON, NEW JERSEY
1995

The Annals of Mathematics Studies are edited by
Luis A. Caffarelli, John N. Mather, and Elias M. Stein

Princeton University Press books are printed on acid-free paper and meet the
guidelines for permanence and durability of the Committee on Production
Guidelines for Book Longevity of the Council on Library Resources

Printed in the United States of America by Princeton Academic Press

10 9 8 7 6 5 4 3 2 1

Library of Congress Cataloging-in-Publication Data

Modern methods in complex analysis : the Princeton conference in honor of
Gunning and Kohn / edited by Thomas Bloom . . . [et al.].
p. cm. — (Annals of mathematics studies : no. 137)
Conference held Mar. 16–20, 1992, at Princeton University.
Includes bibliographical references.
ISBN 0-691-04429-5 (cloth : alk. paper). — ISBN 0-691-04428-7
(pbk. : alk. paper)
1. Mathematical analysis—Congresses. 2. Functions of several complex variables—
Congresses. I. Bloom, Thomas. II. Gunning, R. C. (Robert Clifford), 1931– .
III. Kohn, Joseph John, 1932– . IV. Series.
QA300.M63   1995
515.9—dc20   95-40840

The publisher would like to acknowledge the editors of this volume for providing
the camera-ready copy from which this book was printed

CONTENTS

PREFACE

More than 150 mathematicians honored Robert C. Gunning and Joseph J. Kohn on the occasion of their sixtieth birthdays by participating in a conference on several complex variables. The conference was held in Fine Hall at Princeton University from March 16 to March 20, 1992. This volume contains 15 papers dedicated to the two honorees. Most of the papers are from that conference. Several are by mathematicians unable to attend. The volume also includes articles on the scientific work of both honorees, lists of their Ph.D. students and of the conference participants, and the program for the conference.

The organizers acknowledge financial support from the National Science Foundation and the hospitality of Princeton University for the conference. They also wish to thank Princeton's Mathematics Department manager Scott Kenney for his assistance in organizing the conference.

LIST OF PARTICIPANTS

Marco Abate (Seconda Universita di Roma, Italy)
Khalid Filali Adib (Purdue)
Takao Akahori (Himeji Institute of Technology, Japan)
Solomon Alber (U of Nevada, Reno)
Herbert Alexander (U of Illinois, Chicago)
Eric Amar (Bordeaux)
Salah Baouendi (UC San Diego)
David Barrett (U of Michigan)
Jose Barros-Neto (Rutgers)
Jim Baxter (Washington)
Eric Bedford (Indiana U)
Steve Bell (Purdue)
Shiff Berhanu (Temple)
Carlos Berenstein (U of Maryland)
Craig J. Benham (Mt. Sinai Medical Center)
Bo Berndtsson (Chalmers, Sweden)
Edward Bierstone (U of Toronto)
Thomas Bloom (U of Toronto)
Grigory Bluher (UCLA)
Harold Boas (Texas A & M U)
Raoul Bott (Harvard)
William Browder (Princeton)
Lutz Bungart (U of Washington, Seattle)
Dan Burns (U of Michigan)
David Catlin (Purdue)
Der-Chen Chang (U of Maryland)
P. Charpentier (Bordeaux)
So-chin Chen (SUNY, Albany)
Anne-Marie Chollet (U of Lille, France)
Young-Bok Chung (Purdue)
Salvatore Coen (Universita degli studi di Bologne, Italy)
John P. D'Angelo (U of Illinois)
Maklouf Derridj (Orsay, France)
Klas Diederich (Wuppertal, Germany)
Ricardo Diaz (U of Northern Colorado)
Gilberto Dini (U degli studi di Firenze, Italy)
Pierre Dolbeault (U of Paris VI, France)
Avner Dor (Weizmann Inst., Israel)
Michael Eastwood (U of Adelaide, Australia)
Leon Ehrenpreis (Temple U)

Vladimir Ezhov (Bochum)
Gerd Faltings (Princeton)
Charles Fefferman (Princeton)
Julian Fleron (SUNY, Albany)
John Eric Fornaess (Princeton/U of Michigan)
Gerald Folland (U of Washington, Seattle)
Franc Forstneric (U of Wisconsin)
Edward Frankel (University of Nante, France)
Estela Gavosto (Princeton)
Daryl Geller (SUNY, Stony Brook)
Ian Graham (U of Toronto)
Hans Grauert (Goettingen, Germany)
Sandrine Grellier (France)
Peter C. Greiner (U of Toronto)
Pengfei Guan (McMaster U)
Robert C. Gunning (Princeton)
Nicholas Hanges (IAS)
Adam Harris (Rice U)
Reese Harvey (Rice U)
Zheng-Xu He (Princeton)
G. Henkin (University of Paris VI, France)
Gregor Herbort (Wuppertal, Germany)
C. Denson Hill (SUNY, Stony Brook)
A. Alexandrou Himonas (Notre Dame)
Lop-Hing Ho (Wichita State U)
Soren Illman (Princeton)
Howard Jacobowitz (IAS)
Moonja Jeong (U of Illinois)
Shanyu Ji (U of Houston)
Qin Jing (Harvard)
Leslie Kay (Virginia Polytechnic Institute)
N. Kerzman (U of NC, Chapel Hill)
Joseph J. Kohn (Princeton)
Gen Komatsu (Osaka U Toyonaka, Japan)
Steven Krantz (Washington U, St. Louis)
Oh Nam Kwon (Indiana U)
Christine Laurent-Thiebaut (Institut Fourier, France)
Jurgen Leiterer (Institut Fourier, France)
Laszlo Lempert (Purdue)
Moohyun Lee (Purdue)
David Lieberman (IDA, Princeton)
Haw-Lin Li (Purdue)
Qi-keng Lu (IAS/Academia Sinica, Beijing, China)

Daowei Ma (U of Chicago)
Matei Machedon (Princeton)
Andrew Majda (Princeton/IAS)
Richard Mandelbaum (U of Rochester)
Andrew Markoe (Rider College)
Jeffrey McNeal (Princeton)
Ngaiming Mok (Columbia)
Gerardo Mendoza (Temple)
Adrian Nachman (U of Rochester)
ALan Nadel (IAS)
Alexander Nagel (U of Wisconsin — Madison)
Terrence J. Napier (MIT)
Raghavan Narasimhan (U of Chicago)
W. Whipple Neely (U of Washington, Seattle)
Ed Nelson (Princeton)
Louis Nirenberg (NYU-Courant Institute)
Alan Noell (Oklahoma State U)
Takeo Ohsawa (Nagoya U, Japan)
Masami Okada (Courant Institute, NYU)
Mikeung Park (Purdue)
Mikael Passare (Sweden)
Giorgio Patrizio (U degli studi di Roma, Italy)
Duong H. Phong (Columbia)
Sergey Pinchuk (U of Wisconsin)
Karen R. Pinney (Johns Hopkins)
John C. Polking (Rice U)
Elisa Prato (Princeton)
David Prill (U of Rochester)
Mihai Putinar (UC, Riverside)
Enrique Ramirez (Mexico)
R. Michael Range (SUNY, Albany)
Themistockles Rassias (Greece)
Jean-Pierre Rosay (U of Wisconsin)
Linda Rothschild (UC, San Diego)
Halsey Royden (Stanford)
Weidong Ruan (Harvard)
Norberto Salinas (U of Kansas)
Les Saper (Duke)
Michael Schneider (Bayreuth)
Wolfgang Schwarz (Wuppertal, Germany)
Angela Selvaggi (U degli studi di Firenze, Italy)
Mei-Chi Shaw (Notre Dame)
Bernard Shiffman (Johns Hopkins)

Taka Shiota (Harvard)
Carlos Simpson (Princeton)
Yum-Tong Siu (Harvard)
Zbigniew Slodkowski (U of Illinois, Chicago)
Mikhail Smirnov (Princeton)
Avraham Soffer (Princeton)
Nancy K. Stanton (Notre Dame)
Elias Stein (Princeton)
Wilhelm Stoll (Notre Dame)
Edgar Lee Stout (U of Washington, Seattle)
Emil J. Straube (Texas A & M U)
William Sweeney (Rutgers)
Puqi Tang (Purdue)
David S. Tartakoff (U of Illinois, Chicago)
John Tate (U of Texas, Austin/Princeton)
B.A. Taylor (U of Michigan)
Anthony Thomas (Purdue)
Andrey Todorov (UC, Santa Cruz)
Rodolfo Torres (NYU — Courant Institute)
Marvin Tretkoff (Stevens Inst. of Tech.)
François Treves (IAS)
Alexander Tumanov (Purdue)
Bert G. Wachsmuth (Dartmouth)
Ajith Waidyaratne (Purdue)
Lihe Wang (Princeton)
Sidney Webster (U of Chicago)
John Wermer (Brown)
Arthur Wightman (Princeton)
Andrew Wiles (Princeton)
Bun Wong (UC, Riverside)
Yeren Xu (U of Washington, Seattle)
Stephen Yau (U of Illinois, Chicago)

PROGRAM OF THE CONFERENCE

Monday (March 16, 1992)

9:00 HANS GRAUERT: Meromorphic decomposition in the real case

10:30 LOUIS NIRENBERG: The maximum principle and principal eigenvalue for second order elliptic operators

1:40 LEON EHRENPREIS: Nonlinear Fourier transforms

2:20 LINDA ROTHSCHILD: Analytic discs for real hypersurfaces and generic manifolds in $\mathbf{C}^n$

3:00 MEI-CHI SHAW: Local solvability and estimates for the tangential Cauchy-Riemann operators

4:00 LESLIE SAPER: L^2 cohomology and the theory of weights

4:40 ERIC BEDFORD: Iteration of polynomial automorphisms in $\mathbf{C}^2$

Tuesday (March 17, 1992)

9:00 JOHN ERIC FORNAESS: Complex analytic dynamics in higher dimensions

10:30 FRANÇOIS TREVES: CR structures and complex nonlinear first order ODE

1:40 LASZLO LEMPERT: Stability of embeddings of CR manifolds

2:20 ALEX TUMANOV: Connections and propogations of analyticity for CR functions

3:00 CARLOS SIMPSON: Estimates for singularly perturbed ODE's

4:00 ALAN NADEL: Geometry of Fano varieties

4:40 MICHAEL SCHNEIDER: Compact manifolds with numerically effective tangent bundles

Wednesday (March 18, 1992)

9:00 YUM-TONG SIU: Margulis superrigidity as a result of the nonlinear Matsushima vanishing theorem

10:30 GENNADY HENKIN: Integral representations for analytic varieties with prescribed boundary in complex projective space

1:40 ED BIERSTONE: Resolution of singularities

2:20 REESE HARVEY: Characteristic currents for singular connections

3:00 TAKEO OHSAWA: L^2 cohomology of complex spaces

4:00 KLAS DIEDERICH: A continuity principle for the Bergman kernel

4:40 TAKA SHIOTA: Solitons and algebraic curves

Thursday (March 19, 1992)

9:00 RAOUL BOTT: On invariants of manifolds

10:30 MIKE CHRIST: Analytic hypoellipticity and a nonlinear eigenvalue problem

1:40 MIKE EASTWOOD: A period mapping for conformal four manifolds

2:20 DAVID CATLIN: Embeddings of CR structures

3:00 JOHN D'ANGELO: Spherical space forms and CR mappings

4:00 NGAIMING MOK: A geometric interpretation of some Eichler automorphic forms

4:40 GERALD FOLLAND: A simple proof of the sharp Gårding inequality

Friday (March 20, 1992)

9:00 RAGHAVAN NARASIMHAN: Some remarks on immersions of Stein manifolds

10:30 CHARLES FEFFERMAN: Recent developments in $\overline{\partial}$

1:40 MAKLOUF DERRIDJ: Estimates and real analyticity for the Kohn Laplacian

2:20 JEFF McNEAL: Estimates for the Bergman kernel on convex domains

3:00 QI-KENG LU: Cauchy-Fantappiè formulas for unbounded domains

4:00 TAKAO AKAHORI: A note on the CR analogue of the Tian-Todorov theorem

4:40 VLADIMIR EZHOV: On the problem of equivariant realization of CR automorphisms

Modern Methods in Complex Analysis

Robert C. Gunning

ROBERT C. GUNNING

THE SCIENTIFIC WORK OF
ROBERT C. GUNNING

Robert C. Gunning was born on November 27, 1931 in Longmont, Colorado. After graduation from the University of Colorado in 1952 he went on to graduate study at Princeton University where under the direction of S. Bochner he received his doctorate in 1955.

Gunning's work in mathematics covers a broad range of topics in complex analysis. His early work deals with factors of automorphy and differential operators compatible with them. For a contractible Stein manifold D with a discrete subgroup Γ of automorphisms acting on it, a holomorphic cross section of a holomorphic line bundle over the quotient D/Γ can be regarded as a holomorphic function which gets multiplied by a nowhere zero holomorphic factor, called the *factor of automorphy*, when an element of Γ acts on it. Gunning obtained a classification of factors of automorphy (References 1 and 3). A very important contribution of his in this area is his introduction of the *Eichler cohomology* which has since become a very useful tool in the study of Riemann surfaces, especially in the computation of the dimensions of spaces of automorphic forms and the traces of the Hecke operators on them. G. Bol [B49] observed that the derivative of order $n-1$ of a modular form of degree $n-2$ is a modular form of degree $-n$. Eichler [E57] considered an iterated indefinite integral of order $n-1$ of a modular form of degree $-n$ and defined a bilinear form on modular forms of degree $-n$ by integrating such an indefinite integral against another modular form of degree $-n$. Eichler used such a bilinear form to generalize Abel's theorem to obtain a characterization of a differential on a compact Riemann surface as the logarithmic differential of a meromorphic function. Gunning (Reference 10) introduced a new point of view by considering the short exact sequence over a compact Riemann surface whose middle term is the sheaf of germs of holomorphic automorphic forms of degree $-n$ with the surjective map being the differentiation of order $n+1$. He defined the Eichler cohomology as the cohomology constructed from the global holomorphic sections of the terms of the short exact sequence. From Gunning's point of view the bilinear form which Eichler used can be naturally interpreted as a duality of cohomology groups. Gunning also considered more general differential operators compatible with factors of automorphy and obtained classification results for such differential operators for the Siegel upper half-plane of rank 2 (References 14 and 15).

Another area of Gunning's work is pseudogroups and special coordinate coverings of manifolds. The elements of a pseudogroup are the coordinate transformation functions between different coordinate charts of a manifold. With Bochner (Reference 13) he studied the classification of all pseudogroups of maps $(f_1, \cdots, f_n)$ of the variables $x_1, \cdots, x_n$ on differentiable manifolds which are defined by a system of linear differential equations of degree t with constant coefficients

$$\sum_{s=1}^{t} \sum_{i,j_1,\cdots,j_s=1}^{n} c_{j_1\cdots j_s}^{\nu\ i} \frac{\partial^s f_i}{\partial x_{j_1}\cdots\partial x_{j_s}} = 0 \quad (\nu = 1,2,3,\cdots; \ c_{j_1\cdots j_s}^{\nu\ i} \in \mathbf{C}).$$

An essential property is that, for such a system of linear differential equations, compositions of solutions should again be solutions. A system is irreducible if no nonsingular linear change of coordinates can make the system independent of some of the variables $x_1, \cdots, x_n$. They proved that, for irreducible systems of degree one, the only nontrivial cases are the pseudogroup of differentiable functions, the pseudogroup of holomorphic functions, and the pseudogroup of quaternionic analytic functions (defined analogous to the complex-analytic functions). For irreducible systems of degree two, the only possibilities are the pseudogroup of real affine functions and the pseudogroup of complex affine functions. Their result settled the question whether, besides the well-known classes of manifolds such as differentiable manifolds, complex manifolds, etc., there are other manifolds whose coordinate transformation functions can be defined by such differential equations.

For special coordinate coverings, in a series of papers (References 19, 25, 35, and 37) Gunning studied affine and projective complex structures. For those complex structures the manifold admits a special coordinate covering whose transformation functions are complex affine transformations or complex projective linear transformations. He related the existence of such structures to the existence of special connections, which is equivalent to the vanishing of certain cohomology classes, and to the vanishing of the curvatures of the connections. Let T be the holomorphic tangent bundle of a compact complex manifold M of complex dimension n. For example, in the case of projective structures he introduced the 1-cocyle $\{c_{\alpha\beta}\}$ with coefficient in $T^* \otimes T^* \otimes T$ for a coordinate covering with coordinate charts $\{U_\alpha, z_\alpha\}$ as follows.

$$c_{\alpha\beta} = \sum_{i,j,k} \left(\frac{\partial^2 z_\beta^i}{\partial z_\alpha^j \partial z_\alpha^k} - \delta_j^i \sigma_{\alpha\beta k} - \delta_k^i \sigma_{\alpha\beta j} \right) dz_\alpha^j \otimes dz_\alpha^k \otimes \frac{\partial}{\partial z_\beta^i},$$

where

$$\sigma_{\alpha\beta k} = \frac{1}{n+1} \frac{\partial \ \log \ \det(\partial z_\beta^i / \partial z_\alpha^j)}{\partial z_\alpha^k}$$

and δ_j^i is the Kronecker delta. A holomorphic normal projective connection is a 0-cochain

$$b_\alpha = \sum_{i,j,k} \Gamma_{\alpha jk}^i dz_\alpha^j \otimes dz_\alpha^k \otimes \frac{\partial}{\partial z_\beta^i}$$

whose coboundary is the 1-cocycle $\{c_{\alpha\beta}\}$ with $\Gamma_{\alpha jk}^i$ symmetric in j and k. The transformation functions for a coordinate covering are all projective linear transformations if and only if there is a holomorphic normal projective connection which is flat in the sense that the curvature tensor of the connection vanishes. Gunning linked the existence of holomorphic normal projective connections to some Chern class identities. If M admits a holomorphic normal projective connection, then for every weighted homogeneous polynomial $Q_{n-r} = Q_{n-r}(c_1, \cdots, c_{n-r}) \in H^{2n-2r}(M, \mathbf{C})$ the identity $c_r Q_{n-r} = (n+1)^{-r} \binom{n+r}{r} c_1^r Q_{n-r}$ holds for $1 \leq r \leq n$. In the special case of a Kaehler M the identity $c_r = (n+1)^{-r} \binom{n+r}{r} c_1^r$ holds for $1 \leq r \leq n$. Gunning's work led to the work of Kobayashi-Ochiai [KO80] that a compact complex surface admits a projective structure only when it is the projective plane or a quotient of the ball or when it admits an affine structure. Using Gunning's techniques his student Vitter [V72] classified all possible affine structures on one- and two-dimensional tori, holomorphic families of 1-tori over a 1-torus, Hopf surfaces, and quotients of abelian surfaces.

After a year of post-doctoral work at the University of Chicago, Gunning returned to Princeton as Higgins Lecturer in 1957. He was promoted to full Professor in 1966. This was also the year that he married Wanda S. Holtzinger.

His book with Rossi "Analytic Functions of Several Complex Variables" was published in 1965 (Reference 17). It was among the first books to give a comprehensive treatment of the major topics in the field. To a generation of mathematicians it was the first book that guided them into the subject.

An open problem in several complex variables is to represent a Stein manifold of complex dimension n with trivial tangent bundle as a Riemann domain spread over $\mathbf{C}^n$. In 1967, in a joint paper with Narasimhan (Reference 21) Gunning settled the case $n = 1$.

In 1980-82, Gunning published three papers on Jacobian varieties and theta functions (References 37, 42, and 43). This work provided the most important link in the chain of arguments that culminated in the proof of E. Arbarello and C. de Concini of the Novikov conjecture about the Schottky problem. There are two approaches to the Novikov conjecture. One is the approach of Mulase and Shiota in terms of soliton solutions [Sh86]. Another is the approach of E. Arbarello and C. de Concini [AC87] which depends on Gunning's work. Either approach led to a proof of the conjecture.

Let M be a compact Riemann surface of genus $g > 1$. Let $\omega_1, \cdots, \omega_g$ be a basis over $\mathbf{C}$ of the space of all holomorphic 1-forms on M. Let $\gamma_1, \cdots, \gamma_{2g}$ be loops in M which form a basis in the first homology group of M. The $g \times 2g$ matrix $(\int_{\gamma_\nu} \omega_\mu)$ is the period matrix of M which can be reduced to the form (I_g, Ω) with a suitable choice of the basis $\omega_1, \cdots, \omega_g$, where Ω is a symmetric $g \times g$ matrix with positive definite imaginary part. The abelian variety $J = \mathbf{C}^n/(I_g, \Omega)\mathbf{Z}^{2g}$ is the Jacobian variety of M. Let z_0 be a point of M. The map w which sends z to $(\int_{z_0}^z \omega_\mu) \in \mathbf{C}^g$ modulo $(I_g, \Omega)\mathbf{Z}^{2g}$ embeds M into the Jacobi variety J as a subvariety W_1. The set of all positive divisors $d = z_1 + \cdots + z_r$ of degree r is mapped to the subvariety W_r of J consisting of all $w(d) := \sum_{\mu=1}^r w(z_\mu)$. Let W_r^ν be the subvariety of J consisting of all points of W_r which correspond to divisors $d = z_1 + \cdots + z_r$ with the property that the dimension of the space of meromorphic functions whose divisors $\geq -d$ is $> \nu$. Let $\theta(w; \Omega)$ be the theta function $\sum_{n \in \mathbf{Z}^g} \exp\left(2\pi i(\frac{1}{2}\,{}^t n\Omega n + {}^t nw)\right)$, where ${}^t n$ means the transpose of n. Riemann proved the fundamental vanishing and singularity theorem that W_{g-1}^ν is equal to $c - \Theta^\nu$, where c is the point in J with $2c$ representing the canonical divisor of M and Θ^ν is the set of points in J at which all partial derivatives of θ up to order ν vanish. Gunning proved the analog of Riemann's vanishing and singularity theorem for second-order theta functions which are defined as follows. The theta function θ satisfies the functional equation $\theta(w + \lambda; \Omega) = \xi(\lambda; w)\theta(w, \Omega)$ for $\lambda \in (I_g, \Omega)\mathbf{Z}^{2g}$, where the factor of automorphy is given by $\xi(\lambda; w) = \exp\left(-2\pi i\,{}^t q(w + \frac{1}{2}\Omega q)\right)$ when $\lambda = p + \Omega q$. A second-order theta function is an entire function $f(w)$ satisfying $f(w + \lambda) = \xi(\lambda; w)^2 f(w)$. A basis of the space of all second-order theta functions is given by $f_k(w) = \theta(2w + \Omega k; 2\Omega)\exp\left(2\pi i\,{}^t k(w + \frac{1}{4}\Omega k)\right)$ for $k \in (\mathbf{Z}/2\mathbf{Z})^g$.

Let $\overrightarrow{\theta_2} : \mathbf{C}^g \to \mathbf{C}^{2^g}$ be the map whose components are f_k for $k \in (\mathbf{Z}/2\mathbf{Z})^g$. The analog of Riemann's vanishing and singularity theorem for second-order theta functions which Gunning proved asserts that for any divisor $d = z_1 + \cdots + z_n$ the rank of the $n \times 2^g$ matrix

$$\left(\overrightarrow{\theta_2}\left(w(z_1) + \frac{t - w(d)}{2}\right), \cdots, \overrightarrow{\theta_2}\left(w(z_n) + \frac{t - w(d)}{2}\right)\right)$$

is less than $n - \nu$ if and only if $t \in W^{\nu}_{n-2}$. The special case of $n = 3$ says that $\overrightarrow{\theta_2}\left(\frac{1}{2}(t + w(z_1 - z_2 - z_3))\right), \overrightarrow{\theta_2}\left(\frac{1}{2}(t + w(z_2 - z_3 - z_1))\right), \overrightarrow{\theta_2}\left(\frac{1}{2}(t + w(z_3 - z_1 - z_2))\right)$ are linearly dependent if and only if $t \in W_1 \subset J$. The "if" part is the trisecant formula of Fay [F73]. The "only if" part is Gunning's result and it was a breakthrough in the investigation of the Schottky problem.

The problem of characterizing, among all symmetric $g \times g$ matrices with positive definite imaginary parts, the ones that arise from compact Riemann surfaces, had been posed by Riemann and has been a major preoccupation in the subject of Riemann surfaces ever since. The first significant results beyond Riemann's observations that the matrices are symmetric and have positive definite imaginary parts were obtained by Schottky in 1888 and the problem has subsequently been called the Schottky, or the Riemann-Schottky, problem. Gunning's converse of Fay's trisecant formula implies the following. An irreducible principally polarized abelian variety X is the Jacobian variety of some compact Riemann surface if and only if X contains an irreducible curve Γ such that for any point z on $\frac{1}{2}(\Gamma - \alpha - \beta - \gamma)$ the three points $\overrightarrow{\theta_2}(z + \alpha)$, $\overrightarrow{\theta_2}(z + \beta)$, and $\overrightarrow{\theta_2}(z + \gamma)$ are collinear. He further reduced the existence of Γ to the verification of the positivity of the dimension of a subvariety in X. Let $V_{\alpha,\beta,\gamma}$ be the subvariety in X consisting of the set of all z in X such that $\overrightarrow{\theta_2}(z + \alpha)$, $\overrightarrow{\theta_2}(z + \beta)$, and $\overrightarrow{\theta_2}(z + \gamma)$ are collinear. He then showed that the existence of Γ is equivalent to the following two conditions: (1) The dimension of $V_{\alpha,\beta,\gamma}$ at the point $-\alpha - \beta$ is positive for some α, β, γ in X. (2) There is no complex multiplication on X mapping the points $\beta - \alpha$ and $\gamma - \alpha$ to 0. When these two conditions are satisfied the subvariety $2V_{\alpha,\beta,\gamma}$ will be the curve Γ with X as its Jacobian. Welters [W83] later gave the following infinitesimal version of Gunning's result when the three points α, β, γ approach the same limit point. The abelian variety X is the Jacobian variety of some compact Riemann surface if and only if there exist constant vector fields $D_1 \neq 0$ and D_2 such that the dimension of the subvariety defined by

$$\{z \in X : \overrightarrow{\theta_2}(z) \wedge D_1\overrightarrow{\theta_2}(z) \wedge (D_1^2 + D_2)\overrightarrow{\theta_2}(z) = 0\}$$

at 0 is positive. In that case the subvariety is a smooth curve whose Jacobian is X. Novikov's conjecture which was proved by Shiota and Arbarello-De Concini says that X is the Jacobian variety of some compact Riemann surface if and only if there exist vector fields $D_1 \neq 0$, D_2, D_3, and a complex number d_4 such that the following Kodomcev-Petviashvili equation is satisfied.

$$(D_1^4\theta)\theta - 4(D_1^3\theta)(D_1\theta) + 3(D_2\theta)^2 + 3(D_2^2\theta)\theta$$

$$+3(D_1\theta)(D_3\theta) - (D_1 D_3\theta)\theta + d_4\theta^2 = 0.$$

Gunning obtained his generalization of Riemann's result to second-order theta functions in the framework of generalized theta functions and vector bundles over Riemann surfaces and their Jacobian varieties (References 32, 42). While theta functions are holomorphic sections of line bundles over abelian varieties, generalized theta functions are holomorphic sections of vector bundles over abelian varieties. A vector bundle over the Jacobian variety of a compact Riemann surface can be pulled back to a vector bundle on the Riemann surface. The study of holomorphic vector bundles over abelian varieties or compact Riemann surfaces is a very difficult subject. For compact complex manifolds X and Y, if there is a holomorphic family of holmorphic vector bundles $V(x)$ over Y parametrized by $x \in X$ and if the dimension k of $\Gamma(Y, V(x))$ is independent of x, then one can define a holomorphic vector bundle of rank k over X whose fiber at x is $\Gamma(Y, V(x))$. Every point t of the Jacobian variety J of a compact Riemann surface M of genus g corresponds to a flat line bundle ρ_t on M which is the pullback of a flat line bundle $\tilde{\rho}_t$ over J. For any line bundle ζ of degree 1 over M, the dimension of $\Gamma(M, \rho_t\zeta^{n+g-1})$ is independent of t when $n \geq g$. Mattuck [M61] introduced the vector bundle χ^n over J whose fiber at t is $\Gamma(M, \rho_t\zeta^{n+g-1})$. Such a vector bundle cannot be explicitly described by theta functions. Gunning (Reference 42) considered another vector bundle χ over J closely related to second-order theta functions. Second-order theta functions are holomorphic sections of a line bundle L over J. The fiber at t of the vector bundle χ that Gunning considered is $\Gamma(J, \tilde{\rho}_t L)$. Gunning showed that χ can be explicitly described as a natural extension of χ^{g+1}. From the relation of χ^{g+1} with M and the relation of χ with second-order theta functions, he obtained his analog of Riemann's result for second-order theta functions and a wide class of very useful analytic identities for theta functions.

Besides his original scientific work, Gunning is an exceptionally talented expositor and teacher. He has directed over 25 Ph.D. students. He has authored books on such diverse topics as modular forms, complex analytic varieties, Riemann surfaces, generalized theta functions, and uniformization of complex manifolds. He has recently published, in three volumes, an extensive revision of his 1965 classic book with Rossi.

Gunning is an enthusiastic and hard working member of the Princeton University community, having served as Chairman of the Department of Mathematics from 1976 to 1979 and, since 1989, as Dean of the Faculty.

Through his teaching and writing Gunning has had a broad and indelible influence on a generation of mathematicians.

[AC87] E. Arbarello and C. de Conici, Another proof of a conjecture of S. P. Novikov on periods of abelian integrals on Riemann surfaces, Duke Math. J. 54 (1987), 163-178.

[B49] G. Bol, Invarianten linearer Differentialgleichungen, Abh. Math. Sem. Univ. Hamburg 16 (1949), 1-28.

[E57] M. Eichler, Eine Verallgemeinerung der Abelsche Integrale, Math. Zeitschr. 67 (1957), 267-298.

[F73] J. D. Fay, Theta functions on Riemann surfaces, Lecture Notes in Math. Vol. 352, Springer-Verlag 1973.

[KO80] S. Kobayashi and T. Ochiai, Holomorphic projective structures on compact complex surfaces, Math. Ann. 249 (1980), 75-94.

[M61] A. P. Mattuck, Picard bundles, Illinois J. Math. 5 (1961), 550-564.

[Sh86] T. Shiota, Characterization of Jacobian varieties in terms of soliton equations, Invent. Math. 83 (1986), 333-382.

[V72] A. Vitter, Affine structures on compact complex manifolds, Invent. Math. 17 (1972), 231-244.

[W83] G. E. Welters, A characterization of non-hyperelliptic Jacobi varieties, Invent. Math. 74 (1983), 437-440.

Bibliography of Robert C. Gunning

1. General factors of automorphy, *Proceedings of the National Academy of Sciences*, U.S.A., **41** (1955), 496-498.

2. (with S. Bochner) Existence of functionally independent automorphic functions, *Proceedings National Academy of Sciences, U.S.A.*, **41** (1955), 746-752.

3. The structure of factors of automorphy, *American Journal of Mathematics*, **78** (1956), 357-382.

4. Indices of rank and of singularity of Abelian varieties, *Proceedings of the National Academy of Sciences*, **43** (1957), 167-169.

5. Multipliers on complex homogeneous spaces, *Proceedings of the American Mathematical Society*, **8** (1957), 394-396.

6. Multipliers on complex homogeneous spaces, II, *Seminars on analytic functions, Institute for Advanced Study*, **1** (1957), 103-110.

7. On Vitali's theorem for complex spaces with singularities, *Journal of Mathematics and Mechanics*, **8** (1959), 133-142.

8. Factors of automorphy and other formal cohomology groups for Lie groups, *Annals of Mathematics*, **69** (1959), 314-326.

9. Homogeneous symplectic multipliers, *Illinois Journal of Mathematics*, **4** (1960), 575-583.

10. The Eichler cohomology groups and automorphic forms, *Transactions of the American Math. Society*, **100** (1961), 44-63.

11. On Cartan's theorems A and B in several complex variables, *Annali di Mat.*, **55** (1961), 1-12.

12. *Lectures on modular forms*, Annals of Mathematics Studies, **48** (1962), Princeton University Press, 96 pages.

13. (with S. Bochner) Infinite linear pseudogroups of transformations, *Annals of Mathematics* **75** (1962), 93-104.

14. Generalized symplectic differential forms and differential operators, *Journal of Mathematics and Mechanics*, **11** (1962), 703-724.

15. Differential operators preserving relations of automorphy, *Transactions of the American Mathematical Society*, **108** (1963), 326-352.

16. Connections for a class of pseudogroup structures, *Proceedings of the Conference on Complex Analysis, Minneapolis, 1964*, Springer-Verlag, Berlin, 1965, 186-194.

17. (with H. Rossi) *Analytic functions of several complex variables*, Prentice-Hall, Englewood Cliffs, NJ, 1965, 317 pages, Russian Translation, Mir, Moscow, 1969, 395 pages.

18. *Lectures on Riemann surfaces*, Princeton Mathematical Notes **2**, Princeton University Press, 1966, 254 pages. German translation: Bibliographisches Institut, Mannhein, 1972, 276 pages.

19. Special coordinate coverings of Riemann surfaces, *Mathematische Annalen*, **170** (1967), 67-86.

20. *Lectures on vector bundles over Riemann surfaces*, Princeton Math. Notes **6**, Princeton University Press, 1967, 243 pages.

21. (with Raghavan Narasimhan) Immersion of open Riemann surfaces, *Mathematische Annalen*, **174** (1967), 103-108.

22. Some non-Abelian problems on compact Riemann surfaces, *Proceedings of the Conference on complex analysis, Rice University, 1967, Rice University Studies*, **54** (1968), 39-48.

23. *Lectures on complex analytic varieties: The local parametrization theorem*, Princeton Mathematical Notes **10**, Princeton University Press, 1970, 165 pages.

24. Quadratic periods of hyperelliptic Abelian integrals, *Problems in Analysis*, Princeton University Press, 1970, 239-247.

25. Analytic structures on the space of flat vector bundles over a compact Riemann surface, *Several Complex Variables, II, Maryland, 1970*, Springer-Verlag, 1971, 47-62.

26. Local moduli for complex analytic vector bundles, *Mathematische Annalen*, **195** (1971), 51-78.

27. Some multivariable problems arising from Riemann surfaces, *Actes, Congres Intern. Math.*, **2** (1970), 625-626.

28. Complex analytic varieties, *Lectures in theoretical physics*, New York: Gordon and Breach, 1972, 253-285.

29. *Lectures on Riemann surfaces: Jacobi varieties*, Mathematical Notes **12**, Princeton University Press, 1972, 189 pages.

30. Some special complex vector bundles over Jacobi varieties, *Inventiones Mathematicae*, **22** (1973), 187-210.

31. *Lectures on complex analytic varieties: Finite analytic mappings*, Mathematical Notes **14**, Princeton University Press, 1974, 163 pages.

32. *Rieman surfaces and generalized theta functions*, Springer-Verlag, Berlin and New York, 1976, 165 pages.

33. *Complex numbers and complex variables*, McGraw-Hill Yearbook of Science and Technology (1977), 174-176.

34. On the divisor order of vector bundles of rank two on a Riemann surface, *Bulletin Inst. Math. Academia Sinica*, **6** (1978), 295-303.

35. *On the uniformization of complex manifolds: the role of connections*, Math. Notes *22*, Princeton University Press, 1978, 141 pages.

36. Mathematics, *A Princeton Companion*, A. Leitch, ed., Princeton University Press, 1978, 316-319.

37. Affine and projective structures on Riemann surfaces, *Riemann surfaces and related topics: Proceedings of the 1978 Stony Brook Conference, Annals of Math. Studies*, **97** (1980), 225-244.

38. On the period classes of Prym differentials, II, *J. fur die reine und angew. Math.*, **319** (1980), 153-171.

39. On projective covariant differentiation, *E.B. Christoffel, The influence of his work on mathematics and the physical sciences*, Basel: Birkhauser Verlag, 1981, 584-591.

40. Complex numbers and complex variables, *McGraw-Hill Encyclopedia of Science and Technology* (1982), 466-471.

41. Review of "Families of meromorphic functions on compact Riemann surfaces" by M. Namba, *Bulletin of the American Math. Society*, **4** (1981), 353-357.

42. On generalized theta functions, *American Journal of Mathematics*, **104** (1982), 183-208.

43. Some curves in abelian varieties, *Inventiones Mathematicae*, **66** (1982), 377-389.

44. An identity for Abelian integrals, *Global analysis – Analysis on Manifolds, Teubner Texte zur Math.*, **571**, Teubner, Leipzig (1983), 126-130.

45. Riemann surfaces and their associated Wirtinger varieties, *Bull. Amer. Math. Soc.*, **11** (1984), 287-316.

46. Some identities for Abelian integrals, *Amer. Jour. Math.*, **108** (1986), 39-74.

47. On theta functions for Jacobi varieties, *Algebraic Geometry, Bowdoin 1985, "Proc. Symposia in Pure Math.*, **46**, part 1", Amer. Math. Soc. (1986), 89-98.

48. Analytic identities for theta functions, *Theta Functions, Bowdoin 1987, "Proc. Symposia in Pure Math.*, **49**, part 1", Amer. Math. Soc. (1989), 503-516.

49. *Holomorphic Functions of Several Variables*, Wadsworth and Brooks, Cole Pacific Grove, California, 1990 Vol. I (Function Theory), 203 pp.; Vol. II (Local Theory), 218 pp.; Vol. III (Homological Theory), 194 pp.

50. *The Collected Papers of Salomon Bochner* (editor), Amer. Math. Soc. (1991): Vol. I (762 pp.), Vol. II (790 pp.), Vol. III (732 pp.), Vol. IV (446 pp.).

List of Ph.D. Students of Robert C. Gunning

Andrew Campbell

Craig Benham

Thomas Bloom

Grigory Bluher

Michael Eastwood

Robert Ephraim

Mike Gilmartin

Xavier Gomez-Mont

Richard Hamilton

Eric Jablow

Sheldon Katz

Richard Koch

Henry Laufer

Richard Mandelbaum

J. Peter Matelski

Vernon Alan Norton

Cris Poor

David Prill

John Ries

Martha Katzin Simon

Yum-Tong Siu

John Snively

Charles Stenard

John Stutz

Albert Vitter, III

Bun Wong

David Yuen

Joseph J. Kohn

JOSEPH J. KOHN

THE SCIENTIFIC WORK OF
JOSEPH J. KOHN

The work of J. J. Kohn on the Cauchy-Riemann equations and related operators has fostered an intense interaction between partial differential equations and the theory of functions of several complex variables. It has led to widely applicable analytic techniques and to delicate geometric questions of current interest. This synopsis aims merely to indicate the depth and breadth of Kohn's life and work.

Joseph J. Kohn was born in Prague, Czechoslovakia on May 18, 1932. Seven years later his family moved to a small town in Ecuador and three years after that they moved to Quito, Ecuador. His family came to the United States in 1945 and lived in New York City. Kohn attended high school in New York, received his BS degree from MIT in 1953, and went to Princeton for graduate study. At Princeton he became D. C. Spencer's student and began work on the $\bar{\partial}$-Neumann problem.

Spencer approached complex analysis in several variables by trying to do Hodge theory on domains in complex Euclidean space or in complex manifolds. For simplicity here we consider a smoothly bounded pseudoconvex domain Ω in complex Euclidean space $\mathbf{C}^n$. Suppose that α is a differential (p, q) form with square integrable coefficients that satisfies $\bar{\partial}\alpha = 0$. The $\bar{\partial}$-Neumann problem is to construct the solution u to the Cauchy-Riemann equation $\bar{\partial}u = \alpha$ that is orthogonal to the kernel of $\bar{\partial}$, and to prove regularity results for u in terms of α. This particular solution, now generally known as the Kohn solution to the $\bar{\partial}$-Neumann problem, can be expressed as $\bar{\partial}^* N\alpha$ where N is the inverse to the complex Laplacian $\bar{\partial}^*\bar{\partial} + \bar{\partial}\bar{\partial}^*$. The $\bar{\partial}$-Neumann problem is a boundary value problem, because the condition that a differential form be in the domain of $\bar{\partial}^*$ is a boundary condition resulting from integration by parts. We discuss the technical difficulties of the $\bar{\partial}$-Neumann problem after including some additional biographical information.

Although he had not yet solved the $\bar{\partial}$-Neumann problem, Kohn received his Ph.D. in 1956. He remained in Princeton for two more years, as an instructor at the University and as a visiting member at the Institute for Advanced Study. In 1958 he went to Brandeis University.

In 1962 Kohn solved the $\bar{\partial}$-Neumann problem for strongly pseudoconvex domains. The two papers "Harmonic integrals on strongly pseudoconvex manifolds I, II" (References 7 and 8) were a major achievement; they demonstrate the existence and regularity of solutions and offer applications such as a proof of the Newlander-Nirenberg theorem along the lines forseen by Spencer years before. Kohn later won the Leroy Steele prize for the fundamental contributions to research appearing in these papers.

Kohn's solution to the $\bar{\partial}$-Neumann problem for domains in complex manifolds includes a wealth of information. Kohn's precise formulation of the problem was the first major step. He then realized how to apply the "basic estimate" of Morrey to prove regularity theorems. Under the assumption that the "basic estimate" holds, Kohn obtained both a strong orthogonal Hodge decomposition and precise regularity results. Let H denote the orthogonal projection onto the kernel of the complex Laplacian. Each differential form ϕ of type (p, q) with L^2 coefficients has a Hodge decomposition $\phi = H\phi + \bar{\partial}^*\bar{\partial}N\phi + \bar{\partial}\bar{\partial}^*N\phi$. The operator N is compact and preserves smoothness up to the boundary; in fact for each s there is an estimate $||N\phi||_{s+1}^2 \leq C||\phi||_s^2$ in terms of Sobolev norms. The N operator yields the Kohn solution $u = \bar{\partial}^* N\alpha$ of the inhomogeneous Cauchy-Riemann equation $\bar{\partial}u = \alpha$, and this estimate implies that $||u||_s^2 \leq C||\alpha||_s^2$. Sharper estimates hold for forms supported in the interior, and in general if one considers norms involving tangential derivatives.

We now consider the formidable analytic difficulties of the $\bar{\partial}$-Neumann problem approach to the Cauchy-Riemann equations. The $\bar{\partial}$-Neumann problem is an elliptic system but the boundary conditions are not, so the classical work on elliptic boundary value problems does not apply. Recall that a boundary value problem $Lu = f$ for an operator of order m is called coercive if it is possible to estimate all derivatives of order m of the solution u in terms of Lu and the boundary data (in appropriate norms). The $\bar{\partial}$-Neumann problem is non-coercive. The complex Laplacian is order two; for a strongly pseudoconvex domain it is possible to estimate first derivatives, but this is far from obvious. Kohn discovered this subelliptic nature of the $\bar{\partial}$-Neumann problem and how to use the notion of a $\frac{1}{2}$-estimate (and weaker subelliptic estimates) to prove regularity of solutions. This work indicated the anisotropic behavior of the tangent spaces to the boundary; the estimates for derivatives in complex tangential directions differ from those in the "bad direction". Kohn also introduced, in later work with Rossi, the analog of these notions on the boundary of a domain in a complex manifold. They defined the operator $\bar{\partial}_b$, the boundary Laplacian, and the resulting $\bar{\partial}_b$-Neumann problem. The investigations for $\bar{\partial}$ and for $\bar{\partial}_b$ led to pseudodifferential operators, microlocal methods, the study of weakly pseudoconvex domains, and the notion of CR geometry. The papers in this volume indicate how influential Kohn's work has been.

In 1964 Kohn began the investigation of the boundary operators. Kohn and Rossi wrote the fundamental paper "On the extension of holomorphic functions from the boundary of a complex manifold" (Reference 14). Here the authors thoroughly investigated the tangential Cauchy-Riemann operator $\bar{\partial}_b$, the boundary Laplacian (which is not elliptic in this case), and what is now known as Kohn-Rossi cohomology. In particular they proved that the kernel of the boundary Laplacian is finite dimensional. They also extended Hans Lewy's work by proving that a CR function extends to one side of a hypersurface whenever the Levi form has one positive eigenvalue. In the 1964 paper "Boundaries of complex manifolds" (Reference 17) from the Proceedings of the conference on complex manifolds held in Minneapolis, Kohn generalized the investigation of the boundary operators to abstract real manifolds now known as CR manifolds. The operator $\bar{\partial}_b\bar{\partial}_b^* + \bar{\partial}_b^*\bar{\partial}_b$ is now known as the Kohn Laplacian.

During this time Kohn also began his collaboration with Nirenberg. In the paper "Non-coercive boundary value problems" (Reference 18) they gave an improved proof of the solution of the $\bar{\partial}$-Neumann problem by introducing the method of elliptic regularization. The starting point is the Hermitian quadratic form given by $Q(u,v) = (\bar{\partial}u, \bar{\partial}v) + (\bar{\partial}^* u, \bar{\partial}^* v)$. Elliptic regularization consists of adding epsilon times an elliptic term to this form, proving estimates uniformly in ϵ, and letting ϵ tend to zero to obtain smooth solutions to the original problem. Applications of elliptic regularization to other problems in partial differential equations appear in this paper as well. In particular their methods apply to certain degenerate elliptic and parabolic systems.

The work of Kohn-Nirenberg focuses clearly on consequences of the subelliptic nature of the $\bar{\partial}$-Neumann problem. In particular they proved that an *a priori* estimate of the form

$$|||\phi|||_\epsilon^2 \leq C(||\bar{\partial}\phi||^2 + ||\bar{\partial}^* \phi||^2 + ||\phi||^2)$$

for some positive ϵ and for all smooth and compactly supported (p,q) forms ϕ in the domain of $\bar{\partial}^*$ implied local regularity for the Kohn solution of the $\bar{\partial}$-Neumann problem. The norm on the left is the tangential Sobolev norm; thus the inequality controls tangential derivatives of fractional order in terms of the particular combination of first derivatives on the right hand side. Subelliptic estimates imply the finite dimensionality of the space of harmonic (p, q) forms and also pseudo-locality for the Neumann operator. On $(0, 1)$ forms such estimates yield regularity for the Bergman projection P, because of the formula $P = I - \overline{\partial}^* N \overline{\partial}$. Subelliptic estimates play a critical role in the developments of the seventies and eighties.

By 1965 it was becoming clear that Kohn's methods required obtaining precise estimates for the commutators of operators such as fractional powers of the Laplacian and various derivatives. He and Nirenberg worked out an appropriate general theory; Kohn modestly felt that this belonged in an appendix. Fortunately Nirenberg insisted that they develop the material further, and together they wrote the famous paper "On the algebra of pseudodifferential operators" (Reference 16). There they developed the calculus of pseudodifferential operators using the Fourier transform as the basic tool.

This time period was important for Joe in other ways as well. He and Anna Rosa married in 1966 in Quito, and their son Eduardo was born in 1968. Eduardo is a graduate student in anthropology at the University of Wisconsin; his broad scholarly interests include botany, linguistics, and the oral history of Ecuador. Their daughter Emma was born in 1970. She graduated from Washington University and is now an artist. Their daughter Alicia was born in 1975 and attends Brandeis University.

Kohn returned to Princeton in 1968 as Professor of Mathematics. From 1974-76 and again beginning in 1993 he served as Chairman of the department. Both he and the department flourished during these periods.

In 1972 Kohn published the paper "Boundary behavior of $\overline{\partial}$ on weakly pseudoconvex manifolds of dimension two" in honor of Spencer's 60th birthday (Reference 31). Here Kohn began seeking geometric conditions for subelliptic estimates. He discovered that the behavior of iterated commutators of tangential complex vector fields governed whether a subelliptic estimate held for pseudoconvex domains in $\mathbf{C}^2$. Generalizing this condition to higher dimensions led to considerable work on the geometry of weakly pseudoconvex domains. Subelliptic estimates yield local regularity theorems for the Kohn solution of the $\overline{\partial}$ equation. If, for example, α is a $\overline{\partial}$ closed form with distribution coefficients, and it is known to be smooth on some set, then the Kohn solution to $\overline{\partial}u = \alpha$ must be smooth there whenever a subelliptic estimate holds. This follows from the pseudo-locality for the Neumann operator. Another important consequence is the regularity property known as "condition R", which was used by Bell to extend results on boundary smoothness of biholomorphic and proper mappings to weakly pseudoconvex domains of finite type.

In 1973 Kohn and Nirenberg made another fundamental contribution to the study of weakly pseudoconvex domains. They discovered a pseudoconvex domain (whose defining equation is a polynomial) in $\mathbf{C}^2$ that could not be made linearly convex near a particular boundary point by a local biholomorphism (Reference 32). There is no holomorphic supporting function there; every complex analytic curve through this point intersects both the interior and exterior of the domain. This example reveals a fundamental difficulty in solving the Cauchy-Riemann equations using methods such as integral representation formulas. By 1994 such methods had not yet succeeded on general weakly pseudoconvex domains without holomorphic supporting functions. This makes the success of Kohn's method of L^2 estimates even more striking.

Kohn proposed finding necessary and sufficient conditions for subelliptic estimates on pseudoconvex domains. This problem is difficult for several reasons. One is that the the Levi form does not have constant rank, so bundle techniques from differential geometry do not apply. Another difficulty became clear only later. The $\frac{1}{2}$ estimate arose from commutators. Higher commutators govern the situation in two dimensions, or more generally on $(p, n - 1)$ forms. For $(0, 1)$ forms in higher dimensions, however, such conditions on commutators fail to be non-degeneracy conditions, and hence cannot yield the right geometric condition for subelliptic estimates. Thus it is neither clear what technique to use to derive the estimates, nor what geometric conditions will guarantee that the estimates hold.

In his 1979 Acta paper (Reference 52) Kohn introduced ideals of subelliptic multipliers. He gave a sufficient condition for subelliptic estimates for (p, q) forms on pseudoconvex domains in terms of these ideals. When the boundary is a real analytic manifold, using results of Diederich-Fornaess, the condition is shown to be equivalent to the non-existence of complex analytic varieties of dimension q lying in the boundary. This paper thus revealed a deep connection between the theory of estimates and algebraic-geometric methods from singularity theory. In 1982 and 1987 Kohn's student David Catlin established necessary and sufficient conditions for subelliptic estimates near a boundary point of a smoothly bounded weakly pseudoconvex domain. For $(0, 1)$ forms the condition is that every ambient complex analytic variety has finite order of contact with the boundary at p, a property studied extensively by Kohn's student J. D'Angelo. See D'Angelo's article "Finite type conditions and subelliptic estimates" in this volume for more information about this material.

Kohn's idea of using ideals of subelliptic multipliers for proving estimates has additional applications. Nadel gave several applications of multiplier ideal sheaves to algebraic geometry. He proved a vanishing theorem with applications in Kähler geometry, he proved the existence of Kähler-Einstein metrics for a large class of compact algebraic manifolds with positive first Chern class, and he gave the universal bound for the top power of the first Chern class of Fano manifolds. Demailly, Siu and others used multiplier ideal sheaves for results connected with the Fujita conjecture and the effective Matsusaka big theorem. See Siu's article "Very ampleness criterion of double adjoints of ample line bundles" in this volume.

Kohn's methods have led to many developments in what is now called "microlocal analysis". See Kohn's paper "The range of the tangential Cauchy-Riemann operator" (Reference 61) for a proof that, on a smoothly bounded pseudoconvex domain, $\overline{\partial}_b$ has closed range in L^2 of the boundary. This paper makes extensive use of microlocal methods. See also the article "On Kohn's microlocalization of $\overline{\partial}$-problems" by C. Fefferman in this volume for a discussion of related microlocal methods. Kohn and Fefferman obtain optimal Hölder estimates for $\overline{\partial}$, the Bergman projection operator, and the Szegö projection operator on pseudoconvex domains in two complex dimensions. They also give analogous estimates for the boundary operators on real 3-dimensional CR manifolds. In later work also joint with Machedon they generalize these results to higher dimensions, under the condition of diagonalizable Levi form. The microlocal techniques are related to questions concerning analytic hypoellipticity of the $\overline{\partial}$-Neumann problem and the $\overline{\partial}_b$ problem; many open questions remain about these matters. See the article "Remarks on analytic hypoellipticity of $\overline{\partial}_b$" by M. Christ in this volume for more information.

Many of the problems and techniques discussed so far involve linear generalizations of classical potential theory. The complex Monge-Ampere operator is a non-linear generalization of the Laplacian. In work with Caffarelli, Nirenberg, and Spruck in 1985 (Reference 60), Kohn also studied a Dirichlet problem for the complex Monge-Ampere equation. The authors considered the problem $\mathrm{Det}(u_{z_j \overline{z}_k}) = g$ on Ω and $u = \phi$ on the boundary $b\Omega$. Here u is assumed to be plurisubharmonic on the smoothly bounded strongly pseudoconvex domain Ω and ϕ is assumed positive. Using the method of *a priori* estimates they proved that u must be smooth when g and ϕ are smooth.

Kohn has lectured at conferences all over the world, and has organized conferences on four continents. In 1975 he delivered 5 lectures at the AMS Summer Institute in Williamstown. His article "Methods of partial differential equations in complex analysis" (Reference 44) appears in the proceedings of that meeting and has influenced countless students. The 1972 book by Folland and Kohn entitled "The Neumann problem for the Cauchy-Riemann complex" (Reference 30) remains the basic reference in the strongly pseudoconvex case, and Kohn's article "A survey of the $\overline{\partial}$-Neumann problem" from the Madison conference in 1983 (Reference 58) provides useful information about later developments.

It seems appropriate to indicate that work on the $\overline{\partial}$-Neumann problem continues. Kohn proved in 1973 (Reference 37) that, on a smoothly bounded pseudoconvex domain, the Cauchy-Riemann equation $\overline{\partial}u = \alpha$ always has a smooth solution when α is a globally smooth $(0, 1)$ form with $\overline{\partial}\alpha = 0$. When subelliptic estimates hold, the Kohn solution itself is globally smooth. It remains an open problem whether the Kohn solution always exhibits this global regularity property. As recently as 1991 Boas-Straube gave a general condition ensuring global regularity of the Kohn solution to the Cauchy-Riemann equations.

Joe Kohn speaks many languages and seems comfortable in every social situation. He received an honorary degree from the University of Bologna in Italy. He is a member of the National Academy of Sciences. His mathematical achievements and his international activities have made him one of the most respected mathematicians in the world.

(written by John P. D'Angelo in 1994 at Urbana, IL)

Bibliography of Joseph J. Kohn

1. Linear inequalities and polyhedral convex cones, *ONR Logistics Report*, No. NR 047-002, 1954.

2. Singular integral equations for differential forms on Riemann manifolds, *Proceedings of the National Academy of Sciences*, **42**, 1956, 650-653.

3. (with D.C. Spencer) Complex Neumann problems, *Annals of Mathematics*, **66**, 1957, 89-140.

4. A boundary condition for the vanishing of n holomorphic functions in complex n-space, *Proceedings of the American Math. Soc.*, **9**, 1958, 175-177.

5. Topological methods in the theory of several complex variables, *Annals of the New York Academy of Science*, **86**, 1960, 693-699.

6. Solution of the $\bar{\partial}$-Neumann problem on strongly pseudo-convex manifolds, *Proceedings of the National Academy of Sciences*, **47**, 1961, 1198-1202.

7. Harmonic integrals on strongly pseudo-convex manifolds, I, *Annals of Mathematics*, **78**, 1963, 112-148.

8. Harmonic integrals on strongly pseudo-convex manifolds, II, *Annals of Mathematics*, **79**, 1964, 450-472.

9. Regularity at the boundary of the $\bar{\partial}$-Neumann problem, *Proceedings of the National Academy of Sciences*, **49**, 1963, 206-213.

10. Potential theory and several complex variables, *Annali della Scoula Normale Superiore de Pisa* Series III, **XVII**, Pasc., IV, 1963, 373-386.

11. Introducción a la teoria de integrales harmónicas, Centro de Investigacion del I.P.N., 1963.

12. Non-coercive estimates, Colloques Internationaux deu C.N.R.S., Paris 1963, 47-52.

13. A priori estimates in several complex variables, *Bulletin of the American Mathematical Society*, **70**, 1964, 739-745.

14. (with H. Rossi) On the extension of holomorphic functions on the boundary of a manifold, *Annals of Mathematics*, **81**, 1965, 451-472.

15. Differential operators on manifolds with boundary, Int. Coll. on Diff. Analysis, Tata Institute of Fundamental Research, Bombay, 1964, 57-68.

16. (with L. Nirenberg) An algebra of pseudo-differential operators, *J. Pure and Appl. Math.*, **18**, 1965, 269-305.

17. Boundaries of complex manifolds, *Proc. of the Conference on Complex Analysis*, (Minn., 1964), Springer-Verlag, 1965, 81-94.

18. (with L. Nirenberg) Non-coercive boundary value problems, *Comm. Pure Appl. Math.*, **18**, 1965, 443-492.

19. *Formes integro-differentielles non-coercives*, Les Presses de l'Université de Montréal, January, 1966.

20. Differential complexes, Invited address, *Proc. of the Inter. Congress of Mathematicians*, Moscow, 1966, 402-409.

21. Sub-elliptic complexes, *Proc. of Katata Conference*, December, 1966, 35-44.

22. (with L. Nirenberg) Degenerate elliptic-parabolic equations of second-order, *Comm. P.A.M.*, **20**, 1967, 797-872.

23. (with L. Nirenberg) Degenerate elliptic-parabolic equations, Sem. di Mat. Univ. di Bari, 1967, 1-13.

24. Pseudo-differential operators and non-elliptic problems, *Proc. C. I. M. E. Conference*, Stresa, 1968, 159-165.

25. Harmonic integrals for differential complexes, in *Global Analysis: Papers in honor of K. Kodaira*, Princeton University Press and Tokyo University Press, 1969, 295-308.

26. Complex hypoelliptic operators, *Instituto Naz di Alta Mat.*, **VII**, 1971, 459-468.

27. Integration of complex vector fields, *Bulletin of the American Math. Soc.*, **78**, 1972, 1-12.

28. *Differential complexes*, Les Presses de l'Universié de Montréal, 1972.

29. The $\bar{\partial}$-Neumann problem on (weakly) pseudo-convex two-dimensional manifolds, *Proceedings of the National Academy of Sciences*, **69**, 1972, 1119-1120.

30. (with G. B. Folland) The Neumann problem for the Cauchy-Riemann complex, *Annals of Math. Studies*, **75**, Princeton University Press, 1972.

31. Boundary behavior of $\bar{\partial}$ on weakly pseudo-convex manifolds of dimension two, *Journal of Differential Geometry*, **6**, 1972, 523-542.

32. (with L. Nirenberg) A pseudo-convex domain not admitting a holomorphic support function, *Mathematische Annalen*, **201**, 1973, 265-268.

33. Propagation of singularities for $\bar{\partial}$, *Astérisque, Soc. Math de France*, (**2** and **3**), 1973, 244-251.

34. Introduzione alla teoria degli operatori pseudo-differenziali, mimeographed lecture notes, Instituto di Calcolo delle Probabilita, Fac. di Sci. Staatistiche dell'Univ. di Roma, 1973, 1-49.

35. Pseudo-differential operators and hypoellipticity, *Proceedings of the American Math. Soc. Symp. in Pure Math.*, **XXIII**, 1973, 61-69.

36. Boundary regularity of solutions of the inhomogeneous Cauchy-Riemann equations, *Sem. Goulaouic-Schwartz, 1972-73*, Exposé XXIV, 1-9.

37. Global regularity of $\bar{\partial}$ on weakly pseudo-convex manifolds, *Trans. of the Amer. Math. Soc.*, **181**, 1973, 273-292.

38. Propagation of singularities for the Cauchy-Riemann equations, *Proc. of Complex Analysis*, C. I. M. E. (I ciclo 1973), Ediz. Cremonese, Roma, 1973, 179-280.

39. Subellipticity on pseudo-convex domains with isolated degeneracies, *Proc. of the National Academy of Sciences*, **71**, 1974, 2912-2914.

40. Convexity and pseudo-convexity, *Geom. Diff. Colloque, Santiago de Compostela, Spring 1972*, Lecture Notes in Math., **392**, Springer-Verlag, 195-202.

41. An example of a strange three-dimensional surface in C^2, in *Global Analysis and its Applications*, International Atomic Energy Agency, Vienna, 1974, 323-327.

42. Uno sguardo agli operatori pseudo-differenziali, *Bolletino U.M.I.*, **4**, 1974, 273-297.

43. Subellipticity of the $\bar{\partial}$-Neumann problem on weakly pseudo-convex domains, *Recontre sur L'Analyse Complexe a Plusieurs Variables et els Systemes Indéterminés*, les Presses de L'Université de Montréal, 1975, 105-118.

44. Several complex variables and partial differential equations, *Proceedings of the Amer. Math. Soc. Conference on Several Complex Variables*, Williamstown, MA, 1975.

45. (with P.C. Greiner and E.M. Stein) Necessary and sufficient conditions for solvability of the Lewy equation, *Proc. of the National Academy of sciences*, 1975, 3287-3289.

46. Holomorphic extensions of orthogonal projections into holomorphic functions, *Proceedings of the American Math. Soc.*, **52**, 1975, 333-336.

47. Methods of partial differential equations in complex analysis, *Amer. Math. Soc. Proceedings of Symposium in Pure Math.*, **XXX**, Part 1, 1977, 213-237.

48. Sufficient conditions for subellipticity on weakly pseudo-convex domains, *Proceedings of the National Academy of Sciences*, **74**, No. 6, 1977, 2214-2216.

49. Subelliptic estimates, *Proceedings of Conferences on Several Complex Variables*, Cortona, Italy, 1976-77, 199-204.

50. Degenerate elliptic equations and pseudo-differential operators, *C. I. M. E. Conference*, Bresannone, 1977.

51. Lectures on degenerate elliptic problems, *Proc. of C. I. M. E. Conference on Pseudo-differential Operators with Applications*, Bressanone 1977, pp. 89-151, Liquori, Naples, 1978.

52. Subellipticity of the $\bar{\partial}$-Neumann problem on pseudo-convex domains: sufficient conditions, *Acta Mathematica*, **142**, March 1979.

53. Subelliptic estimates. Harmonic analysis in Euclidean spaces, *Proc. of Symposia in Pure Math.*, Amer. Math. Soc., **XXXV**, Part 2, 1979, 143-152.

54. Several complex variables from the point of view of linear partial differential equations, *Proc. Conference on PDE and Differential Equations*, 1980, Beijing Academy of Sciences, People's Republic of China.

55. Regularity of the $\bar{\partial}$-Neumann problem, *Séminaire Goulaouic-Meyer-Schwartz*, 1980, Exposé No. XIX.

56. Subelliptic estimates for integro-differential forms, *Proc. of the Conference on Linear PDE*, Saint Jean des Monts, 1981.

57. Microlocalization of CR structures, *Proc. Sev. Comp. Var.*, 1981, Hangzhou Conf. Birkhauser, Boston, 1984, 29-36.

58. A Survey of the $\bar{\partial}$-Neumann Problem, *Proc. of Symposia in Pure Math., Amer. Math. Soc.* **41**, 1984, 137-145.

59. Estimates for $\bar{\partial}_b$ on pseudo-convex CR manifolds, *A.M.S. Proc. Symp. Pure Math.*, **43**, 1985, 207-217.

60. (with L. Caffarelli, L. Nirenberg, J. Spruck) The Dirichlet problem for nonlinear second order elliptic equations. II. Complex Monge-Ampère, and uniformly elliptic, equations, *Comm. Pure Appl. Math.*, **38**, 1985, 209-252.

61. The range of the tangential Cauchy-Riemann operator, *Duke Math.*, **53**, 1986, 525-545.

62. (with C.L. Fefferman) Hölder estimates on domains of complex dimension two and on three-dimensional CR manifolds, *Adv. of Math.*, **69**, 1988, 223-303.

63. (with C.L. Fefferman) Estimates of kernels on three-dimensional CR manifolds, *Rev. Mat. Iberoam.*, vol. **4**, no. 3, 1988, 355-405.

64. Analysis on 3-dimensional CR manifolds, *Danish-Swedish Anal. Sem., Report* Ser. No. 3, 1989, 10-11.

65. Microlocal analysis on CR manifolds, *Proc. of Conference on Complex Analysis and Geometry*, Bologna 1989).

List of Ph.D. Students of Joseph J. Kohn

So Chin Chen

David Catlin

John D'Angelo

Ricardo Diaz

Gerald Folland

Marvin Freedman

Pengfei Guan

Lop-Hing Ho

Martin Kolar (1995)

Steven Post

Mei-Chi Shaw

Mikhail Smirnov (1995)

John Stalker

ON INVARIANTS OF MANIFOLDS

RAOUL BOTT

It is a great pleasure to participate in these festive proceedings. Of course from my much advanced vantage point they resemble nothing more than a Bar Mitzvah – a rite of passage of the young. In short, they signify that finally we can welcome you, Joe, and you, Bob, into the community of – shall we say – *maturer* mathematicians. That is as far as I am willing to go, for I doubt that true maturity can ever be attained by mathematicians.

Returning to Princeton after forty years, it is difficult to refrain from reminiscing – especially when one meets such wonderful old comrades in arms as Don Spencer upon entering the hall, and I have therefore chosen to speak about manifold invariants in the context of the forty years that have passed since I first encountered them here.

When our forefathers spoke of an invariant they usually meant a number, and so every discussion of topological invariants must really start with the Euler number of a cell complex:

$$\chi(X) = \Sigma(-1)^i a_i,$$

where a_i denotes the number of cells of dim i in X.

However it is only in the world of manifolds that this invariant really comes into its own and takes on an especially beautiful and geometric meaning associated with the following picture:

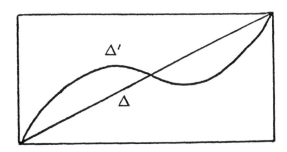

Fig. 1

surely well known to many, if not all, of you. Indeed for manifolds

$$\chi(M) = \text{self-intersection of the diagonal } \Delta \text{ in } M \times M.$$

This antithesis of the factors versus the diagonal of course goes much deeper than the Euler number. First of all, it furnishes the topologist with the right definition of the cup product and in the hands of Steenrod it became the geometric inspiration for his squaring operations. Indeed, any deformation of Δ into a transversal position Δ' necessarily breaks the symmetry of the picture so that the flip of the deformation, say $T\Delta'$, will be distinct from Δ' except at the self intersections. Now, deforming Δ' to $T\Delta'$ as transversally as possible and intersecting this deformation once again with Δ brings us into the realm of the higher Steenrod operations.

I learned these ideas in Steenrod's beautiful and inspiring lectures forty years ago and so it gives me pleasure that in a quite new context I am thinking about this picture once again in 1992.

But, returning to the Euler number $\chi(M)$, note that over the years we have learned to bring it to you in many guises. One, of course, has the relation

$$(*) \qquad\qquad \chi(M) = \sum (-1)^i \dim H^i(M)$$

where the H^i are the cohomology vector spaces, thus establishing $\chi(H)$ as a weaker invariant than the Betti numbers, $\dim H^i(M)$.

But the Euler number $\chi(M)$ can also be given in terms of an integral, and as is usual in so many parts of mathematics, the trail towards such a formula starts with Gauss, whose theorem Egregium is precisely of this genre. Indeed it asserts that if a compact Riemann surface M is endowed with a Riemann structure g, then our Euler number is given by the integral

$$\chi(M) = \frac{1}{2\pi} \int R dv,$$

where R is the *curvature* and v the volume form determined by g.

In the work of Pontryagin, Chern, and Weil we then find a marvelous elaboration of this idea to higher dimensions. For, in dimension n the curvature $R^i_{jk\ell}$ is of course a more formidable tensor. But if properly interpreted – namely as a matrix of 2 forms – then the coefficients of the characteristic polynomial

$$\det(1 + t\frac{i}{2\pi}R) = 1 + t^4 p_1 + t^8 p_2 + \cdots$$

turn out to be closed forms on M whose *cohomology classes are invariants of M.* (These coefficients are identically zero in dimensions not divisible by 4 for symmetry reasons.)

Now, taking monomials in these classes and integrating over M we obtain a whole *new* sequence of numerical invariant

$$p^\alpha(M) = \int_M p_1^{\alpha_1} \cdots p_k^{\alpha_k},$$

the "Pontryagin numbers" of M.

At this stage you may well ask where our Euler number has got to in this scheme, but remember that Gauss's theorem deals with *oriented* manifolds. And to fit $\chi(M)$ into this scheme we need to recall that for oriented even-dimensional manifolds the *Pfaffian* of $\frac{1}{2\pi}R$ makes good sense, and then

$$(**) \qquad \chi(M) = \int_M \text{Pfaff}\, \frac{1}{2\pi}R$$

is a valid but quite *new* formula for our old acquaintance.

The conceptual route from (*) to (**) is best taken via "Poincaré Duality". It comes about from a global (Künneth) and local interpretation of this great principle, which asserts that the linear function from $\Omega^q \to \mathbf{R}$ furnished by integrating a q-form over a q-dimensional submanifold $N \subset M$, i.e., the operation

$$\omega \longrightarrow \int_N \omega,$$

must also be "represented" by a closed differential form $\theta_N \in \Omega^{m-q}$ in the sense that:

$$\int_N \omega = \int_M \omega \wedge \theta_N,$$

and it is then easily seen that the self intersection of a middle dimensional $N \subset M$ is given by $\int_N \theta_N$, that is, the integral of the "Poincaré dual" of N restricted to N.

The question reduces to the problem of constructing the Poincaré dual of N. Now *if* the normal neighborhood of N is trivial, so that near N one can define $r = m - n$ functions $y_1, \cdots, y_r$ whose common zeros precisely describe N, then it is nearly obvious that a Poincaré dual of N can be taken to be

$$\theta_N = \lambda(y) dy_1 \cdots dy_r,$$

where λ has compact support near 0 in $\mathbf{R}^r$ and total volume $= 1$. On the other hand, if the normal tube is not C^∞ trivial, Riemannian geometry is needed to keep track of things – and then the Pfaffian of the curvature of the normal tube turns out to describe the restriction of θ_N to N. Applied to $\Delta \in M \times M$ this construction yeilds (**), – and essentially that is how Chern established this formula in the 50's.

On the other hand, using the Künneth theorem we can construct θ_Δ from any base a_i for $H^*(M)$ of closed forms and for purely functorial reasons find it to be of the form

$$\theta'_N = \sum \pm a_i \otimes a_i^*$$

where the a_j^* are dual to the a_i in the sense that

$$\int_M a_j \wedge a_i^* = \delta_{ij}.$$

Restricting this θ'_N to Δ, yields (*). This second principle is thus based on the two projections of $M \times M$ on its factors – while the first of course works in a small tubular neighborhood of Δ. We see the antithesis of these two at work once more.

In any case, we now have the *Pontryagin* and the *Euler numbers* associated to any C^∞-structure on M; here derived from a diffeogeometric point of view. A decisive step in our understanding of things occurred in the 60's when we for the first time learned to combine the ideas and language of bundles, etc., with the ideas of partial differential equations. The crucial concept for the topologist was the proper understanding of ellipticity, and the corresponding homotopy invariance of the index of an elliptic operator.

I first heard the term "index" at a cocktail party in Stanford while eavesdropping on a serious discussion between Michael Atiyah and Lars Hörmander. With his marvelous instinct for what counts, Michael was clearly very fascinated – while I, as is my wont, was more interested in the excellent gin being served.

But later Michael explained the question to me, and together we soon saw how central these ideas were in all the questions that had occupied us over the years. From this new point of view it was now also very easy to see how numerical invariants of compact manifold could be created out of *analysis* on M.

The principle is simple: every naturally constructed elliptic system D on such a manifold gives rise to an integer invariant of M – namely its index:

$$\chi(D) = \dim(\ker D) - \dim(\operatorname{coker} D).$$

Indeed both ker D and coker D are finite dimensional and as their difference – the index of D – is also a *homotopy invariant*. Thus if the choices involved in creating D can be deformed into each other, $\chi(D)$ is clearly a C^∞ invariant of M. QED

For example, every Riemann structure g on M, yields a natural adjoint d^* to d, and one could manufacture interesting elliptic systems out of these. Indeed by the Hodge theory $d + d^*$ interpreted as

$$d + d^* : \Lambda^{\text{even}} \longrightarrow \Lambda^{\text{odd}}$$

from the even forms to the odd ones, has $\chi(M)$ as its index. On the other hand, if M is orientable one can divide the forms in a *quite different* way in into Λ^+ and Λ^-, as the ones invariant or anti-invariant under the $*$ operator at g. Now, $d \mp d^*$ is seen to map Λ^+ to Λ^- and gives rise to a new elliptic system

$$d_S : \Lambda^+ \longrightarrow \Lambda^-$$

called the *signature* operator of M.

Its index is therefore also a numerical invariant of M. And much more generally, given any representation of $SO(n)$, say on V, we have a corresponding bundle V over M associated to the frame bundle of g, and one can "twist" d_S by V to obtain a new elliptic system

$$d_S \otimes V : \Gamma(\Lambda^+ \otimes V) \longrightarrow \Gamma(\Lambda^- \otimes V).$$

All these constructions, of course, vary smoothly with the Riemann structure, hence index$(d_S \otimes V)$ yields a numerical invariant of M for *every representation V, of $SO(m)$, $m = \dim M$. .*

Over the years we have learned that the invariants created by this procedure are no *greater* – or *less* – than the Pontryagin and Chern numbers discussed earlier, but I want to emphasize that these questions became relevant only after we understood ellipticity in its *most general* form. To explain why, let me presume – for this audience – to "remind you" how ellipticity is defined. Given a local expression for a linear operator

$$D = \sum A^\alpha(x)\frac{\partial^\alpha}{\partial x^\alpha}$$

where α runs over multi-indexes and the $A^\alpha(x)$ are $n \times n$ matrix functions, the "symbol of D" is defined as the expression

$$\sigma_x(D;\xi) = \sum_{|\alpha|=m} A^\alpha(x)\xi_1^\alpha$$

where the ξ's are real variables and the sum is taken over the highest order terms only. Then $\sigma_X(D)$ is called elliptic *if and only if for any system (ξ) with $\xi \neq 0$, $\sigma_X(D;\xi)$ is a nonsingular matrix.*

Hence when D is elliptic, the assignment of

$$\xi \longrightarrow \sigma_X(D;\xi) \qquad |\xi| = 1$$

defines a map of $(n-1)$-sphere

$$S^{n-1} \longrightarrow GL(n),$$

and the *homotopy class* of this element – let's call it [D] – is a fundamental *topological invariant of the elliptic system on every component of M*. I like to call it the "virulence" of D.

Now the only elliptic operator we *all* knew in the 40's and 50's was really the Laplacian of the Hodge theory – and its "virulence" is 0! On the other hand, the signature operator d_S has a *nontrivial* virulence in even dimensions, and it is for that reason that the index problems of $d_S \otimes V$ just described, yield such a rich harvest of invariants as we vary V.

In this connection it has always struck me as quite marvelous – nearly a proof of the existence of God – that in its Euclidean incarnation the famous Dirac operator, describing the behaviour of the electron, has virulence 1 – that is, its symbol $[\not{\partial}]$ generates $\pi_*(GL(N,\mathbf{C}))$, $N \gg 0$, in all dimensions.

But to continue the story of our manifold invariants, let me now take the point of view of the topologist proper, who does not care a fig about analysis and geometry. His main concern is to sort out just how essential the differentiability – so implicit in our constructions – was for the invariants we have created, and from the structure theory of bundles it then soon became apparent that as far as the Pontryagin numbers were concerned, differentiability is used only to the extent that we have to have a tangent bundle TM to M.

Indeed, the classification theory of bundles then guarantees that TM is pulled back from the universal bundle by a map

$$f_T : M \longrightarrow BGL(n),$$

into the *classifying space* of the full linear group, well determined up to homotopy, and one can recover the Pontryagin classes purely from this map! In fact, $H^*(BGL(n))$ is generated *precisely* by the "universal" Pontryagin class $\widetilde{p}_i \in H^{4i}(BGL(n))$ which have the property that

$$f_T^* \widetilde{p}_i = p_i.$$

In short, our geometric constructions represented $f_T^* p_i$. Finally, to show that our "Index invariants" really always amounted to some Pontryagin

and Euler numbers was of course rather more difficult, but is an immediate consequence of the Atiyah-Singer Index theorem.

This then summarizes for you what we have known for a long time,
and I would now like to briefly contrast it with some of the new invariants
which were thrust upon us to a large extent by the physicists – and
in particular by the young man I see characteristically sitting halfway
towards the back in this hall; that is, – Ed Witten.

If we call what I have described so far as an interaction of analysis and
topology, what I am going to describe now is better described as the effect
of "super-analysis" on topology. And here "super" is really meant, not
in its technical sense, but in accordance with its usage in "supernatural"!
So let me describe the "new integrals" of Ed's which we are deciphering
in many ways at this moment.

One starts with an oriented 3-manifold and a compact Lie group G,
with Lie algebra $\mathfrak{g}$. We traditionally write $\mathcal{A}$ for the space of 1-forms on
M, with values in $\mathfrak{g}$, and for simplicity think of $A \in \mathcal{A}$ simply as a matrix
valued 1-form on M. Consider now the function $S_M : \mathcal{A} \to \mathbf{R}$ defined by

$$S_M(A) = \frac{1}{8\pi^2} \int_M Tr(AdA + \frac{2}{3}A^3).$$

It should be clear that under this integral we have defined an ordinary 3-
form and hence $S_M(A)$ is a well-defined number. Actually the expression
under the integral was well known in the mathematical world, and is
called the *Chern-Simons form*, and they constructed it explicitly to have
the following functorial properties.

Let $\mathcal{G}$ be the space of smooth maps of M into G, and consider the
following affine action of $\mathcal{G}$ on $\mathcal{A}$:

$$g \in \mathcal{G} \quad \text{sends} \quad A \quad \text{to} \quad g^{-1}Ag + g^{-1}dg.$$

This is then the famous action of the "group of gauge transformations"
which was found to be so very important in the physics of this century.
Well, the crucial property of the "action", $S_M(A)$, – as the physicists
would think of the function S_M – is that S_M is *nearly invariant under $\mathcal{G}$*.
In fact,

(1) S_M is invariant under the identity component $\mathcal{G}_0$ of $\mathcal{G}$ and

(2) Changes by an *integer* on the different components of $\mathcal{G}$.

It follows then that for every integer k, the function $A \to e^{2\pi i k S_M(A)}$
descends to $\mathcal{A}/\mathcal{G}$. So far so good, but now comes the supernatural part.

Following sound physics folklore, Ed tells us to define an invariant $Z_k^G(M)$ out of these data simply by computing the integral

$$Z_k^G(M) = \int_{\mathcal{A}/\mathcal{G}} e^{2\pi i k S_M(A)} \mathcal{D}A.$$

Q.E.D.

In fact he does not stop there; he tells us also – and indeed, that is how he came upon this recipe in the first place – that this hypothetical measure

$$e^{2\pi i k S(A)} \mathcal{D}(A),$$

on $\mathcal{A}/\mathcal{G}$ can now be used to compute "expectation values of knots" in M. Namely, given an oriented knot $k \subset M$ and a representation V of G, then every $A \in \mathcal{A}$ defines a connection on the trivial bundle over M, so that the closed curve k determines a corresponding holonomy element $h(k)$ in $G/\mathrm{Ad}\, G$.

The trace of $h(k)$ relative to the representation V, is therefore a well-defined function on $\mathcal{A}$, say $\chi_V(k)$, and its expectation value

$$(***) \qquad \langle \chi_V(k) \rangle = \int \chi_V(k) e^{2\pi i k S_M(A)} \mathcal{D}A / Z_k^G(M)$$

should therefore yield invariants of the pair (M, k).

I have run out of time to discuss these formulas in any detail here. But I hope you see how much work will be entailed in bringing these new invariants under control. But there is hope that they can indeed be tamed, and even, that suitably interpreted they lead to a *complete classification* of knots.

The secret weapon of the physicists is of course the "higher loop" expansion procedures of Feynmann integrals. Thus to get a handle on an integral like

$$Z_k^c(M) = \int_{\mathcal{A}/\mathcal{G}} e^{2\pi i k S_M(A)} \mathcal{D}(A),$$

the physicist will start by computing an asymptotic expansion of the right hand side as $k \to \infty$. The rationale for this expansion is derived from the principle of stationary phase for oscillatory integrals of this type over finite dimensional spaces. In that context it is then quite easy to see that an oscillatory integral of the form:

$$\int_M e^{2\pi i k f(x)} \varphi(x) dx,$$

will tend to zero faster than any power of k as $k \to \infty$; *provided* the support of φ avoids the *critical points* of f. Thus the asymptotic expansion of $\int_M e^{2\pi i k f(x)} dx$ is *localized* at the *extrema* of f. If the dimension of M is m, then every nondegenerate extremum, say at x_0, of f contributes a term of the type:

$$\frac{1}{t^{m/2}} \cdot \frac{e^{2\pi i f(x_0) t \pm (i\pi/4) sign Hf}}{\sqrt{|\det Hf|}} \{1 + \frac{a_1}{t} + \frac{a_2}{t^2} + \cdots\}$$

where Hf denotes the Hessian $\frac{\partial^2 f}{\partial x^2 \partial x^j}$ of f at x_0, and $sign Hf$ denotes its signature. Furthermore, the higher order terms $a_\ell; \ell = 1, \cdots$ are derived by universal formulas in terms of the higher order Taylor series of f at x_0, and the entries of the inverse matrix of Hf. The precise Feynmann recipe is quite beautiful and indeed *topological* in flavor. For the contributions to a_ℓ are best described in terms of graphs Γ with Euler number $\chi(\Gamma)$, equal to ℓ.

When this procedure is applied to $Z_k^G(M)$ one obtains a corresponding asymptotic series, now indexed by the *extrema* of S_M. And here the third crucial property of the Chern-Simons expression in dim 3 comes to the fore: that is, if dim $M = 3$, then:

(3) $$\delta S(M)\big|_A = 0 \iff \text{the 2-form } F(A) = dA + A^2$$

vanishes.

Thus the asymptotic series in question is indexed by the *flat connections* on M, which in turn are classified by conjugacy classes of homomorphisms of $\pi_1(M)$ into G.

These remarks lead us to expect that given M, and an isolated homomorphism

$$\mu : \pi_q(M) \to G,$$

there should exist a well defined sequence of numerical invariants

$$a_\ell(M, \mu), \qquad \ell = 0, 1, \cdots$$

corresponding to the ℓth loop approximation to (***) at μ. In attempting to carry out the finite dimensional program in this infinite dimensional context, one encounters all the road blocks which over the years the field theorists have learned to overcome. Indeed the Hessian of S_M at μ turns out to be degenerate and the Fadeev Popof procedures have to be applied. For this purpose an auxiliary Riemann structure, g, on M has to be chosen, and once this choice is made, $a_\ell(M, \mu)$ is seen to make sense, but as a very complicated integral involving the Green's operator of the

Laplacian of g. At this stage one now has to show that this integral is independent of the choice of g, and the physicists have developed formal procedures to show this independence – called "B.R.S. invariance". However it is only the recent work of Dror Bar-Natan (2), and Axelrod & Singer (1) that brings these questions into proper mathematical form. Actually, Dror Bar-Natan's work deals primarily with the higher loop expansions for the expectation functions $\chi_V(k)$ of knots in S^3.

This case is of course simplest because $\pi_1(S^3) = 0$ and so one variable, the homomorphism of π_1 to G, is eliminated. In any case, studying his invariants, and of course influenced by the work of Vasilief, Kontsevich and others, Dror Bar-Natan soon found how to separate the role of the group G and the representation V in the construction of the higher loop invariants $a_\ell(k)$. Indeed he constructed a universal combinatorial chain complex, – in fact, a Hopf algebra "$\mathcal{A}$", in which a given pair (G, V) determine a cycle, and it is this algebra which therefore is now conjectured to be the appropriate combinatorial setting for a sufficient number of knot-invariants, to separate them.

A few days ago I mentioned these matters to my old student and friend, Tom Goodwillie. He thought for a while and then recalled an unpublished construction of his from years ago, meant as a topologist's approximation to the space of imbeddings of an interval in $\mathbf{R}^n$. He had proved convergence for $n > 3(!)$. But when we wrote down his construction for $n = 3$, and then made a few computations we found that the E_2-term of his cosimplicial space was *precisely* the algebra $\mathcal{A}$ mentioned above!

So let me end by writing down the scheme of Goodwillie's "classifying space" , at least for the topologists among you, in the hope that it will prove to be the topological flesh and bones on which we can hang the ℓ'th loop "Feynmann" invariants. Here it is:

$$BK: \quad * \; \overset{\rightarrow}{\underset{\rightarrow}{\rightarrow}} \; C(1) \; \overset{\rightarrow}{\rightarrow} \; C(2) \; \overset{\overset{\textstyle\rightarrow}{\rightarrow}}{\underset{\rightarrow}{\rightarrow}} \; C(3) \; \cdots$$

with $C(r)$ denoting a "jet-version" of the configuration-space of r points in $\mathbf{R}^3$.

Thus at least the constituents $C(r)$ of this space have a very geometric homotopy type, namely just the space of r distinct points in $\mathbf{R}^3$. The "coattaching" maps are more difficult and involve subtle questions involving blowing up the diagonals in $\mathbf{R}^3 \times \cdot \times \mathbf{R}^3$ – in short, precisely a deeper understanding of the dichotomy of "*diagonal*" versus the "*factors*".

For us older and possibly more mature mathematicians, it is of course encouraging to find that our long and adventurous odyssey has brought us back home to old and familiar grounds. But I hope that the youngsters among you, who, I know full well, love nothing more than to teach us *new* things, will not be disappointed by this turn of events. Indeed, via a construction of Kontsevich the Chern-Simons theory is at this moment indispensable in proving that all the differentials in Goodwillie's sequence vanish.

And even if at some future time the topologists will be able to dispense with this construction – the new bridges between analysis and topology which this theory has erected will surely always remain a beautiful achievement of the mathematics of this century.

References

(1) Axelroad & Singer, Chern-Simons perturbation theory, MIT Preprint, October 1991.

(2) Dror Bar-Natan, Perturbative aspects of Chern-Simons topological quantum field theory. Ph.D. Thesis. Princeton Univ., June 1991. Dept. of Math.

(3) Dror Bar-Natan, On Vasiliev Knot Invariants, Harvard Preprint, Oct. 16, 1992.

REMARKS ON
ANALYTIC HYPOELLIPTICITY OF $\bar{\partial}_b$

Michael Christ

1. Introduction

Let there be given a real analytic, pseudoconvex CR structure of finite type on a (small) open three-dimensional manifold M. Let $\bar{\partial}_b$ be the associated Cauchy-Riemann operator, which maps C^∞ functions to sections of a certain bundle B whose fibers have dimension one over $\mathbb{C}$, and let $\bar{\partial}_b{}^*$ be its adjoint, with respect to a nonvanishing real analytic density on M and a real analytic Hermitian inner product structure on B. Denote both the set of all functions real analytic in some open set $U \subset M$, and also the set of all real analytic sections of B over U, by $C^\omega(U)$. $\bar{\partial}_b$ is said to be analytic hypoelliptic modulo its nullspace, in Ω', if for any open set $\Omega \subset \Omega'$, for any distribution u such that $\bar{\partial}_b\bar{\partial}_b{}^*u \in C^\omega(\Omega)$, necessarily $\bar{\partial}_b{}^*u \in C^\omega(\Omega)$. It is a fundamental problem, suggested to this author by J. J. Kohn and to date unsolved, to determine for which such CR structures $\bar{\partial}_b$ is indeed analytic hypoelliptic modulo its nullspace.

Recall that a differential operator L is said to be analytic hypoelliptic in Ω' if for any open set $\Omega \subset \Omega'$, for any distribution u such that $Lu \in C^\omega(\Omega)$, necessarily $u \in C^\omega(\Omega)$. $\bar{\partial}_b$ and $\bar{\partial}_b\bar{\partial}_b{}^*$ are never analytic hypoelliptic in this sense. It is of course also of interest to determine just which operators are analytic hypoelliptic.

In more concrete terms, there are given, in a neighborhood of a point $x_0 \in \mathbb{R}^3$, two real vector fields X, Y, having real analytic coefficients and linearly independent at every point. $\bar{\partial}_b$ is equal to $X + iY$. Fixing a third real vector field T linearly independent of X, Y, pseudoconvexity means that the determinant of the matrix (X, Y, T), computed in some local coordinate system, is either everywhere nonnegative, or everywhere nonpositive. Finite type is the hypothesis of Hörmander, that the Lie algebra generated by X, Y should span the tangent space to $\mathbb{R}^3$ at each point near x_0. The CR structure is weakly pseudoconvex at x if the Lie bracket $[X, Y]$ belongs to the span of X, Y at x, strictly pseudoconvex if $X, Y, [X, Y]$ are linearly independent at x.

Sometimes $\bar{\partial}_b$ is analytic hypoelliptic modulo its nullspace, and sometimes it is not; a necessary condition involves the following geometric notion. To any CR structure is associated a smoothly varying field $\mathcal{T}$ of two-dimensional subspaces of the (real) three-dimensional tangent space, whose complexification is $T^{1,0} \oplus T^{0,1}$; in other words the fiber $\mathcal{T}_x$ is the span of $\{X(x), Y(x)\}$.

1.1. Definitions. *A C^∞ curve $\gamma : (-\varepsilon, \varepsilon) \mapsto \Omega$ for which γ' does not vanish is complex-tangential if for every s, $\gamma'(s) \in \mathcal{T}_{\gamma(s)}$. γ is said to be a weakly pseudoconvex, complex-tangential curve if in addition, $\gamma(s)$ is a weakly pseudoconvex point of the CR structure, for all s.*

1.2. Theorem. *Let there be given a real-analytic, pseudoconvex, three-dimensional CR structure of finite type in an open set Ω. If there exists a weakly pseudoconvex, complex-tangential curve contained in Ω, then $\bar{\partial}_b$ fails to be analytic hypoelliptic modulo its nullspace in Ω.*

This is a special case of a conjecture of Treves [Tr2], which he formulated also for more general operators and in higher dimensions. It is expected that $L = X^2 + Y^2$ and $\bar{\partial}_b$ should share much the same regularity properties. Indeed, retaining the finite type hypothesis but dropping pseudoconvexity, $L = X^2 + Y^2$ fails to be analytic hypoelliptic in Ω whenever there exists a curve γ in Ω whose tangent vector is spanned by X, Y, and such that $X, Y, [X, Y]$ are linearly dependent, at each point of γ. This, and Theorem 1.2, are proved in [C10].

At one point it seemed plausible that the necessary condition of Theorem 1.2, that is, nonexistence of certain curves, should also be sufficient for analytic hypoellipticity. Although this has not yet been proved, the analysis of [C11] now strongly suggests, to the contrary, that analytic hypoellipticity should fail for certain structures having only isolated weakly pseudoconvex points.

Weakly pseudoconvex, complex-tangential curves have arisen in at least two other contexts. Noell [N1] has shown that if $D \Subset \mathbb{C}^2$ is pseudoconvex and C^ω, and if there exist no such curves in ∂D, then any compact subset of ∂D which is locally a peak set for $A^\infty(D)$ is actually a peak set for $A^\infty(D)$, while in the presence of such curves, local peak sets need not be peak sets. Montgomery [Mo] has given examples in which geodesics in sub-Riemannian geometry – curves which minimize distance between two points relative to a degenerate Riemannian metric satisfying a finite type hypothesis – fail to satisfy the "geodesic equations" which are a formal consequence of variational calculations. The underlying geometric structure in these examples is almost identical to the models (1.4) below, and the geodesics in question are weakly pseudoconvex, complex-tangential curves.

In the positive direction, there is one outstanding result: $\bar{\partial}_b$ is analytic hypoelliptic, modulo its nullspace, wherever the CR structure is strictly pseudoconvex. This result and/or closely related ones have been proved by Tartakoff [Ta1],[Ta2], Treves [Tr2], Métivier [M2],[M3], and Geller [G]. Analytic hypoellipticity holds good in other instances, as well. Consider, with coordinates (x, y, t) on $\mathbb{R}^3$,

$$(1.3) \qquad X = \partial_x, \qquad Y = \partial_y + a(x, y)\partial_t$$

where a is a homogeneous polynomial of some degree $m-1$, such that $\partial_x a$ is nonnegative and vanishes only at $x = y = 0$. Since $[X, Y] = \partial_x a \cdot \partial_t$, the CR structure is strictly pseudoconvex where $(x, y) \neq 0$, and weakly pseudoconvex along the curve $\{x = y = 0\}$; nonetheless, $\overline{\partial}_b$ is analytic hypoelliptic, modulo its nullspace. A variety of interesting partial results, for related problems, have been obtained by Derridj and Tartakoff; see [DT1],[DT2],[DT3] and the references therein.

A better result is that of Grigis and Sjöstrand [GS]: with $X = \partial_x$ and $Y = \partial_y + a(x, y)\partial_t$, the operator $L = X^2 + Y^2$ is hypoelliptic when $\partial_x a(x, y) = x^k + y^m$ with k, m even positive integers (and a is analytic), and more generally when $\lambda(x, y) = \partial_x a$ vanishes only at the origin and satisfies $\lambda(r^{1/k}x, r^{1/m}y) \equiv r\lambda(x, y)$.

The first negative result for $\overline{\partial}_b$ was obtained in [CG]. Consider

$$(1.4) \qquad X = \partial_x, \qquad Y = \partial_y + x^{m-1}\partial_t$$

where m is an even, positive integer. Then $\overline{\partial}_b$ fails to be analytic hypoelliptic, modulo its nullspace, in any neighborhood of the origin, in the weakly pseudoconvex case $m > 2$. Earlier, Helffer, Pham The Lai and Robert had shown that for the same vector fields, $L = X^2 + Y^2$ fails to be analytic hypoelliptic in the (non-pseudoconvex) case $m = 3$.

In these examples, any curve $\gamma(s) = (0, s, t_0)$ is weakly pseudoconvex and complex-tangential; the two-dimensional locus of weakly pseudoconvex points is foliated by this one-parameter family of curves. In examples (1.3), when $\partial_x a$ vanishes only at $x = y = 0$, the set of weakly pseudoconvex points is the single curve $\gamma(s) = (0, 0, s)$, which is not complex-tangential.

More generally, when $X = \partial_x$ and $Y = \partial_y + a(x)\partial_t$ with $a(x) = x^m + O(|x|^{m+1})$ near the origin, $\overline{\partial}_b$ fails to be analytic hypoelliptic modulo its nullspace near 0 [C3], for even integers $m > 2$ (and similarly for $X^2 + Y^2$ [C5]). A consequence is the failure of analytic hypoellipticity modulo the nullspace for $\overline{\partial}_b$ on $\partial\Omega$, for any bounded, pseudoconvex, real analytic Reinhardt domain $\Omega \subset \mathbb{C}^2$ whose boundary contains a weakly pseudoconvex point whose orbit, under the rotation group $U(1) \times U(1) \subset U(2)$, is two dimensional. In short, analytic hypoellipticity fails for simple and natural bounded domains. It is to be expected that analytic hypoellipticity fails also for the $\overline{\partial}$–Neumann problem for these same domains, and this should follow from a similar method.

(1.4) are not merely examples, but rather, are models for all CR structures (analytic, pseudoconvex and of finite type) at a point x_0 through which passes a weakly pseudoconvex, complex-tangential curve γ. In this case it is possible to introduce coordinates near x_0 so that along $\gamma = \{(0, y, 0)\}$, $\bar{\partial}_b = X + iY$ where X, Y are as in (1.4), modulo perturbations which are $O(|x|^m + |t|)$.

In the model cases the weakly pseudoconvex locus is two-dimensional, leaving room for doubt as to whether it is correct to regard CR structures in which a single weakly pseudoconvex, complex-tangential curve exists as perturbations of the models. Perhaps the simplest examples of such structures are

$$(1.5) \qquad X = \partial_x, \qquad Y = \partial_y - (x^{m-1} + t^M x)\partial_t$$

where m, M are even integers with $m > 2$, $M \geq 2$; the weakly pseudoconvex locus is the single curve $\{x = t = 0\}$. In §5 a proof of the following special case of Theorem 1.2 is outlined.

1.6. Theorem. *If m, M are both sufficiently large then $\bar{\partial}_b$ is not analytic hypoelliptic, modulo its nullspace, for the CR structure (1.5).*

The proof is of a perturbative character, building on the results for the models (1.4). The hypothesis that M should be large amounts to requiring the structure to be a sufficiently small perturbation of the models.

A related question is that of global analytic hypoellipticity, modulo the nullspace: given a compact three-dimensional CR manifold M without boundary, analytic, pseudoconvex and of finite type, and given both that $\bar{\partial}_b u = f \in C^\omega(M)$, and that u is orthogonal to the nullspace of $\bar{\partial}_b$ in $L^2(M)$, does it follow that $u \in C^\omega(M)$? S.-C. Chen has proved this (or a closely related result) to be true for Reinhardt domains, so that one has simultaneously negative results [C3] for the local problem and positive results for the global one, on the same domain. Chen has further proved global regularity for a large class of circular domains, and Derridj and Tartakoff [DT4] have been able to relax the symmetry hypothesis to the existence of an analytic vector field transverse to the CR field of 2-planes and having certain favorable commutation properties with $\bar{\partial}_b$, $\bar{\partial}_b{}^*$. In [C12] it is shown that global analytic hypoellipticity holds for a wide class of partial differential operators on compact manifolds, given the existence of a group of suitably transverse symmetries.

Other reasons can be advanced in support of the hope that global analytic hypoellipticity should always be valid. For instance, on a torus $\mathbb{T}^n$, convolution with any distribution K defines an operator mapping $C^\omega(\mathbb{T}^n)$ to $C^\omega(\mathbb{T}^n)$, whereas such an operator is analytic pseudolocal if and only if $K \in C^\omega$ on $\mathbb{T}^n$ minus the group identity element. In particular, any C^∞ hypoelliptic, constant-coefficient partial differential operator on a torus is globally analytic hypoelliptic, whereas such an operator is analytic hypoelliptic in the local sense if and only if it is elliptic. Thus global analytic hypoellipticity is a far weaker property than its local analogue.

These heuristics are, regrettably, misleading for the case of variable coefficients [C11].

1.7. Theorem. *There exists a pseudoconvex, bounded domain* $\Omega \subset \mathbb{C}^2$ *with* C^ω *boundary such that* $\overline{\partial}_b$ *fails to be analytic hypelliptic, modulo its nullspace, on* $\partial\Omega$. *Furthermore, the Szegö projection does not map* $C^\omega(\partial\Omega)$ *to* $C^\omega(\partial\Omega)$.

Although this has not yet been proved, the analysis of [C11] suggests that this should be the typical situation for weakly pseudoconvex domains, and that it can happen even for domains having only isolated weakly pseudoconvex points.

2. Analytic hypoellipticity for more general operators

Results known for operators other than $\overline{\partial}_b$, its higher-dimensional analogue $\Box_b$ and sums of squares of vector fields help to place our problem in better perspective. For simplicity we restrict attention to differential operators which have C^ω coefficients, act on scalar-valued functions, are C^∞ hypoelliptic, and whose principal symbols do not vanish identically at any point $x \in \Omega$. Such an operator L is analytic hypoelliptic in Ω if and only if, for every open $\Omega' \subset \Omega$ and every $x_0 \in \Omega'$, there exists $C < \infty$ such that for any $u \in L^2(\Omega')$ satisfying $Lu = 0$ in Ω',

(2.1) $\qquad |\partial^\alpha u(x_0)| \leq C^{1+|\alpha|}|\alpha|^{|\alpha|}\|u\|_{L^2(\Omega')} \qquad$ for all α.

We refer to (2.1) as the Cauchy estimates. If (2.1) is valid and if $Lu = f \in C^\omega$ in a neighborhood of x_0, then (by the assymption on the principal symbol) the Cauchy-Kowalevsky theorem gives an analytic v such that $Lv = f$ near x_0, so that $L(u-v) = 0$ and the growth estimates (2.1) then imply analyticity of $u - v$. The contrapositive of the converse implication may be proved by elementary reasoning.

The elliptic case is simplest. If L is of order n and elliptic, then for any open sets $U \Subset U'$, for any positive integer k, there is an inequality

$$\|u\|_{H^{n+k}(U)} \leq B\|Lu\|_{H^k(U')} + B\|u\|_{H^k(U')}$$

in terms of Sobolev norms, where $B = B(U, U', k)$. Executing the usual bootstrapping argument with an optimally chosen sequence of domains U (and passing from Sobolev to C^k estimates only at the end) yields the Cauchy estimates by induction on $|\alpha|$. However, this argument depends on gaining a full n derivatives at each step of the bootstrap.

An important point is that the theory microlocalizes; the analytic wave front set may be defined in a number of equivalent ways [H1],[H4],[B],[S], and if $Lu \in C^{\omega}$ in a conic subset of phase space where L is elliptic, then u is analytic there. See [H1, Theorem 8.6.1].

L is said to be of principal type, if its principal symbol $\sigma(x, \xi)$ has the property that wherever $\sigma = 0$ and $\xi \neq 0$, the differential forms $d\sigma$ and $\xi \, dx$ are linearly independent. In particular, σ cannot vanish to order greater than one. If such an operator is subelliptic with loss of strictly less than one derivative [H3], then it is not only C^{∞} hypoelliptic, but also analytic hypoelliptic [Tr1],[Trp]; this again microlocalizes. An example is $\bar{\partial}_b$ for pseudoconvex, three-dimensional CR structures of finite type. It is of principal type, and is subelliptic with loss of less than one derivative, in a neighborhood of one half of its characteristic variety Σ (which is a line bundle over $\mathbb{R}^3$). Thus $\bar{\partial}_b$ is analytic hypoelliptic, hence is certainly so modulo its nullspace, microlocally near that half of Σ.

The principal symbols for sums of squares operators $X^2 + Y^2$, and for $\bar{\partial}_b \bar{\partial}_b{}^*$, are sums of squares of real-valued functions, hence vanish to order at least two. Such operators are said to have multiple characteristics, and their C^{∞} and analytic regularity theories diverge markedly.

The outstanding positive result for multiple characteristics was due in various formulations initially to Tartakoff and to Treves, and later to Métivier, Sjöstrand and Geller in other cases and formulations. It states roughly that if L is of second order, subelliptic with loss of one derivative and if its principal symbol vanishes to order exactly two everywhere on its characteristic variety Σ, then L is analytic hypoelliptic, provided that Σ is a *symplectic* submanifold of the cotangent space. Examples are $\Box_b$ on strictly pseudoconvex CR manifolds (of dimension strictly greater than three), $\bar{\partial}_b \bar{\partial}_b{}^*$ on strictly pseudoconvex three-dimensional CR manifolds, and certain, but not all, sums of squares of real vector fields. The result microlocalizes in the usual way, and there are now several distinct methods of proof. An example which satisfies all of these conditions except for being symplectic, and indeed fails to be analytic hypoelliptic, is $\partial_x^2 + x^2 \partial_y^2 + \partial_t^2$ in $\mathbb{R}^3$ [BG],[H1].

As already explained in §1, positive results do hold, for certain non-symplectic operators with multiple characteristics. An easily treated example is the following special case of the subtler result of Grigis and Sjöstrand [GS].

2.2. Proposition. *Let $X = \partial_x$, $Y = \partial_y + a(x, y)\partial_t$ where a is a polynomial homogeneous of degree $m - 1$ for some $m \geq 2$. Suppose that $\partial a / \partial x$ vanishes only at $(0, 0)$. Then $L = X^2 + Y^2$ is analytic hypoelliptic.*

Outline of proof. We take for granted an array of facts belonging to the detailed C^{∞} theory for such operators. First, there exists an operator $Pf(x) = \int_{\mathbb{R}^3} K(u, v) f(v) \, dv$ such that $LP = PL = I$, acting for instance on $C_0^\infty(\mathbb{R}^3)$, whose kernel K is C^∞ off of the diagonal, and may be chosen so as to satisfy certain pointwise bounds of a type first obtained by Nagel, Stein and Wainger [NSW] and Sánchez-Calle [Sa]. K may be taken to be homogeneous: $K(ru, rv) = r^{2-m}K(u, v)$ for all $u \neq v \in \mathbb{R}^3$, $r > 0$. Likewise K may be taken to satisfy $K((x, y, t), (x', y', t')) = K((x, y, t - t'), (x', y', 0))$.

To show that L is analytic hypoelliptic, it suffices to prove K to be analytic with respect to u, at every point off of the diagonal. Write $u = (x, y, t)$. Since $LK = 0$ off the diagonal, where L acts in either variable, K is analytic, with respect to u, off of the diagonal at all strictly pseudoconvex points, which is to say, where $(x, y) \neq 0$.

$f(u) = K(u, 0)$ is analytic where $x = y = 0$, $t \neq 0$, because f satisfies an elliptic system of differential equations; not only is $Lf = 0$ except at 0, but also the dilation symmetry means that f is annihilated by a first-order differential operator, whose characteristic variety is disjoint from the characteristic variety of L in a neighborhood of $\{x = y = 0\}$. Differentiating with respect to t turns the first-order equation into one of second order, so that f is annihilated by a second-order elliptic system near $x = y = 0$, hence is analytic there. Combining this with the result for the strictly pseudoconvex points, $u \to K(u, 0)$ is analytic where $u \neq 0$, and then by the translation symmetry, $u \to K(u, (0, 0, s))$ is analytic where $t \neq s$. The same reasoning, applied to K as a function of $(u, v) \in \mathbb{R}^6$, demonstrates more generally that K is analytic where $t \neq s$ and the first two coordinates of v are close to 0.

$u \mapsto K(u, v)$ is analytic wherever $u \neq v$ and u is a strictly pseudoconvex point, so it remains to treat the case where $v = (x', y', t')$ with $(x', y') \neq 0$ and where $u = (x, y, t)$ with (x, y) in a small neithgborhood of 0. It suffices to prove analyticity microlocally in a conic neighborhood of the characteristic variety of L, and since ∂_t is elliptic in that region, it suffices to obtain the growth estimate

$$\partial_t^k f = O(C^k k^k)$$

for $f(u) = K(u, v)$. This would follow from a bound

(2.3) $$|\hat{f}(x, y, \tau)| = O(\exp(-c|\tau|)) \text{ as } |\tau| \to \infty,$$

where $\hat{f}$ denotes the partial Fourier transform with respect to the variable t.

Setting $w = (x, y)$ and $g(w) = \hat{f}(w, \tau)$ for a fixed τ, g is annihilated by $L_\tau = \partial_x^2 + (\partial_y + ia(w)\tau)^2$ in a fixed neighborhood U, at a positive distance from (x', y'), of 0. Various elementary arguments yield a preliminary bound $g = O(1)$ in U, as $|\tau| \to \infty$. Fixing smaller neighborhoods $0 \in$

$U'' \Subset U' \Subset U$, another estimate proper to the C^∞ theory [C8],[C9] for solutions of L_τ is

$$(2.4) \qquad \sup_{U''} |g| \leq C_\tau \sup_{U \setminus U'} |g| \qquad \text{where } C_\tau = O(\tau^{-N}) \text{ for all } N.$$

In fact, $C_\tau = O(\exp(-c\tau^{1/m}))$ as $\tau \to \infty$ (and this does not require the origin to be an isolated zero of $\partial_x a$). This bound alone is related to Gevrey regularity of order m, and is wholly inadequate for our purpose.

However, the fundamental result on analytic hypoellipticity at strictly pseudoconvex points yields a bound

$$(2.5) \qquad\qquad\qquad \sup_{U \setminus U'} |g| = O(e^{-c|\tau|}).$$

This is seen for instance by considering $h(u) = e^{it\tau} g(x,y)$, which is annihilated by L. Since $\partial_t^k h = (i\tau)^k h$, the Cauchy estimates (2.1) together with the bound $g = O(1)$ yield (2.5). Combining (2.5) with (2.4) yields (2.3), completing the proof. $\qquad\qquad\qquad\qquad\qquad\qquad\qquad\qquad\square$

$\bar{\partial}_b \bar{\partial}_b{}^*$ is likewise analytic hypoelliptic modulo its nullspace, in the situation of Proposition 2.2, and the same method of proof applies. Another interesting analytic hypoelliptic operator is $L = (X + iY)(X - iY) + c = \bar{\partial}_b \bar{\partial}_b{}^* + c$ where c is any nonzero constant, and $X = \partial_x$, $Y = \partial_y + x\partial_t$ [St]. It satisfies no subelliptic estimate, and in fact there exist other examples [H] which actually lose derivatives, in the sense that in the scale of Sobolev spaces, u is in general less regular than Lu.

Further perspective is provided by the scale of Gevrey classes G^s. A function u belongs to G^s in an open set Ω if there exists $C < \infty$ such that

$$|\partial^\alpha u| \leq C^{1+|\alpha|} |\alpha|^{s|\alpha|}$$

in Ω. Thus G^1 is the analytic class, and $G^s \subset G^{s'}$ if $s < s'$. For $\bar{\partial}_b$, the best that can be proved by an appropriate generalization of the bootstrapping argument indicated above for the elliptic case, is that for an otherwise arbitrary (real analytic, as always) CR structure of type at most m, if $\bar{\partial}_b \bar{\partial}_b{}^* u = 0$ in some open set, then $\bar{\partial}_b{}^* u \in G^m$ in that set. The same goes for sums of squares of vector fields [DZ]. This result is in fact optimal, for in the model situation (1.4), there exists a solution of $\bar{\partial}_b \bar{\partial}_b{}^* u = 0$ such that $\bar{\partial}_b{}^* u$ does not belong to G^s for any $s < m$ [CG], [C3]. This is related to questions about the size of the reproducing Bergman kernel for weighted L^2 spaces of entire functions on $\mathbb{C}^1$ [C9].

3. Hypoellipticity and group representations

As shown by Folland, Rothschild and Stein [FS],[RS] and others, the analysis of wide classes of differential operators with multiple character-istics is closely linked to the study of invariant operators on nilpotent Lie groups. C^∞ hypoellipticity of such an operator L is linked to injectivity of $d\pi(L)$ for all irreducible unitary representations π [RS],[HN1],[HN2]. Helffer [He] outlined a connection between analytic hypoellipticity and the non-unitary representations. Before considering the non-abelian case, we review the situation for translation-invariant operators on Euclidean space.

Let L be a constant-coefficient differential operator on $\mathbb{R}^n$ and let σ be its symbol. Define $\Gamma = \{\xi \in \mathbb{C}^n : \sigma(\xi) = 0\}$. Then L is C^∞ hypoelliptic if and only if $|\Im(\xi)| \to \infty$ as $|\Re(\xi)| \to \infty$ for $\xi \in \Gamma$. There exists an equivalent formulation involving only $\xi \in \mathbb{R}^n$ [H2]. Similarly, L is analytic hypoelliptic if and only if $|\Im(\xi)|/|\xi| \geq c > 0$ as $|\Re(\xi)| \to \infty$ for $\xi \in \Gamma$. Taking the Fourier transform amounts to decomposing the regular representation of $\mathbb{R}^n$ on $L^2(\mathbb{R}^n)$ into its irreducible components; to each representation $\pi \leftrightarrow \xi \in \mathbb{R}^n$ corresponds the operator $d\pi(L) = \sigma(\xi)$ on the one-dimensional Hilbert space $\mathbb{C}$.

If σ is a homogeneous polynomial which is nonvanishing in $\mathbb{R}^n \backslash \{0\}$, then L is elliptic, hence C^∞ and analytic hypoelliptic. However, there is also a negative result. Fix some $0 \neq \xi \in \mathbb{C}^n$ such that $\sigma(\xi) = 0$. The functions $f_\tau(x) = e^{i\tau x \cdot \xi}$ are annihilated by L, for all $\tau \in \mathbb{R}^+$. These satisfy the Cauchy estimates, but fail to satisfy a strengthened version: if ε is sufficiently small, then there exists no finite C_ε such that

$$|\partial_x^\alpha f_\tau(0)| \leq C_\varepsilon \, \varepsilon^{|\alpha|} |\alpha|^{|\alpha|} \sup_{|x|<1} |f_\tau(x)| \qquad \text{for all } \tau, \alpha.$$

The failure of analytic hypoellipticity for $\bar{\partial}_b$ in the model cases (1.4) results ultimately from a generalization of this trivial remark.

In the situations of the last two paragraphs, negative results on the regularity of solutions of L are derived from failure of the operators $d\pi(L)$ to be injective. Consider next the model vector fields $X = \partial_x$, $Y = \partial_y - x^{m-1}\partial_t$. Letting $\mathbb{R}^2$ act on $L^2(\mathbb{R}^3)$ by translation in the second and third coordinates, these remain invariant. Taking the Fourier transform in these variables decomposes $L^2(\mathbb{R}^3)$ into a continuous direct sum of copies of $L^2(\mathbb{R}^1)$, indexed by variables $\eta, \tau \in \mathbb{R}$ dual to y, t. $L = \bar{\partial}_b \bar{\partial}_b{}^* = (X + iY) \circ (X - iY)$ is transformed into the two-parameter family of ordinary differential operators $d\pi(L) = (d/dx - (\eta - \tau x^{m-1})) \circ (d/dx + (\eta - \tau x^{m-1}))$. Taking into account the dilation symmetry, all the information is contained in the one-parameter family

$$L_\zeta = \left(\frac{d}{dx} - (\zeta - x^{m-1})\right) \circ \left(\frac{d}{dx} + (\zeta - x^{m-1})\right), \qquad \zeta \in \mathbb{R}$$

together with the variant family obtained by replacing $\zeta - x^{m-1}$ by $\zeta + x^{m-1}$.

Since X, Y generate a nilpotent Lie algebra, they may be regarded as images of generators of the Lie algebras of certain nilpotent Lie groups G, under appropriate representations of the groups as unitary operators on $L^2(\mathbb{R}^3)$, and what we have actually done is to decompose those representations into their irreducible unitary components; each such component is a representation of G on the Hilbert space $L^2(\mathbb{R}^1)$. This point of view is discussed further in [RS],[He],[HN3],[C4],[C6].

If g is a solution, on all of $\mathbb{R}$, of L_ζ then

$$f_\tau(x, y, t) = e^{it\tau} e^{iy\zeta\tau^{1/m}} g(\tau^{1/m}x)$$

defines a one-parameter family of solutions of $\bar{\partial}_b\bar{\partial}_b{}^*$. If there were to exist $\zeta \in \mathbb{R}$ such that L_ζ admitted a solution g which was globally bounded but not identically zero, then $\{f_\tau\}$ would violate those *a priori* bounds, analogous to the Cauchy estimates, which are implied by C^∞ hypoellipticity in a neighborhood of the origin: assuming for simplicity that it happened that $g(0) \neq 0$, then for any finite C, N and neighborhood U of 0, the bound $|\partial_t^k \bar{\partial}_b{}^* f_\tau(0)| \leq C_k \|f_\tau\|_{C^N(U)}$ would be violated for $k = N + 1$, for all sufficiently large $\tau \in \mathbb{R}^+$. No such $\zeta \in \mathbb{R}$ exists, as follows from an examination of $\langle L_\zeta g, g \rangle$ and integration by parts. While $\bar{\partial}_b\bar{\partial}_b{}^*$ is indeed not C^∞ hypoelliptic, no obstruction arises in this way, essentially because for large $\tau > 0$, the supports of the f_τ, in the microlocal sense, are concentrated primarily along that half of Σ where $\bar{\partial}_b\bar{\partial}_b{}^*$ does happen to be hypoelliptic.

Helffer, Pham The Lai and Robert [He],[PR] showed that $L = X^2 + Y^2$ is not analytic hypoelliptic, with the same definition of X, Y and for $m = 3$, by proving the existence of $\zeta \in \mathbb{C}$ (here again, no such real ζ exists) such that the corresponding variant $\tilde{L}_\zeta$ of L_ζ admits a globally bounded, nonzero solution g. Defining f_τ in terms of g as before, fixing any neighborhood U of 0, and supposing for simplicity of exposition that $g(0) \neq 0$, then the Cauchy estimates $|\partial_t^k f_\tau(0)| \leq C^{k+1} k^k \sup_U |f_\tau|$ fail for $\tau \sim k^m$, since the left-hand side is $\gtrsim k^{mk}$, while the right side is $O(k^k e^{ck})$. The factor e^{ck} on the right-hand side comes from $\exp(i\tau^{1/m}\zeta y)$, which grows since $\zeta \notin \mathbb{R}$.

More precisely, this analysis shows that there exist solutions to $Lu = 0$ which do not belong to any Gevrey class G^s with $s < m$. This may be proved either by arguing that Gevrey regularity would imply *a priori* estimates generalizing the Cauchy estimates (2.1), with $|\alpha|^{|\alpha|}$ replaced by $|\alpha|^{s|\alpha|}$, or equivalently by forming a series $\sum c_j f_{\tau_j}(x, y, t)$ and choosing the sequences c_j, τ_j so as to obtain a convergent series whose sum does not belong to G^s.

Given the existence of a ζ with the required property, the negative result for analytic hypoellipticity emerges directly from the underlying anisotropic dilation structure; the factor $\exp(i\tau t)$ oscillates much more rapidly than the factor $\exp(i\tau^{1/m}\zeta y)$ grows. This is almost completely analogous to the remark several paragraphs above concerning homogeneous, elliptic operators with constant coefficients and the stronger form of the Cauchy estimates, but the dilation groups are different, so the implication of non-injectivity is different. There is however no fundamental theorem of algebra to guarantee the existence of such ζ, here.

Note that even if U is small, the growth properties of $g(x)$ for large x come into play in estimating the right-hand side. It is not essential that g should be bounded, but for the argument to succeed, $g(x)$ must grow less rapidly than $\exp(cx^m)$ as $|x| \to \infty$, for all $c > 0$. Now it happens that there is a sharp dichotomy: for any given $\zeta \in \mathbb{C}$, either every solution grows like $\exp(cx^m)$, or there exists a solution which decays exponentially as $|x| \to \infty$, so in particular is in L^2. In terms of representation theory, then, the issue is whether $d\pi(L) = \tilde{L}_\zeta$ fails to be injective on a non-unitary representation π, corresponding to some $\zeta \in \mathbb{C}\backslash\mathbb{R}$.

The remarkable method used by Pham The Lai and Robert to prove, indirectly, the existence of such a ζ was in part rather general, but relied at one step on the oddness of m. The equations $L_\lambda g = 0$, $\tilde{L}_\lambda g = 0$ are examples of nonlinear eigenvalue problems, in which the usual eigenvalue equation $Ag = \lambda Bg$ is generalized to $\sum_k \lambda^k A_k g = 0$. There is a substantial theory devoted to this problem, but the general results [Ke],[FrS] require hypotheses not satisfied by L_ζ, $\tilde{L}_\zeta$. In §4 we will outline a different approach which applies to L_ζ, $\tilde{L}_\zeta$ and a family of related operators.

There is the following heuristic connection between local and global analytic hypoellipticity. Suppose that the only obstruction to local analytic hypoellipticity were to arise from the functions f_τ and variants of them. Fix a compact CR manifold M without boundary, satisfying the usual hypotheses. Then any non-analytic solution $\overline{\partial}_b{}^* u$ of $\overline{\partial}_b\overline{\partial}_b{}^* u = 0$ would be synthesized out of functions resembling $\overline{\partial}_b{}^* f_\tau$, hence its high-frequency Fourier components would exhibit growth like $\exp(c\tau^{1/m}y)$ along weakly pseudoconvex, complex-tangential curves (parametrized by y). But $\overline{\partial}_b{}^* u$ must be globally bounded on the compact manifold M, so can not exhibit such growth. Thus any obstruction to global analyticity should be different in nature.

4. Analysis of the models

Throughout this section the vector fields

$$X = \partial_x, \qquad Y = \partial_y - x^{m-1}\partial_t$$

are fixed, and $m \geq 2$ is assumed to be even. Set $\bar{\partial}_b = X + iY$, $\bar{\partial}_b{}^* = X - iY$, and

$$L_z = \left(\frac{d}{dx} - (z - x^{m-1})\right) \circ \left(\frac{d}{dx} + (z - x^{m-1})\right), \qquad z \in \mathbb{C}.$$

The problem is to prove the existence of z and a solution $g \in L^\infty(\mathbb{R})$, not identically vanishing, of $L_z g \equiv 0$.

One solution is $\exp(-zx + m^{-1}x^m)$, but it grows too rapidly. Two solutions are

$$f_z^{\pm}(x) = e^{-zx + m^{-1}x^m} \int_x^{\pm\infty} e^{2(zs - m^{-1}s^m)}\, ds.$$

For all z, f_z^+ remains bounded, in fact decays rapidly, as $x \to +\infty$, while f_z^- behaves similarly at $-\infty$. Up to scalar multiples, these are the only solutions bounded at $\pm\infty$, respectively. Therefore given z, there exists a nontrivial globally bounded solution to L_z, if and only if $f_z^{\pm}$ are linearly dependent. This is equivalent to the vanishing of their Wronskian, which up to a constant factor equals

$$N(z) = \int_{-\infty}^{\infty} e^{2(zs - m^{-1}s^m)}\, ds.$$

In the strictly pseudoconvex case $m = 2$, N is a Gaussian $c_1 \exp(c_2 z^2)$ and assuredly has no zeroes.

For $m \geq 4$, at least three elementary arguments establish the existence of zeroes of N. First, the asymptotics of N for real z, coupled with simple upper bounds, show that N is an entire holomorphic function of order exactly $m/(m-1) \in (1,2)$, hence N has infinitely many zeroes [CG],[C3]; this property of entire functions of finite order replaces the fundamental theorem of algebra. Alternatively, for $\lambda \in \mathbb{R}$, $N(i\lambda)$ is real-valued, and as $\lambda \to \infty$ satisfies for some exponent $\sigma > 0$

$$N(i\lambda) = c_1 \lambda^{-\sigma} \exp(-c_2 \lambda^{m/(m-1)}) \cdot \cos(c_3 \lambda^{m/(m-1)} + c_4)$$

$$+ O\left(\lambda^{-2\sigma} \exp(-c_2 \lambda^{m/(m-1)})\right).$$

This may be deduced either from a change of the contour of integration and the method of stationary phase, or by observing that N itself satisfies an ordinary differential equation, the asymptotics of whose solutions follow from the theory of equations with irregular singular points at infinity, or equivalently from the WKB approximation. In any event, one obtains the existence of infinitely many imaginary zeroes along with their asymptotic locations.

G. Pólya devoted substantial effort to the study of entire functions all of whose zeroes are imaginary (presumably motivated by the Riemann hypothesis), and one of his results [P] was that N has this property[1]! This has an interpretation in quantum field theory [Si]. All eigenvalues of a self-adjoint linear operator must be real; it might be interesting if the confinement of all nonlinear eigenvalues for the one-parameter family L_z to the imaginary axis were due to some variant of self-adjointness. (It is possible to view the nonlinear eigenvalues for $\{L_z\}$ as ordinary, linear eigenvalues for a system of ordinary differential operators – but that system is not self-adjoint.)

For other, related one-parameter families of ordinary differential equations such as

$$\tilde{L}_z = -d^2/dx^2 \,+\, (z - x^{m-1})^2,$$

or more generally $P(i\,d/dx, (z - x^{m-1}))$ where P is a homogeneous polynomial in two non-commuting variables, no explicit formulas such as those for $f_z^{\pm}$ and for N are available. In fact, some such operator families have no nonlinear eigenvalues at all, while for others, every $z \in \mathbb{C}$ is a nonlinear eigenvalue. The following general result does hold [C4],[C6]:

4.1. **Theorem.** *Let $m \geq 3$ be an integer. If P is a generic, homogeneous polynomial then either the set of all nonlinear eigenvalues of $\{P(i\,d/dx, (z - x^{m-1}))\}$ equals the set of all zeroes of an entire holomorphic function of order precisely $m/(m-1)$, or there exists a nontrivial solution g of at least one of the two equations $P(i\,d/dx, \pm x)g = 0$ such that $g(x) = O(|x|^N)$ as $|x| \to \infty$, for some finite N.*

In the former case, constructing a family of solutions f_r as in §3 leads to a contradiction with the Cauchy estimates, hence implies that $L = P(X, Y)$ is not analytic hypoelliptic. In the latter case, an auxiliary argument establishes the existence of solutions which are not even C^∞. The definition of "generic" may be found in [C4],[C6].

The first step in the proof is the existence of solutions $f_{z,j}^{\pm}$ to $L_z f = P(i\,d/dx, (z - x^{m-1}))f = 0$, whose asymptotic behavior as $x \to \pm\infty$ may be computed precisely, and which depend holomorphically on $z \in \mathbb{C}$. The Wronskian of an appropriate subcollection of these plays the role of N and is $O(\exp(C|z|^{m/(m-1)}))$ as $|z| \to \infty$. The operator $P(i\,d/dx, -x)$ may be realized as the limit, in an appropriate sense, of L_z as $z \to \infty$. If the Wronskian of the $f_{z,j}^{\pm}$ does not grow like $\exp(cz^{m/(m-1)})$ as $\mathbb{R}^+ \ni z \to \infty$,

---

[1] I am indebted to R. Askey, G. Gasper and D. Geller for this information.

then the functions $f_{z,j}^{\pm}$ become closer to being linearly dependent than they ought to be, and in the limit $z \to \infty$ become dependent, resulting in a solution of $P(i\,d/dx, x)g = 0$ with mild growth properties.

5. Non-analyticity in a perturbed case

In this section we outline the proof[2] of Theorem 1.6; details are in [C7]. Fix $X = \partial_x$, $Y = \partial_y - (x^{m-1} + xt^M)\partial_t$. By p we denote always an exponent which depends on m, M and which may be taken to be as large as we please, by choosing m, M to be sufficiently large. The value of p may vary from one occurrence to the next, and m, M are always implicitly assumed to be large.

One additional fact about the model case (1.4) will be required: there exists $z_0 \in \mathbb{C}$ such that $N(z_0) = 0$ but $N'(z_0) \neq 0$. Indeed, $N'(i\lambda)$ satisfies asymptotics similar to those indicated in §4 for $N(i\lambda)$, and these imply that all but finitely many of the imaginary zeroes of N are simple. Fix for the remainder of the discussion such a z_0.

Set $\bar{\partial}_b = X + iY$. The formal adjoint, with respect to Lebesgue measure, is $-X + iY + iMxt^{M-1}$, but in order to simplify the exposition we will define $\bar{\partial}_b{}^* = -X + iY$; the necessary corrections are in [C7]. The aim is to construct a one-parameter family of solutions u_τ to $\bar{\partial}_b\bar{\partial}_b{}^*$, so that $\bar{\partial}_b{}^* u_\tau$ violate the Cauchy estimates. (In fact $\bar{\partial}_b\bar{\partial}_b{}^* u_\tau$ will not be zero, but will be small for large τ.) These will take the form

$$u_\tau = e^{i\tau t} e^{-\tau t^2} e^{\phi(y,t)} f(x,y,t)$$

where ϕ, f are to be determined; both of these depend on the large parameter $\tau \in \mathbb{R}^+$.

As compared to the model case, the factor $\exp(\phi)$ plays the role of $\exp(i\tau^{1/m}\zeta y)$, f that of the solution of the ordinary differential equation. The factor $\exp(-\tau t^2)$ is new, and serves to localize our solutions near $t = 0$. Where $t \neq 0$, the CR structure is strictly pseudoconvex, and hence our construction must fail, which in the model cases is reflected in the absence of zeroes for N when $m = 2$. Without the factor of $\exp(-\tau t^2)$, the procedure outlined below would still produce a good trial solution u_τ, but only in a small neighborhood of $\{t = 0\}$ shrinking to $\{t = 0\}$ as $\tau \to \infty$. In order to violate the Cauchy estimates, solutions, or near-solutions, are required in a fixed open set. Multiplication by cutoff functions would extend u_τ to a fixed domain — but would destroy the equation $\bar{\partial}_b\bar{\partial}_b{}^* u_\tau = 0$. If, however, u_τ is sufficiently small on the

---

[2]Since preparation of this article a more powerful and simpler variant [C10] has supplanted the argument outlined here.

support of the gradient of such a cutoff function, then $\bar{\partial}_b \bar{\partial}_b^* u_\tau$ is still approximately zero, and this suffices.

Set
$$\lambda = \tau^{1/m}$$

and $x_0 = \lambda^{-3/4}$, $t_0 = \lambda^{-m/4}$. r denotes always an element of $[\frac{1}{2}, 1]$.

The fundamental quantity in the construction is a function $\theta(y, t)$, depending also on τ, related to ϕ by

$$\partial_y \phi = i\lambda z_0 + \theta, \qquad \phi(0, t) \equiv 0.$$

It is convenient to ask that θ, ϕ, f be holomorphic functions in certain domains: Set

$$D_r = \{(y, t) \in \mathbb{C}^2 : |t| < r t_0, \; |y| < r\lambda^{m/8}\},$$
$$\omega_r = \{x \in \mathbb{C} : -r\lambda^{-1} < \Re(x) < 3rx_0 \text{ and } |\Im(x)| < r(\lambda + \tau|x|^{m-1})^{-1}\},$$
$$\Omega_r = \omega_r \times D_r.$$

For any function $h(y, t)$ holomorphic in D_r, we write

$$Jh(y, t) = \int_0^y h(u, t)\, du.$$

Set
$$\psi(x, y, t) = \lambda z_0 x - i\theta x - (m^{-1}x^m + 2^{-1}x^2 t^M)(\tau + 2i\tau t - iJ\partial_t \theta).$$

Then
$$\bar{\partial}_b u_\tau = e^{i\tau t - \tau t^2} e^\phi (\mathcal{D} + E) f$$

where

$$\mathcal{D} = e^\psi \circ \partial_x \circ e^{-\psi} \quad \text{and} \quad E = iY = i\partial_y - i(x^{m-1} + xt^M).$$

Similarly $\bar{\partial}_b^*$ is conjugated to $\mathcal{D}_* + E$ where $\mathcal{D}_* = -e^{-\psi}\partial_x e^\psi$.

Set $\mathcal{L} = \mathcal{D}\mathcal{D}_*$. $\mathcal{D}$ is inverted by

$$\mathcal{D}^{-1}g(x, y, t) = e^{\psi(x,y,t)} \int_{x_0}^x e^{-\psi(s,y,t)} g(s, y, t)\, ds.$$

A similar formula defines $\mathcal{D}_*^{-1}$, and we set $\mathcal{L}^{-1} = \mathcal{D}_*^{-1} \circ \mathcal{D}^{-1}$. Then $\mathcal{D} \circ \mathcal{D}^{-1}$ and $\mathcal{L} \circ \mathcal{L}^{-1}$ equal the identity. We have

$$(\mathcal{D} + E)(\mathcal{D}_* + E) = \mathcal{L} + \mathcal{E}, \qquad \mathcal{E} = \mathcal{D}E + E\mathcal{D}_* + E^2.$$

Suppose for the moment that $\theta \in H^\infty(D_r)$ were given. To obtain a solution f to $(\mathcal{L} + \mathcal{E})f = 0$, define the initial approximation

$$f_0(x, y, t) = e^{-\psi(x,y,t)} \int_{x_0}^x e^{2\psi(s,y,t)}\, ds.$$

For $j \geq 1$ set
$$f_j = (-1)^j (\mathcal{L}^{-1}\mathcal{E})^j f_0.$$

Fix large exponents p, q. Given r, set $r' = r - \lambda^{-q}$.

5.1. Lemma. *Assume that $\|\theta\|_{H^\infty(D_r)} = O(1)$. If m, M are sufficiently large, then for all $(x, y, t) \in \Omega_{r'}$, for all $j \geq 1$,*

$$|f_j(x, y, t)| \leq C^j \lambda^{-jp} |\exp(-\tau m^{-1} x^m)|.$$

Thus the series $f^+ = \sum_0^\infty f_j$ converges, and it can be shown to define a solution to $\mathcal{L} + \mathcal{E}$ in $\Omega_{r'}$. However, this solution is defined only, essentially, for $\Re(x) \geq 0$.

In the model case, we saw that growth properties of the solutions of the ordinary differential equations were of the essence, and only a discrete set of ζ led to adequate growth properties. The situation here is the same; for most θ, our solution f^+ must be expected to be very large for $x < 0$ (assuming it exists there at all). Instead we repeat the construction for $\Re(x) < 0$, obtaining a solution f^-, and attempt to choose θ so that $f^\pm$ match across the hypersurface $x = 0$, defining together a single solution having good growth properties on both sides.

$\mathcal{L}$ is a second-order ODE, so both $f^\pm$ and $\partial_x f^\pm$ ought to be matched at $x = 0$; so a second unknown function is needed. To that end, f^- is defined by taking as the initial approximation

$$f_0^- = (1 + \beta(y, t)) e^{-\psi} \int_{-x_0}^x e^{2\psi(s, y, t)} ds$$

where $\beta \in H^\infty(D_r)$. If $\|\beta\|_{H^\infty} = O(1)$, then the analogue of Lemma 5.1 holds for $(-x, y, t) \in \Omega_{r'}$.

The matching equations $f^+(x = 0) = f^-(x = 0)$, $\partial_x f^+(x = 0) = \partial_x f^-(x = 0)$ amount to a system of nonlinear integro-differential equations of infinite order and are solved by Newton's method. Define

$$\mathcal{W}(\theta, \beta, y, t) = \det \begin{pmatrix} f^+ & f^- \\ \partial_x f^+ & \partial_x f^- \end{pmatrix} (0, y, t)$$

$$\mathcal{M}(\theta, \beta, y, t) = (f^+ - f^-)(0, y, t)$$

and

$$\mathcal{G}(\theta, \beta)(y, t) = \begin{pmatrix} \mathcal{W} \\ \mathcal{M} \end{pmatrix}.$$

We aim to solve the system $\mathcal{G}(\theta, \beta) = 0$ for the two unknowns $\theta, \beta \in H^\infty$. Assume always that $\|\theta\| < 1$, $\|\beta\| < 1$.

5.2. Lemma. *There exists a nonzero constant c_0 such that*

$$\|\partial_\theta \mathcal{W}(\theta, \beta, \Delta\theta) - c_0(1 + \beta)\lambda^{-2}\Delta\theta\|_{H^\infty(D_{r'})}$$

$$\leq C\lambda^{-p}\|\Delta\theta\|_{H^\infty(D_r)} + C\lambda^{-3}\|\theta\| \cdot \|\Delta\theta\|$$

for all sufficiently large λ, where $r' = r - \lambda^{-q}$, and $p, q \to \infty$ as $m, M \to \infty$.

In fact, c_0 is a normalizing factor times $N'(z_0)$, so that a simple zero of N for the model case is needed for Newton's method to be applicable. The other first-order derivatives may also be estimated:

5.3. Lemma. *If* $\|\beta\|_{H^\infty(D_r)} + \|\theta\|_{H^\infty(D_r)} \lesssim \lambda^{-p_0}$ *then in* $D_{r'}$,

$$\frac{\partial \mathcal{G}}{\partial(\theta,\beta)}(\theta,\beta,\Delta\theta,\Delta\beta) = \mathcal{A}\begin{pmatrix} \Delta\theta \\ \Delta\beta \end{pmatrix} + \mathcal{R}(\theta,\beta,\Delta\theta,\Delta\beta),$$

where

$$\mathcal{A} = \begin{pmatrix} c_0\lambda^{-2} & 0 \\ c_2\lambda^{-2} & c_1\lambda^{-1} \end{pmatrix}$$

with $c_0, c_1 \neq 0$, *and*

$$\|\mathcal{R}(\theta,\beta,\Delta\theta,\Delta\beta)\|_{(H^\infty\oplus H^\infty)(D_{r'})} \lesssim \lambda^{-p}\|\Delta\theta\| + \lambda^{-p}\|\Delta\beta\|$$

where $p \to \infty$ *as* $\min(m, M, p_0) \to \infty$.

Upper bounds may be given for the second derivatives of $\mathcal{G}$ with respect to θ, β, in the same spirit.

This is the crux of the construction. We are unable to invert the differential of $\mathcal{G}$, only its leading-order term. Even when m, M are small, $\mathcal{R}$ will be small relative to $\mathcal{A}$, provided that $r - r'$ is not too small. But Newton's method will converge relatively slowly since $\partial\mathcal{G}$ will not be inverted exactly, necessitating a large number of iterations in order to make $\mathcal{G}$ sufficiently close to zero. The assumption that m, M are large permits sufficiently many iterations, for large τ, that $\mathcal{G}$ may be made sufficiently close to zero for the Cauchy estimates ultimately to be violated.

The upshot of Newton's method is as follows.

5.4. Proposition. *There exist* θ, β *holomorphic in*

$$D = \{|y| \leq \frac{1}{2}\lambda^{m/8}, \ |t| \leq \frac{1}{2}\lambda^{-m/4}\}$$

and satisfying $\|\theta\|_{H^\infty} + \|\beta\|_{H^\infty} \leq \lambda^{-p}$, *such that*

$$|\mathcal{W}(\theta,\beta)(y,t)| + |\mathcal{M}(\theta,\beta)(y,t)| \leq Ce^{-\lambda^p}$$

for all $(y,t) \in D$, *where* $p \to \infty$ *as* $\min(m, M) \to \infty$.

The construction may be thought of as an analogue of the method of geometric optics. θ satisfies a nonlinear "eiconal" equation, while the f_j are determined successively by solving linear "transport" equations, which are determined by θ. However, here it is necessary to solve the infinitely many transport equations, before the eiconal equation is even known; we have an infinite coupled system.

Fix such θ, β, and let $f^\pm$ be the functions constructed in terms of θ, β by the procedure already outlined. Fix a sequence of cutoff functions

$\eta_k \in C^\infty(\mathbb{R})$ satisfying $0 \le \eta_k \le 1$, $\eta_k(x) \equiv 0$ for $x \le -1$ and $\eta_k(x) \equiv 1$ for $x \ge 1$, such that

$$\|\frac{d^j}{dx^j}\eta_k\|_{C^0} \le C^{j+1}k^j \quad \text{for all } 0 \le j \le k.$$

Such functions exist [H4]. Set

$$\eta_\lambda^+(x) = \eta_K(c^{-1}\lambda x), \qquad \eta_\lambda^-(x) = 1 - \eta_\lambda^+(x)$$

where $\lambda \le K < 1 + \lambda$ and c is a small constant. Define

$$g(x,y,t) = \eta_\lambda^+(x)f^+(x,y,t) + \eta_\lambda^-(x)f^-(x,y,t).$$

5.5. Lemma. *For $|y| < 1/4$ and $|t| < \frac{1}{4}\lambda^{-m/4}$, for all $x \in \mathbb{R}$,*

$$(\mathcal{L}+\mathcal{E})g \equiv 0 \qquad \text{for } |x| \ge \lambda^{-1},$$
$$(\mathcal{L}+\mathcal{E})g(x,y,t) = O(\exp(-\lambda^p)) \qquad \text{for } |x| \le \lambda^{-1},$$
$$|\partial_x^\alpha \partial_y^\beta \partial_t^\gamma g(x,y,t)| \lesssim C^{\alpha+\beta+\gamma}\lambda^{-1}(\alpha+\beta+\gamma)!$$
$$\times (\lambda+\tau|x|^{m-1})^\alpha \lambda^{m\gamma/4}e^{-m^{-1}\tau x^m}$$

for $|x| < \lambda^{-3/4}$, and

$$|\partial_x^\alpha \partial_y^\beta \partial_t^\gamma (\mathcal{L}+\mathcal{E})g(x,y,t)| \lesssim C^{\alpha+\beta+\gamma}(\alpha+\beta+\gamma)!\,\lambda^{m\gamma/4}\exp(-\lambda^p)$$

for all $0 \le \alpha,\beta,\gamma$ with $\alpha < \lambda+1$.

Define

$$\tilde{u} = \tilde{u}_\tau(x,y,t) = e^{i\tau t - \tau t^2}e^{\phi(y,t)}g(x,y,t)$$

and $u(x,y,t) = \tilde{u}(x,y,t) \cdot \eta_K(t_0^{-1}t) \cdot \eta_K(\lambda^{3/4}x)$, where the new cutoff functions $\eta_K \in C_0^\infty(\mathbb{R})$ are supported in $(-1,1)$, are identically equal to one on $[-\frac{1}{2},\frac{1}{2}]$ and satisfy

$$\|\frac{d^j}{dx^j}\eta_K\|_{C^0} \le C^j\lambda^j \qquad \text{for all } 0 \le j \le K,$$

where $\lambda \le K < \lambda+1$.

5.6. Lemma. *There exist $c > 0$ and $A < \infty$ such that for all sufficiently large τ, for all $k \ge 1$,*

$$|\partial_t^k \bar{\partial}_b{}^* u_\tau(0)| \ge c\lambda^{mk}.$$

Also

$$\|\partial_{x,y,t}^\alpha \bar{\partial}_b \bar{\partial}_b{}^* u_\tau\|_{L^2(\Omega)} \le C\exp(-\lambda^p)$$

provided that $|\alpha| < \lambda+1$.

Let Γ_0 denote a sufficiently small conic neighborhood of that half of the characteristic variety of $\bar{\partial}_b\bar{\partial}_b{}^*$ in $T^*(\mathbb{R}^3)$ on which $\bar{\partial}_b$ is not hypoelliptic. Denote by $H^s(\Gamma_0)$ the usual L^2 Sobolev space of order s, but microlocalized to Γ_0.

5.7. Lemma. *There exists $c > 0$ such that for all sufficiently large τ,*

$$\|\bar{\partial}_b{}^* u_\tau\|_{H^\lambda(\Gamma_0)} \geq c^\lambda \lambda^{m\lambda}.$$

The last sticking point is that $\bar{\partial}_b \bar{\partial}_b{}^* u_\tau$ is not exactly zero, nor is it even real analytic, because of the imperfect matching of $f^\pm$. Consequently there is no hope of using $\{u_\tau\}$ directly to contradict the Cauchy estimates.

The same difficulty was encountered and surmounted by Métivier in earlier work [M1]. Building on model examples such as $\partial_x^2 + x^2 \partial_y^2 + \partial_t^2$, he obtained the following negative results for sums of squares of vector fields. Let there be given n analytic real vector fields X_j in $\mathbb{R}^{n+1}$, which are linearly independent at every point. Firstly, if n is odd, then $L = \sum X_j^2$ is never analytic hypoelliptic. Secondly, if n is even, and if the "Levi form" associated to $\{X_j\}$ is degenerate at *every* point x in some open set U, then L is not analytic hypoelliptic[3].

Métivier proved the abstract result that if L is analytic hypoelliptic, and if in addition L is locally solvable in L^2, then a certain sequence of inequalities hold, which are phrased in terms of the usual Sobolev norms. These are suitable for testing with trial functions u_τ which are C^∞, but not necessarily analytic. The natural analogues of Métivier's Cauchy/Sobolev estimates, for $\bar{\partial}_b$, are as follows. For positive integers k, consider the following norms on the usual L^2 Sobolev spaces:

$$\|g\|_{k,\Omega}^2 = \sum_{0 \leq |\alpha| \leq k} k^{2(k-|\alpha|)} \|\partial^\alpha g\|_{L^2(\Omega)}^2.$$

Denote by σ the Fourier variable dual to t.

5.8. Lemma. *Suppose that $\bar{\partial}_b$ were analytic hypoelliptic modulo its nullspace, in some neighborhood of the origin. If Ω, Γ are a sufficiently small neighborhood of 0 and a sufficiently narrow conic neighborhood of $\{\sigma > 0,\ \xi = \eta = 0\}$, then there would exist $B < \infty$ such that for any $u \in C^\infty(\Omega)$ and any $k \in \mathbb{Z}^+$,*

$$\|\bar{\partial}_b{}^* u\|_{H^k(\Gamma)} \leq B^k \|\bar{\partial}_b \bar{\partial}_b{}^* u\|_{k,\Omega} + B^k k^k \|u\|_{L^2(\Omega)}.$$

The proof of Lemma 5.8 follows Métivier's argument (which belongs to the subject of real interpolation) in outline, relying on two properties of $\bar{\partial}_b$: microlocally in Γ it is solvable in an appropriate sense, even though it is not truly locally solvable, and it is analytic hypoelliptic on the complement of Γ. We omit the precise definition of the norm for $H^k(\Gamma)$, which involves the cutoff functions of Andersson and Hörmander [H4].

Finally, Lemmas 5.6 and 5.7 imply that u_τ fails to satisfy the bounds of Lemma 5.8 for large τ, once B is fixed, and the proof is complete.

---

[3]This is only of significance in higher dimensions, since when $n = 2$, the hypothesis and the Frobenius theorem imply that the field of 2-planes span (X_1, X_2) is integrable, whence for trivial reasons, there exist solutions of $Lu = 0$ which are not even continuous.

REFERENCES

[BG]   M. S. Baouendi and C. Goulaouic, *Nonanalytic-hypoellipticity for some degenerate elliptic operators*, Bulletin AMS **78** (1972), 483-486.

[B]   J. M. Bony, *Equivalence des diverses notions de spectre singulier analytique*, Exposé III, Séminaire Goulaouic-Schwartz (1976-7).

[C1]   M. Christ, *Some non-analytic-hypoelliptic sums of squares of vector fields*, Bulletin AMS **16** (1992), 137-140.

[C2]   _____, *Certain sums of squares of vector fields fail to be analytic hypoelliptic*, Comm. Partial Differential Equations **16** (1991), 1695-1707.

[C3]   _____, *Analytic hypoellipticity breaks down for weakly pseudoconvex Reinhardt domains*, International Math. Research Notices **1** (1991), 31-40.

[C4]   _____, *Nonexistence of invariant analytic hypoelliptic differential operators on nilpotent groups of step greater than two*, Proceedings of symposium in honor of E. M. Stein, Princeton University Press (in press).

[C5]   _____, *A class of hypoelliptic PDE admitting nonanalytic solutions*, The Madison Symposium on Complex Analysis, Contemporary Mathematics **137** (1992), 155-168.

[C6]   _____, *Analytic hypoellipticity, representations of nilpotent groups, and a nonlinear eigenvalue problem*, Duke Math. J. **72** (1993), 595-639.

[C7]   _____, *Examples of analytic nonhypoellipticity of $\bar{\partial}_b$*, Comm. Partial Differential Equations **19** (1994), 911-941.

[C8]   _____, *Embedding compact three-dimensional CR manifolds of finite type in $\mathbb{C}^n$*, Annals of Math. **129** (1989), 195-213.

[C9]   _____, *On the $\bar{\partial}$ equation in weighted L^2 norms in $\mathbb{C}^1$*, J. Geometric Analysis **1** (1991), 193-230.

[C10]   _____, *A necessary condition for analytic hypoellipticity*, Math. Research Letters **1** (1994), 241-248.

[C11]   _____, *The Szegö projection need not preserve global analyticity*, submitted.

[C12]   _____, *Global analytic regularity in the presence of symmetry*, Math. Research Letters **1** (1994), 559-564.

[CG]   M. Christ and D. Geller, *Counterexamples to analytic hypoellipticity for domains of finite type*, Annals of Math. **235** (1992), 551-566.

[CL]   E. Coddington and N. Levinson, *Theory of Ordinary Differential Equations*, McGraw-Hill, New York, 1955.

[DT1]   M. Derridj and D. S. Tartakoff, *Local analyticity for $\Box_b$ and the $\bar{\partial}$-Neumann problem at certain weakly pseudo-convex points*, Comm. Partial Differential Equations **13** (1988), 1521-1600.

[DT2]   _____, *Local analyticity in $\Box_b$ and the $\bar{\partial}$-Neumann problem – some model domains without maximal estimates*, Duke Math. J. **64** (1991), 377-402.

[DT3]   _____, *Local analyticity in the $\bar{\partial}$-Neumann problem for a class of totally decoupled weakly pseudoconvex domains*, J. Geometric Analysis (to appear).

[DT4]   _____, *Global analyticity for $\Box_b$ on three dimensional pseudoconvex CR manifolds*, preprint.

[DZ]   M. Derridj and C. Zuily, *Régularité analytique et Gevrey pour des classes d'opérateurs elliptiques paraboliques dégénérés du second ordre*, Astérisque **2,3** (1973), 371-381.

[FS]   G. B. Folland and E. M. Stein, *Estimates for the $\bar{\partial}_b$ complex and analysis on the Heisenberg group*, Comm. Pure Appl. Math. **27** (1974), 429-522.

[FrS]   A. Friedman and M. Shinbrot, *Nonlinear eigenvalue problems*, Acta Math. **121** (1968), 77-128.

[G]   D. Geller, *Analytic Pseudodifferential Operators For The Heisenberg Group And Local Solvability*, Mathematical Notes 37, Princeton University Press, Princeton, NJ, 1990.

[GS]   A. Grigis and J. Sjöstrand, *Front d'onde analytique et sommes de carrés de champs de vecteurs*, Duke Math. J. **52** (1985), 35-51.

[HH]   N. Hanges and A. A. Himonas, *Singular solutions for sums of squares of vector fields*, Comm. Partial Differential Equations **16** (1991), 1503-1511.

[He]   B. Helffer, *Conditions nécessaires d'hypoanalyticité pour des opérateurs invariants à gauche homogènes sur un groupe nilpotent gradué*, J. Diff. Eq. **44** (1982), 460-481.

[HN1]   B. Helffer and J. Nourrigat, *Hypoellipticité pour des groupes nilpotents de rang 3*, Comm. Partial Differential Equations **3** (1978), 643-743.

[HN2]   _____ , *Caractérisation des opérateurs hypoelliptiques homogènes invariants à gauche sur un groupe nilpotent gradué*, Comm. Partial Differential Equations **4** (1979), 899-958.

[HN3]   _____ , *Hypoellipticité Maximal Pour des Opérateurs Polynômes de Champs de Vecteurs*, Prog. Math. vol. 58, Birkhäuser, Boston, 1985.

[H]   P. Heller, *Analyticity and regularity for nonhomogeneous operators on the Heisenberg group*, PhD dissertation, Princeton University, 1986.

[H1]   L. Hörmander, *The Analysis of Linear Partial Differential Operators I*, Springer Verlag, Berlin, 1983.

[H2]   _____ , *The Analysis of Linear Partial Differential Operators II*, Springer-Verlag, Berlin, 1983.

[H3]   _____ , *The Analysis of Linear Partial Differential Operators IV*, Springer-Verlag, Berlin, 1985.

[H4]   _____ , *Uniqueness theorems and wave front sets for solutions of linear differential equations with analytic coefficients*, Comm. Pure Appl. Math. **24** (1971), 671-704.

[Ke]   M. V. Keldysh, *On the completeness of the eigenfunctions of classes of non-selfadjoint linear operators*, Russian Math. Surveys **26** (1971), 15-44.

[Ki]   A. A. Kirillov, *Unitary representations of nilpotent Lie groups*, Russian Math. Surveys **17** (1962), 53-104.

[K]   J. J. Kohn, *Estimates for $\overline{\partial}_b$ on pseudoconvex CR manifolds*, Proc. Symp. Pure Math. **43** (1985), 207-217.

[M1]   G. Métivier, *Une classe d'opérateurs non hypoélliptiques analytiques*, Indiana Math. J. **29** (1980), 823-860.

[M2]   _____ , *Analytic hypoellipticity for operators with multiple characteristics*, Comm. Partial Differential Equations **6** (1981), 1-90.

[M3]   _____ , *Hypoellipticité analytique sur des groupes nilpotents de rang 2*, Duke Math. J. **47** (1980), 195-221.

[Mo]   R. Montgomery, *Geodesics which do not satisfy the geodesic equations*, preprint.

[NSW]   A. Nagel, E. M. Stein and S. Wainger, *Balls and metrics defined by vector fields, I: Basic properties*, Acta Math. **155** (1985), 103-147.

[N1]   A. V. Noell, *Properties of peak sets in weakly pseudoconvex boundaries in $\mathbb{C}^2$*, Math. Zeitschrift **186** (1984), 99-116.

[N2]   _____ , *Local versus global convexity of pseudoconvex domains*, preprint.

[Pe]   I. G. Petrowsky, *Sur l'analyticité des solutions des systèmes d'équations différentielles*, Mat Sb. **4** (1939), 3-70.

[PR]   Pham The Lai and D. Robert, *Sur un problème aux valeurs propres non linéaire*, Israel J. Math. **36** (1980), 169-186.

[P]   G. Pólya, *Über trigonometrische Integrale mit nur reelen Nullstellen*, J. Reine
        Angew. Math. **58** (1927), 6-18.

[RS]   L. P. Rothschild and E. M. Stein, *Hypoelliptic differential operators and nilpo-
        tent groups*, Acta Math. **137** (1976), 247-320.

[Sa]   A. Sánchez-Calle, *Fundamental solutions and geometry of the sum of squares
        of vector fields*, Invent. Math. **78** (1984), 143-160.

[Si]   B. Simon, *The $P(\phi)_2$ Euclidean (Quantum) Field Theory*, Princeton Univer-
        sity Press, Princeton, NJ, 1974.

[S]   J. Sjöstrand, *Analytic wavefront sets and operators with multiple characteris-
        tics*, Hokkaido Math. J. **12** (1983), 392-433.

[Sm]   H. Smith, *A calculus for three-dimensional CR manifolds of finite type*, J.
        Functional Analysis (to appear).

[St]   E. M. Stein, *An example on the Heisenberg group related to the Lewy operator*,
        Invent. Math. **69** (1982), 209-216.

[Ta1]   D. Tartakoff, *Local analytic hypoellipticity for $\Box_b$ on non-degenerate Cauchy-
        Riemann manifolds*, Proc. Nat. Acad. Sci. USA **75** (1978), 3027-3028.

[Ta2]   _____, *On the local real analyticity of solutions to $\Box_b$ and the $\bar{\partial}$-Neumann
        problem*, Acta Math. **145** (1980), 117-204.

[Trp]   J-M. Trepreau, *Sur l'hypoellipticité analytique microlocale des opérateurs de
        type principal*, Comm. PDE **9** (1984), 1119-1146.

[Tr1]   F. Treves, *Analytic-hypoelliptic partial differential equations of principal type*,
        Comm. Pure Appl Math. **24** (1971), 537-570.

[Tr2]   _____, *Analytic hypo-ellipticity of a class of pseudodifferential operators with
        double characteristics and applications to the $\bar{\partial}$-Neumann problem*, Comm.
        Partial Differential Equations **3** (1978), 475-642.

Department of Mathematics, UCLA, Los Angeles, Ca. 90024

FINITE TYPE CONDITIONS AND SUBELLIPTIC ESTIMATES

JOHN P. D'ANGELO

Introduction

The present paper has two purposes. The primary purpose is to survey some work on finite-type conditions that arose because of Kohn's work on subelliptic estimates. We also introduce a class of weakly pseudoconvex domains for which it is possible to understand Kohn's ideals of subelliptic multipliers in a rather concrete and precise manner. Although these domains are rather special, for them it is possible to see clearly the relationship between analysis and commutative algebra anticipated by Kohn's work.

We discuss the geometry of the boundary of a domain in complex Euclidean space and its relationship to the function theory on the domain. There are many relevant geometric conditions; here we emphasize those finite-type conditions that arose (directly or indirectly) as consequences of the work of Kohn. These notions apply when one considers subelliptic estimates for the $\bar{\partial}$-Neumann problem; the author believes that they will be useful in many other analytic problems. We consider several different generalizations of strong pseudoconvexity for a point on a real hypersurface. The language of commutative algebra unifies these ideas.

The only new result in this paper is Proposition 5, where we apply Kohn's algorithm for finding subelliptic multipliers to an interesting class of domains. We call these domains 'regular coordinate domains' because of an analogy with the use of regular systems of coordinates for ideals of germs of holomorphic functions. For these domains we compute the Kohn ideals of subelliptic multipliers, even though the Levi form is not generally diagonalizable. We obtain a value for the parameter epsilon in such estimates in terms of the given exponents.

This paper is not intended to be complete. Our discussion of subelliptic estimates is restricted to $(0, 1)$-forms. We do not consider important recent results about estimates in function spaces other than L^2 and Sobolev spaces. Most important there are many theorems known for strongly pseudoconvex domains that generalize to appropriate classes of weakly pseudoconvex domains. The final forms of such theorems will involve different finite-type conditions. Thus there is no single notion of finite-type applicable to all problems. The author acknowledges Dave Catlin for helpful criticisms of a preliminary version of this article.

1. The Levi form

We begin by considering a domain $\Omega \subset \mathbf{C}^n$ whose boundary is a smooth real hypersurface M. The complexified tangent bundle $\mathbf{C}TM$ contains an integrable subbundle $T^{1,0}M$ whose local sections are complex vector fields of type $(1,0)$. We denote by η a purely imaginary non-vanishing 1-form that annihilates both $T^{1,0}M$ and its conjugate bundle. In terms of a local defining function r for M, we may put $\eta = \frac{1}{2}\left(\partial - \overline{\partial}\right)r$. The Levi form λ is the Hermitian form on $T^{1,0}M$ defined (up to a multiple) by

$$\lambda\left(L, \overline{K}\right) = \left\langle \eta, [L, \overline{K}]\right\rangle. \tag{1}$$

The hypersurface (or the domain on one side of it) is called strongly pseudoconvex when λ is positive semi-definite, and is called weakly pseudoconvex when λ is semi-definite but not definite. We note immediately that strong pseudoconvexity is a non-degeneracy condition: if λ is positive definite at a point $p \in M$, then it is positive definite in a neighborhood. Furthermore strong pseudoconvexity is 'finitely determined': if M' is another hypersurface containing p, and M' osculates M to second order there, then M' is also strongly pseudoconvex at p. We seek generalizations of strong pseudoconvexity that have applications in analytic problems; we also want these two properties to remain valid.

For later use we compute the Levi form for domains in $\mathbf{C}^{n+1}$ defined locally by the equation

$$r\left(z, \overline{z}\right) = 2\operatorname{Re}\left(z_0\right) + \sum_{k=1}^{N}\left|f^k\left(z\right)\right|^2 < 0. \tag{2}$$

Here the functions f_k are holomorphic near the origin, vanish there, and depend only on the variables $z_1, z_2, ..., z_n$. The domain defined by (2) is pseudoconvex. Its Levi form near the origin has a nice expression:

$$\left(\lambda_{i\overline{j}}\right) = \left(\sum_{k=1}^{N} f^k_{z_i}\overline{f^k_{z_j}}\right) = (\partial f)^*\,(\partial f). \tag{3}$$

It follows immediately from (3) that the origin will be a weakly pseudoconvex point if and only if the rank of (∂f) (as a mapping on $\mathbf{C}^n$) is less than full there. It is a point of finite-type in the sense of this paper if and only if the germs of the functions f_k define a trivial variety. See the last paragraph of Section 2. This simple example allows us to glimpse the role of commutative algebra in later discussions.

2. Finite type conditions and subelliptic estimates

The study of the inhomogeneous Cauchy-Riemann equations motivates much of this work. Suppose that $\Omega \subset\subset \mathbf{C}^n$ is smoothly bounded and pseudoconvex. Let α be a differential $(0\text{-}1)$-form α with square integrable coefficients that satisfies $\bar{\partial}\alpha = 0$. We wish to solve the inhomogeneous Cauchy-Riemann equations $\bar{\partial}u = \alpha$ and discuss the regularity of the solution. See [K1-K5]. Kohn's solution of the $\bar{\partial}$-Neumann problem constructs the particular solution $u = \bar{\partial}^* N\alpha$ that is orthogonal to the holomorphic functions. Suppose that the $\bar{\partial}$-Neumann operator is pseudo-local; that is, $N\alpha$ must be smooth wherever α is smooth. Then $u = \bar{\partial}^* N\alpha$ is also smooth wherever α is. The pseudolocality property of N follows from certain a priori estimates called subelliptic estimates. The finite-type condition emphasized here arises from Kohn's program of seeking necessary and sufficient conditions for subelliptic estimates.

Definition 1. Suppose that $\Omega \subset\subset \mathbf{C}^n$ is smoothly bounded and pseudoconvex. Let $p \in \overline{\Omega}$ be any point in the closure of the domain. The $\bar{\partial}$-Neumann problem satisfies a subelliptic estimate at p on $(0,1)$-forms if there is a neighborhood $U \ni p$ and positive constants C, ϵ such that

$$|||\phi|||_\epsilon^2 \leq C \left(\left\| \bar{\partial}\phi \right\|^2 + \left\| \bar{\partial}^*\phi \right\|^2 + \|\phi\|^2 \right) \tag{4}$$

for every $(0,1)$-form ϕ that is smooth, compactly supported in U, and in the domain of the operators on the right hand side.

In (4) $|||\phi|||_\epsilon$ denotes the tangential Sobolev norm of order ϵ. Also, the statement that ϕ be in the domain of $\bar{\partial}^*$ is a boundary condition on its components. We abbreviate the right side by $Q(\phi, \phi)$.

A basic theorem of Kohn-Nirenberg [KN] implies that pseudolocality for N on $(0,1)$-forms follows from the estimate (4). In fact, this estimate has, in terms of local Sobolev norms, the following consequences:

$$\alpha \in H_s \Rightarrow N\alpha \in H_{s+2\epsilon};$$

$$\alpha \in H_s \Rightarrow \bar{\partial}^* N\alpha \in H_{s+\epsilon}. \tag{5}$$

Therefore it is significant to determine when there is a subelliptic estimate at a given point $p \in \overline{\Omega}$. Observe that the set of points for which an estimate of the form (4) holds must be an open subset of the closed domain.

For interior points, the estimate (4) is elliptic, and holds with $\epsilon = 1$. At strongly pseudoconvex boundary points, the estimate holds for $\epsilon = \frac{1}{2}$. Catlin has found necessary and sufficient conditions for a subelliptic estimate of some order to hold. See Theorem 4. The precise value of the parameter ϵ is not known in general.

To gain some feeling for the estimates, we recall the basic formula for $Q(\phi, \phi)$ on $(0,1)$-forms. We assume that r is a defining function for $b\Omega$.

Lemma 1. *The quadratic form Q satisfies*

$$Q(\phi, \phi) = \sum_{i,j=1}^{n} \int_{\Omega} \left| (\phi_i)_{\overline{z}_j} \right|^2 dV + \sum_{i,j=1}^{n} \int_{b\Omega} r_{z_i \overline{z}_j} \phi_i \overline{\phi_j} dS + \int_{\Omega} \sum_{i=1}^{n} |\phi_i|^2 dV$$

$$= \|\phi\|_{\overline{z}}^2 + \int_{b\Omega} \lambda(\phi, \phi) dS + \|\phi\|^2. \tag{6}$$

This formula reveals an asymmetry between the barred and unbarred derivatives. Observe also that the integral of the Levi form appears. This term is non-negative when Ω is pseudoconvex. Since an estimate holds at strongly pseudoconvex points, and no estimate holds at points where the Levi form vanishes identically, we see that the existence of an estimate is a non-degeneracy condition of the type we seek.

We consider the case of $(0, 1)$-forms in this paper and recall some of the history. The first result was in two dimensions. Let us say that a $(1,0)$ vector field L on a real hypersurface M is of finite-type at p if there is an integer k such that $\langle \eta, [... [L_1, L_2], ..., L_k] \rangle_p \neq 0$. In this definition each L_j equals either L or $\overline{L}$. The type of L at p is the smallest such integer. It is evident that the type of every non-vanishing $(1,0)$ vector field at p equals two precisely when M is strongly pseudoconvex there. For hypersurfaces in $\mathbf{C}^2$, $T_p^{1,0} M$ is one dimensional; it follows easily that the type of each non-zero $(1,0)$ vector field L at p is the same. Kohn proved in [K3], a paper dedicated to Spencer on the occasion of his 60th birthday, the sufficiency condition in the following result. Greiner [Gr] established the converse. The precise relationship between the type of the vector field and the value of epsilon is due to Rothschild-Stein. [RS].

Theorem 1. *Suppose that $\Omega \subset\subset \mathbf{C}^2$ is pseudoconvex, and that $p \in M = b\Omega$. Let L be a non-vanishing $(1,0)$ vector field at p. If L is of finite-type m at p, then there is a subelliptic estimate. The estimate holds for $\epsilon = \frac{1}{m}$, but for no larger value. Conversely, if there is a subelliptic estimate for some ϵ, then every such vector field is of finite-type.*

Kohn then conjectured that a necessary and sufficient condition for subellipticity on pseudoconvex domains in higher dimensions would be that every $(1,0)$ vector field is of finite-type. This turns out to be false, because this latter condition is not an open condition. The following simple example reveals this:

Example. The polynomial $r(z, \overline{z}) = 2\mathrm{Re}(z_3) + \left| z_1^2 - z_2^3 \right|^2$ defines a pseudoconvex hypersurface containing the origin, and each non-vanishing $(1,0)$ vector field is of type 4 or 6 there. Let $L_j = \frac{\partial}{\partial z_j} - r_j \frac{\partial}{\partial z_3}$ for $j = 1, 2$ be the usual commuting basis for $(1,0)$ vector fields. Let V be the complex analytic 1-dimensional variety given by $z_1^2 = z_2^3$ and $z_3 = 0$. The vector field $L = \frac{3z_2^2}{2z_1} L_1 + L_2$ is then smooth along V, but it cannot be smoothly extended to a neighborhood of the origin. This difficulty occurs because V is not a normal variety. We cannot detect the existence of this complex-analytic curve by studying the types of smooth vector fields.

From this example the author realized that the finite-type condition should consider the contact with all holomorphic curves, including those with singularities. This led to the following definition of point of finite-type. [D2] We discuss also a recent refinement of this idea.

Let M be the germ at p of a real hypersurface in $\mathbf{C}^n$, and let r be a defining function. (The resulting notion will not depend on the choice of defining function). We denote by $v_p(g) = v(g)$ the order of vanishing of the function $g - g(p)$ at the point p. Consider the collection of non-constant germs of holomorphic mappings $z : (\mathbf{C}, 0) \to (\mathbf{C}^n, p)$. We say that M is of finite-type at p if there is a constant T so that $v(z^* r) \leq T v(z)$ for every such holomorphic mapping. The infimum of all such constants is called the type of M at p, and is written $\Delta^1(M, p)$. Observe that $\Delta^1(M, p) = \sup_z \left(\frac{v(z^* r)}{v(z)} \right)$. This number is therefore a measurement of the maximum order of contact of complex analytic one-dimensional curves with M at p. More generally, if J is an ideal in some local ring at p, we define $\mathbf{T}(J) = \sup_z \inf_g \left(\frac{v(z^* g)}{v(z)} \right)$. Here the infimum is taken over the local ring. Thus $\mathbf{T}(J) = \Delta^1(M, p)$ when J is the ideal in the ring of germs of smooth functions at p of functions vanishing on M. One of the main points in [D2] is to relate this number to $\mathbf{T}(I)$ where I is an ideal in the ring O of germs of holomorphic functions.

We now give a refined equivalent definition of point of finite-type. This uses the idea of scheme, but no results from the theory of schemes. The idea is the following.

For each integer k, we consider the hypersurface M_k defined by the k-th order Taylor polynomial $j_{k,p} r$. Thus M_k osculates M at p to order k. We consider $j_{k,p} r$ as an Hermitian form on the finite-dimensional vector space of holomorphic polynomials of degree at most k. After diagonalizing this form, we can write $j_{k,p} r = ||F^p||^2 - ||G^p||^2$, where F^p, G^p are holomorphic vector-valued polynomials. This construction leads to a family of ideals of holomorphic polynomials, generated by the components of $F^p - UG^p$, where U is unitary. We call this family $I(U, k, p)$. The point of this construction is that the complex-analytic varieties defined by $\mathbf{V}(I(U, k, p))$ lie in M_k, and conversely, any irreducible complex-analytic variety lying in M_k must be a subvariety of one of these. Thus the order of contact of complex-analytic varieties with M can be completely analyzed by letting the degree of osculation tend to infinity. For each k we have the estimate

$$\sup_U \mathbf{T}(I(U, k, p)) + 1 \leq \Delta^1(M_k, p) \leq 2\sup_U \mathbf{T}(I(U, k, p)). \quad (7)$$

For pseudoconvex M_k we have an equality

$$\Delta^1 \left(M_k, p \right) = 2\mathrm{sup}_U \mathbf{T} \left(I \left(U, k, p \right) \right). \tag{8}$$

The construction assigns a family of ideals of germs of holomorphic functions to each point on a smooth hypersurface. The ideals depend on the degree of osculation and on a unitary matrix. The varieties of the ideals have high order of contact with the hypersurface. As in the theory of schemes, the ideals themselves are more important than the sets they define. Even when all the varieties are trivial, essentially the finite-type case, we obtain numerical information on the geometry of the hypersurface by computing numerical invariants of the ideals. In particular the codimension $\mathbf{D} \left(I \right) = \mathrm{dim}_{\mathbf{C}} \left(O/I \right)$ is particularly useful. The author used its upper semicontinuity properties and its relationship with $\mathbf{T} \left(I \right)$ to establish the following theorem.

Theorem 2. *Let M be a smooth real hypersurface in $\mathbf{C}^n$. Suppose that $p_0 \in M$ and that $\Delta^1 \left(M, p_0 \right) < \infty$. Then there is a neighborhood of p_0 on which*

$$\Delta^1 \left(M, p \right) \leq 2 \left(\Delta^1 \left(M, p_0 \right) \right)^{n-1}. \tag{9}$$

If M is pseudoconvex, then we can improve (9) to the following sharp bound:

$$\Delta^1 \left(M, p \right) \leq 2 \left(\frac{\Delta^1 \left(M, p_0 \right)}{2} \right)^{n-1}. \tag{10}$$

Furthermore, there is an integer k_0 that 'finitely determines' the type. That is, whenever $k \geq k_0$, $\Delta^1 \left(M, p_0 \right) = \Delta^1 \left(M_k, p_0 \right)$; also $\mathbf{V} \left(I \left(U, k, p_0 \right) \right) = \{p_0\}$ for every unitary U, and in fact $2\mathrm{sup}_U \mathbf{T} \left(I \left(U, k, p_0 \right) \right) \leq k$.

This theorem establishes the two fundamental non-degeneracy properties of finite-type. Observe that the ideals $I \left(U, k, p \right)$ all equal the maximal ideal in the strongly pseudoconvex case. The generalization to points of finite-type parallels the generalization from the maximal ideal to ideals primary to it.

In case the hypersurface is real-analytic, we do not need to osculate the hypersurface by algebraic hypersurfaces. We can define the ideals $I \left(U, p \right)$ directly. Finite type for a real analytic hypersurface M is then equivalent to the statement that there is no complex-analytic variety of positive dimension passing through p and lying in M. This statement is also equivalent to the statement that there is a constant k such that $\mathbf{D} \left(I \left(U, p \right) \right) < k \ \forall U$. See [D1] for these results. For domains defined by (2) we do not need to consider unitary matrices. For such domains the origin is a point of finite-type if and only if the functions f_k together define a trivial variety.

3. The Methods of Kohn and Catlin

We return to the discussion of subelliptic estimates. We assume that Ω is a smoothly bounded pseudoconvex domain. In [K5] Kohn first developed the method of subelliptic multipliers. Let x be a boundary point of Ω. The idea is to consider the set of all germs of functions f such that

$$|||f\phi|||_\epsilon^2 \leq C \left(||\bar{\partial}\phi||^2 + \left||\bar{\partial}^*\phi\right||^2 + ||\phi||^2 \right). \tag{11}$$

Here both constants may depend on f. Let J_x denote the collection of all such germs at x; its elements are called subelliptic multipliers. It is easy to see that the defining equation is a subelliptic multiplier, with $\epsilon = 1$, essentially because the estimate is elliptic in the interior. It is considerably harder to prove that the determinant of the Levi form is also a multiplier, with $\epsilon = \frac{1}{2}$. From this starting point, Kohn gives an algorithmic procedure for constructing new multipliers, for which the value of epsilon is typically smaller. [K5]. We summarize the procedure after stating some properties of the multipliers.

Proposition 1. *Let x be a boundary point of the psuedoconvex domain Ω. Then the collection of subelliptic multipliers J_x on $(0,1)$-forms is a radical ideal. In particular,*

$$g \in J, \; |f|^N \leq |g| \; \Rightarrow f \in J. \tag{12}$$

We also have the estimate

$$|||g\phi|||_\epsilon^2 \leq c \, |||g^m \phi|||_{m\epsilon}^2 + c \, ||\phi||^2 \quad m\epsilon \leq 1. \tag{13}$$

Proposition 2. *Suppose that g is a subelliptic multiplier, and that*

$$|||g\phi|||_{2\epsilon}^2 \leq cQ(\phi, \phi) \tag{14}$$

for all appropriate ϕ and for $0 < \epsilon \leq \frac{1}{2}$. Then there is a constant $c > 0$ so that

$$\left|\left|\left|\sum_{j=1}^n \frac{\partial g}{\partial z_j} \phi_j\right|\right|\right|_\epsilon^2 \leq cQ(\phi, \phi). \tag{15}$$

We say that $\sum_{j=1}^n a_j dz_j$ is an allowable row if, for some positive epsilon,

$$\left|\left|\left|\sum_{j=1}^n a_j \phi_j\right|\right|\right|_\epsilon^2 \leq cQ(\phi, \phi). \tag{16}$$

Proposition 2 states that ∂g is an allowable row whenever g is a subelliptic multiplier.

Proposition 3. *The rows of the Levi form are allowable rows; we may take $\epsilon = \frac{1}{2}$ in (16).*

Proposition 4. *The determinant of any square matrix of allowable rows is a subelliptic multiplier. We may take epsilon equal to the minimum over the rows of the epsilons guaranteed by (16).*

Kohn's algorithm begins with the statement that the Levi form consists of allowable rows. By Proposition 4, the determinant of the Levi form is a subelliptic multiplier. We consider the ideal generated by this determinant and also the defining function. By applying Proposition 1 we may extract roots. Then we apply Proposition 2 to such roots of determinants, obtaining new allowable rows. We continue the process by taking all possible determinants, and taking radicals. The first of Kohn's theorems is that a subelliptic estimate holds if the process yields a non-zero constant function after finitely many steps.

Kohn proved that, for real-analytic pseudoconvex domains, the process must terminate after finitely many steps. Either a non-zero constant results, or the process uncovers a real-analytic real subvariety in the boundary of 'positive holomorphic dimension'. Diederich-Fornaess then proved that such varieties can exist only when there are complex-analytic varieties in the boundary passing through points arbitrarily close by. [DF] This is equivalent to the statement that there are no complex-analytic varieties in the boundary passing through p. See [D6,D1] for a proof of this last equivalence that applies without the hypothesis of pseudoconvexity. Conversely the estimate cannot hold when there is a complex-analytic variety passing through p and lying in the boundary.

This gives the result in the pseudoconvex real-analytic case.

Theorem 3. *Let $\Omega \subset\subset \mathbf{C}^n$ be pseudoconvex, and suppose that its boundary is real analytic near p. Then there is a subelliptic estimate at p if and only if there is no germ of a complex analytic variety lying in $\mathrm{b}\Omega$ and passing through p.*

Finally Catlin generalized this result to the smooth case. In the papers [C1,C2,C3] he established that finite-type is a necessary and sufficient condition for subellipticity on pseudoconvex domains.

Theorem 4. *Let $\Omega \subset\subset \mathbf{C}^n$ be a pseudoconvex domain with smooth boundary. Then there is a subelliptic estimate at p if and only if $\Delta^1(\mathrm{b}\Omega, p) < \infty$. In this case, the parameter epsilon must satisfy $\epsilon \leq \frac{1}{\Delta^1(\mathrm{b}\Omega,p)}$.*

Catlin applies the method of weight functions used earlier by Hörmander. [H]. Rather than working in $L^2(\Omega)$ with respect to Lebesgue measure dV, consider the measure $e^{-\Phi}dV$ where Φ will be chosen according to the needs of the problem. After this choice is properly made, one employs, as a substitute for Lemma 1, the inequality

$$\int_\Omega \sum_{i,j=1}^n \Phi_{z_i \bar{z}_j} a_i \bar{a}_j \, dV + \sum_{j,k=1}^n \left|\left|\overline{L}_j a_k\right|\right|^2 \leq C Q\left(a, a\right) \tag{17}$$

where $|\Phi| \leq 1$. Here $\overline{L}_j$ are vector fields of type $(0, 1)$ on $\mathbf{C}^n$. There could be also a term on the left side involving the boundary integral of the Levi form, but such a term does not need to be used in this approach to the estimates. Instead, one needs to choose Φ with a large Hessian. One step in Catlin's proof is the following reduction:

Theorem 5. *Suppose that $\Omega \subset\subset \mathbf{C}^n$ is a pseudoconvex domain defined by $\Omega = \{r < 0\}$, and that $p \in b\Omega$. Let U be a neighborhood of p. Suppose finally that for all $\delta > 0$ there is a smooth real-valued function Φ_δ satisfying the following properties:*

$$|\Phi_\delta| \leq 1 \text{ on } U$$

$$(\Phi_\delta)_{z_i \bar{z}_j} \geq 0 \text{ on } U$$

$$\sum_{i,j=1}^n (\Phi_\delta)_{z_i \bar{z}_j} a_i \bar{a}_j \geq c \frac{||a||^2}{\delta^{2\epsilon}} \text{ on } U \cap \{-\delta < r \leq 0\}. \tag{18}$$

Then there is a subelliptic estimate of order ϵ at p.

Theorem 5 reduces the problem to constructing such bounded smooth plurisubharmonic functions whose Hessians are at least as large as $\delta^{-2\epsilon}$. Catlin accomplishes this by a considerable refinement of the ideas in [D2], [DF], and [K5].

We briefly discuss the necessity result. An early example of the author showed that one cannot in general choose epsilon as large as the reciprocal of the order of contact. [D5,D1]. The result is very simple. The function $p \to \Delta^1 (b\Omega, p)$ is not in general upper semicontinuous, so its reciprocal is not lower semicontinuous. Definition 1 reveals that, if there is a subelliptic estimate of order epsilon at one point, then there also is one at nearby points. Catlin has extended this example to show that the parameter value cannot be determined by information based at one point alone. [C1]. Nevertheless Theorem 2 on openness shows that the condition of finite-type does propagate to nearby points. This suggests that one can always choose epsilon as large as $\epsilon = \frac{2^{n-2}}{(\Delta^1(b\Omega,p))^{n-1}}$. A more precise conjecture is that we may always choose epsilon as large as $\epsilon = \frac{1}{B(b\Omega,p)}$. The denominator is the 'multiplicity' of the point, defined by the author to be

$$B\left(M, p\right) = \limsup_{k \to \infty} 2 \sup_U \mathbf{D}\left(I\left(U, k, p\right)\right). \tag{19}$$

The function $p \to B\left(M, p\right)$ is upper semicontinuous.

4. Regular coordinate domains

We continue this paper by discussing the best value of epsilon that we can get from the methods of Kohn, for a class of domains we call 'regular coordinate domains'. We use this term because the Weierstrass polynomials involved exhibit the usual coordinates as a regular system of coordinates for an ideal. [Gu]. There are several reasons for considering the Kohn ideals of subelliptic multipliers. One is that they can be defined on CR manifolds without reference to an embedding into $\mathbf{C}^n$. This has applications to the boundary operator $\overline{\partial}_b$. Furthermore it remains an open problem whether, in the smooth case, Kohn's condition on subelliptic multipliers is equivalent to finite-type. By understanding subelliptic multipliers better, even in the very special case considered here, one hopes to obtain the deeper understanding required to solve this problem.

Put

$$r\left(z, \overline{z}\right) = 2\mathrm{Re}\left(z_0\right) + \sum_{j=1}^{n} \left|f_j\left(z\right)\right|^2 \tag{20}$$

where $z = \left(z_0, z_1, ..., z_n\right)$ and the f_j are Weierstrass polynomials of the following form:

$$f_1\left(z\right) = z_1^{m_1}.$$
$$f_2\left(z\right) = z_2^{m_2} + \cdots.$$
$$f_3\left(z\right) = z_3^{m_3} + \cdots. \tag{21}$$

More precisely, each f_j is a Weierstrass polynomial in z_j whose coefficients depend only on $\left(z_1, ..., z_{j-1}\right)$. We assume that $m_j \geq 2$ for all j. Our earlier computation shows that the Levi form is the matrix product $\left(\partial f\right)^* \left(\partial f\right)$. The especially nice property following from (21) is that the matrix $\left(\partial f\right)$ is lower triangular, so it is easy to compute determinants.

We assume that Ω is a smoothly bounded pseudoconvex domain for which (20) is a defining equation in some neighborhood of the origin. The origin is a point of finite-type. Although the number $\Delta^1\left(M, p\right)$ cannot be computed from the given information alone, we can compute the multiplicity easily. From standard commutative algebra [D1] it follows that $B\left(M, 0\right) = 2 \prod_{j=1}^{n} m_j$. By work of the author, we have (in dimension $n+1$) that

$$\Delta^1\left(M, 0\right) \leq B\left(M, 0\right) \leq \frac{\left(\Delta^1\left(M, 0\right)\right)^n}{2^{n-1}}. \tag{22}$$

As a consequence of (22) the type satisfies

$$2\left(\prod_{j=1}^{n} m_j\right)^{\frac{1}{n}} \leq \Delta^1\left(M, 0\right) \leq 2\prod_{j=1}^{n} m_j \tag{23}$$

for these domains; thus the origin is a point of finite-type, but we cannot compute the maximum order of contact from the information given. Simple examples show that each extreme equality in (23) is possible.

The following lemma is not stated in [K5], but follows from the work there; see the section in [K5] called 'some special domains'.

Lemma 2. *(Holomorphic version of Proposition 3). Suppose that (20) is the defining equation near the origin for a pseudoconvex domain. Then the rows of (∂f) are allowable rows; we may take $\epsilon = \frac{1}{4}$ in (16).*

Kohn's algorithm thus reduces to the situation where the starting collection of allowable rows is given by $\{\partial f_j\}$, and we may stay within the category of holomorphic functions. This information could be useful for general holomorphic functions; for maps defined by (21) things are much simpler yet. By following Kohn's procedure we determine an epsilon for which a subelliptic estimate holds at the origin. The result is not best possible. The author and Catlin believe that the maximum value of epsilon in this case is equal (generically) to $\frac{1}{2\prod m_j}$. In simple cases it is larger. One interesting aspect of this example is that calculation of the value of epsilon is independent of everything except the exponents. Although the order of contact depends on the rest of the terms in the Weierstrass polynomials, our calculations do not.

In order to facilitate the understanding of the computation, we first perform it in the simplest cases. We suppose that (20) is the defining equation, and we assume that $m_j \geq 2$. The simplest case is when $n = 1$. Put $m = m_1$. Then z_1^{m-1} is a subelliptic multiplier with $\epsilon = \frac{1}{4}$ according to Lemma 2. Applying Proposition 1 we see that z_1 is a multiplier with $\epsilon = \frac{1}{m-1}\frac{1}{4}$. Applying Proposition 2 to differentiate and then Proposition 4, reveals that 1 is a subelliptic multiplier with

$$\epsilon = \frac{1}{(m-1)}\frac{1}{8}. \tag{24}$$

The case in more variables is considerably more difficult. To simplify notation we put $E_j = \frac{\partial f_j}{\partial z_j}$. We write DE_j to denote $\frac{\partial f_j}{\partial z_j}$, and $D^k E_j$ for the derivative of order k. For these domains we never need to know the derivatives of E_j with respect to any other variables. The determinant of the Levi form is the product $E_1 E_2 ... E_n$. The technique of proof involves "peeling off" the E_k until we obtain E_1 as a subelliptic multiplier. Since this is a (constant times) a power of a coordinate function, we then take radicals to discover that z_1 is a multiplier, and hence that $(1, 0, \cdots, 0)$ is an allowable row. Working modulo the first variable, we obtain a similar problem in fewer variables.

We begin with the case $n = 2$. The starting point is then that $E_1 E_2$ is a subelliptic multiplier. Its gradient is an allowable row. Taking determinants again reveals that $E_1^2 DE_2$ is a multiplier. We repeat this process $m_2 - 1$ times, obtaining $E_1^{m_2}$ as a multiplier, because $D^{m_2-1}E_2$ is a non-zero constant. Up to this point we have multiplied the originial epsilon by $2^{-(m_2-1)}$. Taking radicals shows that E_1 is a multiplier and multiplies the epsilon by a factor of $\frac{1}{m_2}$. Therefore $z_1^{m_1-1}$ is a multiplier with the original epsilon now muliplied by $\frac{1}{m_2}2^{-(m_2-1)}$. Taking radicals again shows that z_1 is a multiplier, and applying Proposition 2 reveals that $(1,0)$ is an allowable row. These last two steps multiply epsilon by $\frac{1}{m_1-1}\frac{1}{2}$. Since z_1 is a multiplier, and the set of multipliers is an ideal, we may now set z_1 equal to zero. Using the same notation we see that E_2 is now (a constant times) $z_2^{(m_2-1)}$. Hence, we take radicals and one gradient to obtain the allowable row $(0,1)$. The final value of epsilon is

$$\frac{1}{2}\frac{1}{m_2-1}\frac{1}{2}\frac{1}{m_1-1}\frac{1}{m_2}\frac{1}{2}^{(m_2-1)}. \tag{25}$$

Now we set up some notation to handle any number of dimensions. For $k = n+1, n, n-1, ..., 2$ we define A_k by $A_{n+1} = 1$, $A_n = \frac{1}{m_n}\frac{1}{2}^{m_n-1}$, and

$$A_k = \frac{1}{m_k}\frac{1}{2}^{m_k-1}\left(A_{k+1}\right)^{m_k}. \tag{26}$$

The idea is that the process of peeling off the last factor from the product $E_1 E_2 ... E_k$ multiplies epsilon by the value A_k. We begin with $E_1 E_2 ... E_n$. As in the case $n = 2$ we can "peel off" E_n to see that $E_1 E_2 ... E_{n-1}$ is a multiplier. Now we use its gradient in the penultimate row and take determinants to obtain $E_1 ... E_{n-2}DE_{n-1}E_n$ as a multiplier. Again "peel off" the E_n. Use the gradient again in the penultimate row. Take determinants. Apply this process until $E_1 ... E_{n-2}$ is a multiplier. Now use its gradient in row $n-2$. Eventually we "peel off" everything but E_1. As above we take radicals and gradients to obtain that $(1,0,...,0)$ is an allowable row. Working modulo z_1 we obtain a similar situation in $n-1$ variables. We obtain the following result.

Proposition 5. *Let Ω be a pseudoconvex domain for which (20) is a defining equation in some neighborhood of the origin. Then the*

$$\prod_{j=1}^{n}\left(\frac{1}{2}\frac{1}{m_j-1}\right)\prod_{j=2}^{n}A_j. \tag{27}$$

The reader should observe using (26) how complicated the term $\prod A_j$ is.

5. Other uses of finite-type conditions

Other finite-type conditions arise in various analytic problems. We consider but a few of these. Many of them involve preventing complex analytic objects of various dimensions from osculating the boundary to high order. In particular one can define numbers $\Delta^q(M,p)$, $\Delta^q_{\text{reg}}(M,p)$ that measure the contact of q-dimensional complex-analytic varieties or manifolds with a hypersurface. [D4,D1]. Related numbers arise in Catlin's proof of subelliptic estimates for $(0,q)$-forms. The special case where $q = n - 1$ arises very often, for two reasons. First, it arises in the extension of CR functions to one side of a hypersurface. Second, results on estimates for $\overline{\partial}$ in other function spaces seem to be very difficult. These results are known in two dimensions; the two dimensional proofs can be extended to the case of $(0, n - 1)$-forms.

We recall the notions of CR submanifolds of $\mathbf{C}^n$ and CR functions. Let M be a real submanifold of $\mathbf{C}^n$. Suppose that the $(1,0)$ vectors tangent to M define a bundle $T^{1,0}M$, that is, the fibres have constant rank. Then $T^{1,0}M$ is integrable, and $T^{1,0}M \cap T^{0,1}M = 0$. The codimension of the CR structure is the codimension of $T^{1,0}M + T^{0,1}M$ in the complexified tangent bundle $TM \otimes \mathbf{C}$. A real hypersurface therefore has CR codimension equal to unity. A CR function f is a continuous function such that $\overline{L}f = 0$ (in the distribution sense) for all smooth local sections $\overline{L}$ of $T^{0,1}M$.

A basic question concerns the possible holomorphic extension of a CR function from a given CR manifold. The theorem of Trepreau [T] answers the question completely for hypersurfaces.

Theorem 6. *Suppose that M is a smooth real hypersurface in $\mathbf{C}^n$ and that $p \in M$ is an arbitrary point. Then every germ at p of a CR function on M extends (locally) to be holomorphic on one side of M if and only if there is no germ of a complex analytic hypersurface passing through p and lying in M.*

For a real-analytic hypersurface, the nonexistence of a complex hypersurface germ is equivalent to the condition that every complex hypersurface must osculate it to at most finite order. The extension result had been proved earlier in the real-analytic case by Baouendi-Treves. [BT]. The condition on finite osculation is called finite $(n - 1)$-type, and can be expressed by $\Delta^{n-1}(M,p) < \infty$. It is equivalent to the existence of at least one vector field of type $(1,0)$ being of finite-type. Another way to say this is that the Lie algebra generated by local sections of $T^{1,0}M$ and $T^{0,1}M$ spans $TM \otimes \mathbf{C}$ at each nearby point. Thus this generalization of Kohn's condition for hypersurfaces in $\mathbf{C}^2$ arises both in subelliptic estimates for $(n - 1)$-forms and in the extension of CR functions.

Work of Tumanov [Tu] and Baouendi-Rothschild completely solves the problem of holomorphic extension of a CR function from a CR manifold of higher codimension to a wedge. See [BR3] for references and historical discussion. In the real-analytic case, the condition can be expressed again by saying that the Lie algebra generated by local sections of $T^{1,0}M$ and $T^{0,1}M$ spans $TM \otimes \mathbf{C}$. Tumanov discovered the notion of minimality and proved its sufficiency for holomorphic extension to some wedge. Baouendi and Rothschild proved the necessity of minimality. Other than giving the definition we omit any discussion of this beautiful work. A CR submanifold in $\mathbf{C}^n$ is minimal at $p \in M$ if there is no smooth submanifold $N \subset M$ of lower dimension for which all the sections of $T^{1,0}M$ and $T^{0,1}M$ are tangent to N.

Finally we consider briefly another finiteness condition that involves multiplicities of holomorphic ideals. Suppose that $f : (M', p') \to (M, p)$ is a smooth CR diffeomorphism between (germs of) real-analytic real hypersurfaces. A regularity theorem of Baouendi-Jacobowitz-Treves [BJT] states that such a mapping must be itself real analytic in case the germ (M, p) is 'essentially finite' This finiteness condition is that a certain ideal of holomorphic functions constructed from the defining function must define a trivial variety. Assume that p is the origin and write the power series defining the target hypersurface germ as

$$r(z, \overline{z}) = \sum c_{\alpha\beta} z^\alpha \overline{z}^\beta$$

$$= \operatorname{Re} \sum_{|\alpha| \geq 1} c_{\alpha 0} z^\alpha + \sum_{|\beta| \geq 1} \sum_{|\alpha| \geq 1} c_{\alpha\beta} z^\alpha \overline{z}^\beta$$

$$= \operatorname{Re}(h(z)) + \sum_\beta h_\beta(z) \overline{z}^\beta + \operatorname{Im}(h(z)) \Phi(z, \overline{z}). \qquad (28)$$

Here the second term in the last line is independent of h, the functions h_β are holomorphic, and Φ is defined by the equation. The germ $(M, 0)$ is essentially finite if the ideal

$$(h, h_\beta) \qquad (29)$$

(including all the h_β) in O defines the trivial variety consisting of the origin alone. Note that we may always take h to be a coordinate function. Any numerical invariant such as $\mathbf{D}, \mathbf{T}$ of a holomorphic ideal can be applied to give a measurement of the extent of 'essential finiteness'. Baouendi-Rothschild [BR1,BR2] have given many uses of this concept and its generalization to formal power series germs. See [D1,D7] for the relationship between these ideals and the ideals used to consider points of finite-type.

The author finds it amazing that two deep theorems, subellipticity for the $\overline{\partial}$-Neumann problem and the real analyticity of CR diffeomorphisms, each involve checking whether certain ideals of germs of holomorphic functions define trivial varieties. This conclusion evinces a deep connection between partial differential equations and commutative algebra that was anticipated by the work of Kohn.

References

[BR1] Baouendi, M.S. and Rothschild, L.P., Germs of CR maps between real analytic hypersurfaces, Inventiones Mathematicae 93(1988), 481-500.

[BR2] ____, Geometric properties of mappings between hypersurfaces in complex space, Journal of Differential Geometry 31(1990), 473-499.

[BR3] ____, Minimality and the extension of functions from generic manifolds, Pp. 1-13 in Proceedings of Symposia in Pure Mathematics, Vol. 52, Part 3, Several Complex Variables and Complex Geometry, (1991) American Math. Society, Providence, Rhode Island.

[BR4] ____, Cauchy Riemann functions on manifolds of higher codimension in complex space, Inventiones Mathematicae 101(1990), 45-56.

[BJT] Baouendi, M.S., Jacobowitz, H., and Treves, F., On the analyticity of CR mappings, Annals of Math. 122(1985), 365-400.

[BT] Baouendi, M. S. and Treves, F. , About the holomorphic extension of CR functions on real hypersurfaces in complex space, Duke Math J. 51(1984), 77-107.

[C1] Catlin, D., Necessary conditions for subellipticity of the $\bar{\partial}$-Neumann problem, Annals of Math 117(1983), 147-171.

[C2] ____, Boundary invariants of pseudoconvex domains, Annals of Math. 120(1984), 529-586.

[C3] ____, Subelliptic estimates for the $\bar{\partial}$-Neumann problem on pseudoconvex domains, Annals of Math 126(1987), 131-191.

[D1] D'Angelo, J. P., Several Complex Variables and the Geometry of Real Hypersurfaces, CRC Press, Boca Raton, 1992.

[D2] ____, Real hypersurfaces, orders of contact, and applications, Annals of Math 115(1982), 615-637.

[D3] ____, Intersection theory and the $\bar{\partial}$-Neumann problem, Proc. of Symposia in Pure Math. 41(1984), 51-58.

[D4] ____, Finite type conditions for real hypersurfaces, in Lecture Notes in Math. 1268, Springer, Berlin, 1987.

[D5] ____, Subelliptic estimates and failure of semi-continuity for orders of contact, Duke Math. J., Vol. 47(1980), 955-957.

[D6] ____, Finite type and the intersections of real and complex varieties, Pp. 103-117 in Proceedings of Symposia in Pure Mathematics, Vol. 52, Part 3, Several Complex Variables and Complex Geometry, (1991) American Math. Society, Providence, Rhode Island.

[D7] ____, The notion of formal essential finiteness for smooth real hypersurfaces, Indiana Univ. Math. J. 36(1987), 897-903.

[DF] Diederich K., and Fornaess, J.E., Pseudoconvex domains with real analytic boundary, Annals of Math (2) 107(1978), 371-384.

[Gr] Greiner, P., Subellipticity estimates of the $\bar{\partial}$-Neumann problem, J. Differential Geometry 9 (1974), 239-260.

[Gu] Gunning, R.C., Lectures on complex analytic varieties: the local parameterization theorem, Princeton Univ. Press, Princeton, New Jersey, 1970.

[H] Hörmander, L. An introduction to complex analysis in several variables, Van Nostrand, Princeton 1966.

[K1] Kohn, J., Harmonic integrals on strongly pseudoconvex manifolds I,II, Annals of Math (2) (1963), 112-148; ibid 79 (1964), 450-472.

[K2]____, Boundary behavior of $\bar{\partial}$ on weakly pseudo-convex manifolds of dimension two, J. Diff. Geom., Vol 6 (1972), 523-542.

[K3] ____, Global regularity for $\bar{\partial}$ on weakly pseudoconvex manifolds, Transactions of the American Math Society 181(1973), 273-292.

[K4] ____, A survey of the $\bar{\partial}$-Neumann problem, Proceedings Symposia Pure Math, No. 41, 137-145, Providence, RI, 1984.

[K5] ____, Subellipticity of the $\bar{\partial}$-Neumann problem on pseudoconvex domains: Sufficient conditions, Acta Math. 142(1979), 79-122.

[KN] Kohn, J. and Nirenberg, L. Non-coercive boundary value problems, Comm. Pure Appl. Math 18(1965), 443-492.

[RS] Rothschild, L. P. , and Stein, E. M. , Hypoelliptic operators and nilpotent Lie groups, Acta Math. 137(1976), 247-320.

[T] Trepreau, J. M., Sur le prolongement holomorphe des fonctions CR definies sur une hypersurface reele de classe C^2 dans C^n, Inventiones Mathematicae 83(1986), 583-592.

[Tu] Tumanov, A. E. , Extension of CR-Functions into a wedge, Math. USSR Sbornik Vol. 70(1991), No. 2, 385- 398.

Department of Mathematics, University of Illinois, Urbana, IL 61801

CHARACTERIZATION OF CERTAIN HOLOMORPHIC GEODESIC CYCLES ON HERMITIAN LOCALLY SYMMETRIC MANIFOLDS OF THE NONCOMPACT TYPE

PHILIPPE EYSSIDIEUX AND NGAIMING MOK

Contents

Introduction

In the study of quotients of bounded symmetric domains a main theme is that of rigidity. In this article we are interested in compact complex submanifolds of quotients of bounded symmetric domains. In this respect the compact holomorphic totally-geodesic cycles, which are themselves quotients of bounded symmetric domains, are rigid except for trivial deformations (as can be obtained in the case of product domains). Here we explore the question of characterizing compact holomorphic totally-geodesic cycles of quotients Ω/Γ of bounded symmetric domains by a discrete properly discontinuous subgroup Γ of $\text{Aut}(\Omega)$ and study a stronger rigidity phenomenon for such geodesic cycles, which we term the gap phenomenon.

Our point of departure is the study of sufficiently pinched compact complex submanifolds of Ω/Γ. Given a bounded symmetric domain Ω, there are only a finite number of isomorphism classes of totally-geodesic complex submanifolds Σ, which are themselves bounded symmetric domains. Let $S \hookrightarrow \Omega/\Gamma$ be an immersed compact complex submanifold such that the second fundamental form is everywhere of norm $< \epsilon$. By an application of Bishop's Theorem of subconvergence of complex-analytic subvarieties, we show that when the pinching constant ϵ is sufficiently small, S can be locally approximated by a unique isomorphism class of of totally-geodesic complex submanifold of Ω, represented by some $i : \Sigma \hookrightarrow \Omega$. We will say that S is modelled on $(\Omega, \Sigma; i)$ or simply on (Ω, Σ). We say that the gap phenomenon holds for (Ω, Σ) if there exists an ϵ such that for every discrete subgroup $\Gamma \subset \mathrm{Aut}(\Omega)$ acting without fixed points, any compact ϵ-pinched complex submanifold $S \hookrightarrow \Omega/\Gamma$ modelled on (Ω, Σ) is necessarily totally geodesic. In this regard Mikhaïl Gromov has suggested that a sufficiently pinched compact complex submanifold of Ω/Γ should be totally-geodesic. This amounts to saying that the gap phenomenon holds for every bounded symmetric domain Ω and every totally-geodesic bounded symmetric subdomain $\Sigma \subset \Omega$. While we do not have supporting evidence that the gap phenomenon holds for all (Ω, Σ), our investigation confirms that it indeed holds in certain special cases. The significance of the gap phenomenon is that, as opposed to most pinching theorems in Riemannian geometry, we do not impose any bound on the *volume* of the submanifold S. As a matter of fact, if we impose an upper bound on the volume of S, the gap phenomenon is an easy consequence of the rigidity of compact totally-geodesic holomorphic cycles (modulo trivial deformations), at least if we fix Γ and assume that Ω/Γ is compact (*cf.* (1.5)). From this point of view the gap phenomenon may be regarded as a strengthened version of rigidity of compact totally-geodesic holomorphic cycles on Ω/Γ.

A second motivation for our investigation is to find *optimal* inequalities on Chern numbers for $S \hookrightarrow \Omega/\Gamma$. Here an inequality on Chern numbers is said to be optimal if and only if equality is attained when S is itself totally geodesic and modelled on some (Ω, Σ) for some bounded symmetric subdomain Σ. From this perspective there is first of all an inequality on Chern numbers, due to Arakelov, for the special case of compact holomorphic curves in quotients of the Siegel upper half-plane, where equality is attained precisely for compact totally-geodesic holomorphic curves of maximal Gauss curvature. We are interested to generalize such inequalities on Chern numbers for compact complex submanifolds on bounded symmetric domains.

The two perspectives are related to each other in the proof of the following pinching theorem for the Siegel upper half-plane $\mathcal{H}_n$.

Theorem. (Mok[Mok2]) Normalize the canonical metric on $\mathcal{H}_n$ so that the maximum holomorphic sectional curvature is -1. Let $C \hookrightarrow X = \mathcal{H}_n/\Gamma$ be an immersed compact holomorphic curve on X verifying the curvature condition

$$-\left(1 + \frac{1}{4n}\right) \leq \text{Gauss curvatures of } C \leq -1$$

Then, $C \hookrightarrow X$ is totally geodesic in X.

The Siegel upper half-plane is here interpreted as the parameter space of polarized abelian varieties. As a consequence, the curve C can be interpreted as the base space of a variation of polarized Hodge structures of weight one. The immersion $C \hookrightarrow X$ gives rise to a linear representation $\Phi : \pi_1(C) \to \pi_1(X) \hookrightarrow Sp(n, \mathbf{R})$. Let $L_{\mathbf{R}}$ be the associated real 2n-dimensional locally constant bundle over C and $L_{\mathbf{C}} = L_{\mathbf{R}} \otimes_{\mathbf{R}} \mathbf{C}$ be its complexification. The variation of polarized Hodge structures leads to a theory of harmonic forms and to Hodge decomposition. In Mok [Mok2] the proof was geometric and uses the Gauss-Bonnet formula. Here we give a variation of the proof in terms of harmonic forms. The proof consists of of an analytic and a topological part. The analytic part consists of a *vanishing* theorem arising from a Bochner formula. We prove that for a curve satisfying the pinching hypothesis of the theorem, a certain component of the Hodge decomposition of $H^1(C, L_{\mathbf{C}})$ is always zero. The topological part is on the other hand a *non-vanishing* theorem, which asserts that for *any* immersed compact curve $C \hookrightarrow X$, that particular component (which is $H^1(C, L_{\mathbf{C}})^{1,1}$ in standard notations) is non-zero unless C is totally-geodesic and of maximum Gaussian curvature. The topological part comes from a calculation of an Euler-Poincaré characteristic. Let V denote the universal seminegative holomorphic vector bundle over C. The non-vanishing theorem follows from Arakelov's inequality $-\deg(V) \leq n(g(C) - 1)$, where $g(C)$ denotes the genus of the compact curve C, and the assertion that equality is attained if and only if C is totally-geodesic and of maximum Gauss curvature. The Siegel upper half-plane $\mathcal{H}_n$ consists of n-by-n symmetric matrices with positive-definite imgainary part. The subdomain of diagonal matrices is an n-fold Cartesian product $\mathcal{H}^n$ of the upper half-plane $\mathcal{H}$. The pinching theorem above then verifies the gap phenomenon for $(\mathcal{H}_n, \mathcal{H}; i)$ where $i : \mathcal{H} \hookrightarrow \mathcal{H}_n$ is the composition of the diagonal embedding $\delta : \mathcal{H} \hookrightarrow \mathcal{H}^n$ with the inclusion $\mathcal{H}^n \subset \mathcal{H}_n$.

This new formulation of the proof of the pinching theorem above opens up an approach to verify the gap phenomenon. According to an idea of Pierre Deligne's we interpret bounded symmetric domains Ω as parameter space for polarized Hodge structures with some additional structure on an underlying real vector space W. More precisely, let p be a representation of $\mathrm{Aut}(\Omega)$ in the linear group of a real finite-dimensional vector space W. The flat vector bundle $\Omega \times W$ on Ω is endowed with a compatible action of $\mathrm{Aut}(\Omega)$ by the formula $g \cdot (x, w) = (g \cdot \times, \rho(g) \cdot w)$. Zucker [Zu] constructs $\mathrm{Aut}(\Omega)$- equivariant Hodge structures on this flat vector bundle, which form the basis of our interpretation. This flat vector bundle can be descended to Ω/T and pulled back to any m-dimensional compact complex submanifold S of $X = \Omega/\Gamma$. Denote by $W_{\mathbf{C}}$ the resulting complex local system on S. Following Gromov [Gro2] by proving asymptotic vanishing theorems for $H^r(S, W_{\mathbf{C}})$ for $r \neq m$, S not necessarily pinched, with respect to a tower of unramified coverings S_k, we conclude that the asymptotic dimension of each component $H^m(S_k, W_{\mathbf{C}})^{p,q}$, normalized by $\mathrm{Volume}(S_k)$, is computable in terms of Euler-Poincaré characteristics of holomorphic vector bundles over S. We obtain in particular a universal inequality $\nu_{p,q}(S) \geq 0$ where $\nu_{p,q}(S)$ is up to sign the Euler-Poincaré characteristic pertaining to the particular Hodge component $H^m(S_k, W_{\mathbf{C}})^{p,q}$. We will call such an inequality on Chern numbers an Arakelov inequality. In certain special situations there exist Hodge components for which $\nu(S_o) = 0$ (ν meaning $\nu_{p,q}$) when S_o is totally-geodesic, modelled on (Ω, Σ), say. We will then call the inequality $\nu(S) \geq 0$ an optimal Arakelov inequality on Chern numbers for the pair (Ω, Σ). Given an optimal Arakelov inequality $\nu(S) \geq 0$ we can look for Bochner-Kodaira formulas to explain the vanishing of $\nu(S)$. Such Bochner-Kodaira formulas can then be used to prove the vanishing $\nu(S) = 0$ for sufficiently pinched compact holomorphic cycles S modelled on (Ω, Σ). Finally to verify the gap phenomenon for (Ω, Σ) it is sufficient to prove the topological non-vanishing theorem $\nu(S) > 0$ for all compact S which are not totally geodesic and modelled on (Ω, Σ). By the Chern-Weil theory of characteristic forms, $\nu(S)$ is represented by an integral of an (m, m)-form ν_S. To prove the nonvanishing theorem desired it suffices to show that the inequality $\nu(S) \geq 0$ is *local* in the sense that $\nu_S \geq 0$ for all germs S of complex submanifolds and that $\nu_S \equiv 0$ implies that S is totally geodesic. As in the case of curves on quotients of the Siegel upper half-plane, this will lead to a verification of the gap phenomenon for (Ω, Σ). We have thus reduced the gap phenomenon to finding local optimal Arakelov inequalities with accompanying Bochner-Kodaira formulas. In this context it is also natural to formulate the gap phenomenon for period domains of Griffiths as parameter spaces for polarized Hodge structures. The gap phenomenon is then defined for $(\mathcal{D}, \Sigma; i)$, where $\mathcal{D}$ is a period domain of Griffiths, Σ is a bounded symmetric domain, and $i : \Sigma \hookrightarrow \mathcal{D}$ is a totally-geodesic period map (where $\mathcal{D}$ is equipped with a canonical pseudo-Kähler metric).

The article is organized as follows. In §1 we prove a local approximation theorem in order to formulate the gap phenomenon. In §2 we give the first examples when the gap phenomenon holds in the situation of a product domain $\Omega^n = \Omega \times \cdots \times \Omega$, showing that the gap phenomenon holds for $(\Omega^n, \Omega; \delta)$, δ denoting the diagonal embedding, as a consequence of the uniqueness of Kähler-Einstein metrics. In §3 we give a proof of the pinching theorem of Mok [Mok2] for the Siegel upper half-plane $\mathcal{H}_n$ by making use of harmonic forms in place of the Gauss-Bonnet formula. The resulting verification of the gap phenomenon for $(\mathcal{H}_n, \mathcal{H}; i)$ then strengthens the gap phenomenon for $(\Delta^n, \Delta; \delta)$ as obtained from the elementary method of §2. In §4 we obtain an optimal Arakelov inequality and an accompanying Bochner-Kodaira formula for $(B^2 \times B^2, B^2; \delta)$ by using the adjoint representation. This inequality is a special case of optimal Arakelov inequalities obtained by Eyssidieux [Eys1] in the context of Hodge theory. In §5 we show that the optimal Arakelov inequality for $(B^2 \times B^2, B^2; \delta)$ is local. Together with the Bochner-Kodaira formula this gives an elaborate proof that the gap phenomenon holds for $(B^2 \times B^2, B^2; \delta)$. While this is a special case of results in §2 the optimal Arakelov inequality obtained in [Eys1] applies also to the situation of $(\mathcal{D}, B^2; i)$ for some Griffith domain $\mathcal{D}$ and may serve as a basis for verifying the gap phenomenon for $(\mathcal{D}, B^2; i)$.

This article is expository in nature. Although the most natural context of our investigation is that of Hodge theory, our purpose here is to illustrate the principle behind the approach in some simple classical cases. In §3 for the case of curves on the Siegel upper half-plane we will give a direct approach by using the geometric description of the Siegel upper half-plane as the parameter space for polarized abelian varieties. The equivalent formulation in terms of polarized Hodge structure of weight 1 motivates then a general formulation of our approach in terms of Hodge theory. In §4 we will state facts from Hodge theory necessary for the general approach and explain how this can be applied to the special case of $B^2 \times B^2$ by regarding each individual factor B_i^2 as the base space for polarized Hodge structures of weight 2 arising from the adjoint representation of SU(2,1).

In the conference held in March 1992 in Princeton University in honor of the sixtieth birthdays of Professor Robert Gunning and Professor Joseph J. Kohn, the second author gave a talk in relation to the differential geometry of Kuga families of abelian varieties. This article is in part an outgrowth from the circle of ideas expounded in the talk, especially the idea that elliptic operators and eigensection equations are useful in such a context. We would like to thank Mikhaïl Gromov for discussions in relation to the gap phenomenon and Wing-Keung To for comments concerning asymptotic vanishing theorems for cohomology groups in the context of Hodge theory.

1. Local approximation of almost-geodesic complex submanifolds on bounded symmetric domains

(1.1) Let Ω be a bounded symmetric domain, Γ be a torsion-free discrete subgroup of holomorphic isometries and $S \subset \Omega/\Gamma = X$ be an immersed compact complex submanifold such that the second fundamental form is everywhere of norm $< \epsilon$. We say that S is ϵ-pinched. We are going to approximate S locally by totally-geodesic complex submanifolds. For an Ω fixed, there exists a finite set of totally-geodesic submanifolds $\{\Sigma_1, ..., \Sigma_m\}$ such that any totally-geodesic complex submanifold $\Sigma \subset \Omega$ is of the form $\varphi(\Sigma_j)$ for some k, $1 \leq k \leq m$, and for some $\varphi \in \mathrm{Aut}(\Omega)$. It follows readily that, if we can approximate S locally by totally-geodesic complex submanifolds sufficiently well all such local approximations are isomorphic to open pieces of some $\Sigma_k \to \Omega$ for a fixed k (*cf.* (1.2), (1.3)). We will then say that the ϵ-pinched complex submanifold S is modelled on (Ω, Σ_k).

Proposition 1. Let Ω be a bounded symmetric domain of dimension n and let B be a geodesic ball of radius 1 (with respect to the Kähler-Einstein metric with Einstein constant -1) centered at $x_o \in \Omega$. Let $p < n$ be a positive integer and $\{V_i\}$ be a sequence of connected p-dimensional complex-analytic submanifolds of B containing x_o such that the second fundamental forms $\sigma_i = \sigma_{V_i|\Omega}$ are pointwise of norm $\leq \epsilon_i$ with $\epsilon_i \to 0$. Then, some subsequence of $\{V_i\}$ converges to a totally-geodesic complex submanifold.

Proof: By Bishop's theorem on subconvergence of complex-analytic sub-varieties, in order to extract a convergent subsequence of $\{V_i\}$, whose limit is to be denoted by V, it is enough to prove that $\{V_i\}$ is of uniformly bounded volume on any relatively compact domain of B. For the rest of this paragraph we will change notations and use $\{V_i\}$ to denote an extracted convergent subsequence. We will further prove that the limit V is unreduced. For any smooth point x of V, there is an open neighborhood U_x of x in B with the following property: $V \cap U_x$ is the graph of some vector-valued holomorphic functions $(f_1, ..., f_p)$ over a p-dimensional polydisk D; $V_i \cap U_x$ is the graph of some vector-valued holomorphic functions $(f_1^i ..., f_p^{(i)})$ for i sufficiently large such that $f_k^{(i)}$ converges uniformly to f_k for $1 \leq k \leq p$. By expressing the second fundamental forms σ_i of V_i in U_x in terms of the vector-valued functions $(f_1^i ..., f_p^{(i)})$ it follows from the pointwise bound $|\sigma_i| \leq \epsilon_i \to 0$ that V is totally-geodesic where it is smooth. By considering exponential maps this implies that V is actually non-singular and everywhere totally geodesic.

We proceed to prove that the given sequence $\{V_i\}$ is of uniformly bounded volume on any $B(R)$, $0 < R < 1$, where $B(R)$ denotes the geodesic ball of radius R centered at x_o. Fix such an R. We assert that

(*) There exists R', $R < R' < 1$, such that for i sufficiently large, any geodesic $\gamma_i(t)$ on V_i emanating from x_o and parametrized by arc-length can be continued for $0 \leq t \leq R'$ with $d(\gamma_i(R'), x_o) > R$.

This results from standard estimates for ordinary differential equations. To be more precise, we describe this in normal geodesic coordinates.

Write $(x_1, ..., x_{2n})$ for normal geodesic coordinates on the geodesic ball B centered at x_o. (The exponential map at x_o is a diffeomorphism because Ω is complete and of nonpositive Riemannian sectional curvature.) We compare geodesics on V_i with geodesics on B, as follows. Write (Γ^ℓ_{jk}) for the Riemann-Christoffel symbols for B. A geodesic $\gamma = \gamma(t)$ on B is a solution of $\nabla_{\dot\gamma}\dot\gamma = 0$, i.e., writing $\gamma(t) = (u^1, ..., u^{2n})$, a solution of

$$\frac{\partial^2 u^j}{\partial t^2} + \Sigma_{k,\ell}\frac{\partial u^k}{\partial t}\frac{\partial u^\ell}{\partial t}\Gamma^j_{k\ell}(\gamma(t)) = 0, \tag{1}$$

for $1 \leq j \leq 2n$. A geodesic $\gamma_i(t) = (u_i^1, ..., u_i^{2n})$ on V_i is on the other hand a solution to $\nabla^i_{\dot\gamma_i}\dot\gamma_i = 0$, where ∇^i is the Levi-Civita connection of V_i with respect to the restriction of the canonical metric on B. In other words, $\nabla^i_{\dot\gamma_i}\dot\gamma_i$ is normal to the submanifold V_i. But the component of $\nabla_{\dot\gamma_i}\dot\gamma_i$ normal to S is by definition $\sigma_i(\dot\gamma_i, \dot\gamma_i)$. For geodesics parametrized by arc-length $\dot\gamma_i$ is of length one. Together with the assumption on second fundamental forms σ_i this means that $\gamma_i(t) = (u_i^1, ..., u_i^{2n})$ satisfies the same equation as in (1) except that the right-hand side should be replaced by a function pointwise of norm $\leq \epsilon_i$. If we fix the same initial conditions $\gamma_i(o) = \gamma(o) = o$ and $\dot\gamma_i(o) = \dot\gamma(o)$ by Gronwall's lemma we have $|u_i(t) - u(t)| \leq \epsilon_i' \to 0$ for $0 \leq t \leq \frac{1}{2}(1+R)$. This proves (*).

As a consequence of (*) for any R such that $0 < R < 1$, $V_i \cap B(R)$ is contained in the geodesic ball $B_{V_i}(x_o, R')$ on V_i centered at x_o and of radius R'. From the assumption on the second fundamental forms the submanifolds V_i are of uniformly bounded Riemannian sectional curvatures. It follows readily from the comparison theorem in Riemannian geometry for volumes of geodesic balls that $V_i \cap B(R) \subset B_{V_i}(x_o, R')$ are of uniformly bounded volume, from which we conclude by Bishop's theorem that there is a convergent subseqence. Denote such a limit by V, which is irreducible since V_i are connected. It remains to prove that V is unreduced. It suffices to show that V is of multiplicity 1 at x_o. This will follow if for $r > 0$ and i sufficiently large, $\mathrm{Vol}(V_i \cap B(r)) < \frac{3}{2}\mathrm{Vol}(B_E^p(r))$, where $B_E^p(r)$ is the p-dimensional Euclidean ball of radius r. But this is again obvious from uniform boundedness of Riemannian sectional curvatures and the standard volume-comparison theorem.

(1.2) We consider now the structure of the set S of all totally-geodesic complex submanifolds of a bounded symmetric domain. Let Ω be a bounded symmetric domain and $\Sigma \subset \Omega$ be a totally-geodesic complex submanifold. Since involutions on a Riemannian symmetric manifolds can be defined by reflection along geodesics, it follows that Σ is invariant under the involution $\imath_x$ of Ω at any point $x \in \Sigma$. Thus, the complex submanifold $\Sigma \subset \Omega$ is itself a bounded symmetric domain. Regarding the structure of S we only need the following simple fact.

Proposition 2.   Denoting $\mathrm{Aut}_o(\Omega)$ by G, we say that two totally-geodesic complex submanifolds Σ, $\Sigma' \subset \Omega$ are G-equivalent if and only if $\Sigma' = \gamma(\Sigma)$ for some $\gamma \in G = \mathrm{Aut}_o(G)$. Then, there are only a finite number of equivalence classes of totally-geodesic complex submanifolds $\Sigma \subset \Omega$.

Proof: Proposition 2 is in fact a special case of the statement that for any Riemannian symmetric manifold $X = G/K$ with G semisimple, there are only a finite number of equivalence classes of totally-geodesic submanifolds. Write g resp. k for the Lie algebras of G resp. K. Denote by θ the Cartan involution on g with fixed point set k. Any totally-geodesic submanifold S of X is itself symmetric with respect to the metric induced from X. To classify totally-geodesic submanifolds of X it is equivalent to classify embeddings $(h, \tau) \hookrightarrow (g, \theta)$ up to conjugation, where (h, τ) is a semisimple Lie algebra h together with a Cartan involution τ. From the classical theory of semisimple Lie algebras there are only a finite number of solutions up to conjugation by G.

For a general discussion on totally-geodesic embeddings between bounded symmetric domains (*cf.* Satake [Sa, pp.47-51]. For a complete classification we refer the reader to Ihara [Iha].

(1.3) In what follows to simplify notations we will assume that the ϵ-pinched immersed compact complex submanifold $S \hookrightarrow X = \Omega/\Gamma$ is of injectivity radius ≥ 1, so that we can apply [(1.1), Proposition 1]. This is not essential since one can always argue by lifting to the universal covering. By Proposition 2, there are only a finite number of equivalence classes of totally-geodesic complex submanifolds on Ω, represented by $\Sigma_k \hookrightarrow \Omega$, $1 \leq k \leq m$. By Proposition 1, given any $\delta > 0$ there exists an $\epsilon > 0$ such that the following holds: For any ϵ-pinched complex submanifold $S \hookrightarrow X = \Omega/\Gamma$ of injectivity radius ≥ 1 there is a covering of S by open sets U_ℓ such that U_ℓ is an irreducible component of $S \cap B_\ell$, where B_ℓ is a geodesic ball of radius $\frac{1}{2}$ on X. ($S \cap B_\ell$ is itself the union of a finite number of ϵ-pinched submanifolds which may intersect each other).

The covering $\{U_\ell\}$ has furthermore the property that over B_ℓ there exists a totally-geodesic complex submanifold Ξ_ℓ with the property that Ξ_ℓ is of Hausdorff distance $< \delta$ from U_ℓ.

Given a point $x \in \Omega$, the set of germs of p-dimensional complex totally-geodesic submanifolds passing through x is parametrized by a compact subset E of the Grassmannian $Gr(T_x(\Omega), p)$ of complex p-planes on the holomorphic tangent space $T_x(X)$. By [(1.2), Proposition 2] $E \subset Gr(T_x(\Omega), p)$ is a finite union of K_x-orbits, where $K_x \subset G$ is the isotropy subgroup at x and acts on $Gr(T_x(X), p)$ by isometries. It follows easily that for $\delta > 0$ sufficiently small (and hence for $\epsilon > 0$ sufficiently small), the totally-geodesic complex submanifolds Ξ_ℓ can all be identified as open pieces of the same totally-geodesic complex submanifold $\Sigma = \Sigma_k \subset \Omega$. We will say that the ϵ-pinched complex submanifold $S \hookrightarrow X$ is modelled on $\Sigma \subset \Omega$. The preceding discussion can be formulated as

Proposition 3. Let $\Omega \subset\subset \mathbf{C}^N$ be a bounded symmetric domain. Fix $x_o \in \Omega$ and let $B(r) \subset \Omega$ denote the geodesic ball (with respect to the Bergman metric) of radius r and centered at x_o. For $\delta > 0$ sufficiently small ($\delta < \delta_o$) there exists $\epsilon > 0$ such that the following holds: For any ϵ-pinched connected complex submanifold $V \subset B(x_o; 1)$, $x_o \in V$, there exists a *unique* equivalence class of totally-geodesic complex submanifold on Ω, to be represented by $i : \Sigma \hookrightarrow \Omega$, and a totally-geodesic complex submanifold $\Xi \subset B(1)$ modelled on $(\Omega, \Sigma; i)$ such that the Hausdorff distance between $\Xi \cap B(\frac{1}{2})$ and $V \cap B(\frac{1}{2})$ is less than δ.

Proposition 3 allows us to define the gap phenomenon, as follows.

Definition 1. Let $\Omega \subset\subset \mathbf{C}^N$ be a bounded symmetric domain and $i : \Sigma \hookrightarrow \Omega$ be a totally-geodesic complex submanifold. We say that the gap phenomenon holds for $(\Omega, \Sigma; i)$ if and only if there exists $\epsilon < \epsilon(\delta_o)$ (δ_o as in Proposition 3) for which the following holds: For any torsion-free discrete group $\Gamma \subset \mathrm{Aut}(\Omega)$ of automorphisms and any ϵ-pinched immersed compact complex submanifold $S \hookrightarrow \Omega/\Gamma$ modelled on $(\Omega, \Sigma; i)$, S is necessarily totally geodesic.

(1.4) In most pinching theorems in Riemannian geometry one considers compact Riemannian manifolds whose volumes are by hypothesis bounded, such as in Gromov [Gro] and Min-Oo/Ruh [MR], or else there is an *a-priori* bound on the volume as a consequence of hypotheses on the curvature, such as in the $\frac{1}{4}$-Sphere Theorem (*cf.* Klingenberg [Kli]) or in Ros [Ros].

We want to explain in our situation that with the additional hypothesis of a bound on the *volume* of the submanifold S, the gap phenomenon for a fixed and compact $X = \Omega/\Gamma$ will follow from the local rigidity (modulo deformation obtained from geodesic translations) of compact totally-geodesic holomorphic cycles. Fix any positive real number M and a positive integer $p < \dim_{\mathbf{C}} \Omega$ and consider the space $\mathcal{D}_M$ of p-dimensional complex submanifolds S of volume not exceeding M. Then $\mathcal{D}_M$ is a union of a finite number of compact irreducible branches of the Douady space of X. We claim that there exists a constant $\epsilon_M > 0$ such that any ϵ_M-pinched complex submanifold $[S] \in \mathcal{D}_M$ is totally geodesic. We argue by contradiction. Since $\mathcal{D}_M$ is compact the absence of such an ϵ_M would imply by Bishop's theorem on subconvergence of complex subvarieties that there exists a sequence of distinct $\{S_i\}$ of non-geodesic ϵ_i-pinched complex submanifolds of X in $\mathcal{D}_M$, $\epsilon_i \to 0$, which converges as a subvariety to some $S \in \mathcal{D}_M$. By the argument of [(1.1),Proposition 1] S is necessarily totally geodesic and reduced.

Since the submanifolds S_i are distinct and they converge to S it follows that S is not an isolated point in $\mathcal{D}_M$. Passing to a subsequence we may assume that S_i belongs to the same irreducible branch $\mathcal{D}_M^o$. There is a canonical holomorphic map from the germ of $\mathcal{D}_M^o$ at $[S]$ to the germ of the semi-universal Kuranishi deformation space $\mathcal{K}$ of S at $[S]$. We write o for the point correponding to $[S]$ and write the canonical map as $i : (\mathcal{D}_M^o, o) \to (\mathcal{K}, o)$. We claim that i is constant. Otherwise there exists a holomorphic 1-paramenter family $\{S_t : t \in \mathbf{C}, |t| < 1\}$ of complex submanifolds of X centered at $S_o = S$ such that the complex structures are not identical on any neighborhood of o. Suppose for the time being that the infinitesimal deformation of the abstract family $\{S_t\}$ at $t = 0$ is non-trivial, *i.e.*, given by a non-trivial element in $\theta \in H^1(S, T_S)$. Write $T_X|_S$ for the restriction of the holomorphic tangent bundle of X to S and $N_{S|X}$ for the holomorphic normal bundle of S in X. The infinitesimal deformation at o of $\{S_t\}$ as a family of complex *submanifolds* in X is given by an element $\mu \in H^o(S, N_{S|X})$. From the short exact sequence $0 \to T_S \to T_X|_S \to N_{S|X} \to 0$ we obtain a connecting homomorphism $\delta : H^o(S, N_{S|X}) \to H^1(S, T_S)$ such that $\theta = \delta(\mu)$. As $S \subset X$ is totally geodesic, $T_S \subset T_X|_S$ are Hermitian locally homogeneous holomorphic vector bundles on S, so that $T_X|_S \cong T_S \oplus N_{S|X}$ holomorphically and isometrically. It follows that $H^o(S, T_X|_S) \to H^o(S, N_{S|X})$ is surjective and hence that the image of the connecting homomorphism $\delta : H^o(S, N_{S|X}) \to H^1(S, T_S)$ is trivial, so that $\theta = \delta(\mu) = 0$, contradicting with the assumption $\theta \neq 0$. In general the family $\{S_t\}$ of complex manifolds may be infinitesimally rigid at o up to exactly order k. We have then to consider the $(k+1) - st$ infinitesimal deformation θ_{k+1} and higher order tangent maps between the Douady space and the Kuranishi space. A modification of the same argument will lead to a contradition to the non-vanishing of θ_{k+1}.

We have thus proved that $i : (\mathcal{D}_M^o, o) \to (\mathcal{K}, o)$ is constant. It means that in deforming $S \subset X$ as a complex submanifold the complex structure does not change for small deformations. From the uniqueness of harmonic maps up to geodesic translations (Hartman [Har]) it follows that small deformations of $S_o = S$ are obtained by geodesic translations. This contradicts with the existence of non-geodesic S_i converging to S and proves by argument by contradiction that there indeed exists an $\epsilon_M > 0$ such that any ϵ_M-pinched p-dimensional complex submanifold of X of volume $\leq M$ is necessarily totally geodesic.

We can thus conclude that, with $X = \Omega/\Gamma$ fixed and compact and under the additional hypothesis of a bound on the volume of the submanifold $S \hookrightarrow X$, the gap phenomenon for (Ω, Σ) on X follows readily from Bishop's theorem and the local rigidity (modulo deformations obtained by geodesic translations) of totally-geodesic holomorphic cycles $S_o \subset X$. The gap phenomenon for (Ω, Σ), if verified, can thus be regarded as a strengthened version of local rigidity for compact totally-geodesic holomorphic cycles modelled on (Ω, Σ). In the next section we will verify one very simple case of the gap phenomenon in the case when Ω is a product domain with isomorphic direct factors.

2. The gap phenomenon on a product symmetric domain as a consequence of the uniqueness of Kähler-Einstein metrics

(2.1) In this section we prove a very simple case of the gap phenomenon, resulting in the following theorem.

Theorem 1. Let Ω be an bounded symmetric domain and $\Omega^n = \Omega \times \cdots \times \Omega$ be the n-fold Cartesian product of Ω. Let $\delta : \Omega \to \Omega^n$ be the diagonal embedding defined by $\delta(x) = (x, ..., x)$. Then, the gap phenomenon holds for $(\Omega^n, \Omega; \delta)$.

This result serves two purposes. On the one hand, it gives the first instance where one has a pinching theorem on quotients of bounded symmetric domains (hence of negative Ricci curvature) without assumptions on the volume of the complex submanifolds. On the other hand, it serves as a motivation for the formulation of an *effective* pinching theorem on the Siegel upper half-plane by suggesting the correct growth order of the pinching constants (as a function of n).

In Theorem 1 we consider immersed compact complex submanifolds $S \hookrightarrow \Omega^n/\Gamma$. Since $\mathrm{Aut}(\Omega^n)$ contains $\mathrm{Aut}(\Omega)^n$ as a normal subgroup of finite index, replacing Γ by a normal subgroup of finite index if necessary, we may assume without loss of generality that $\Gamma \subset \text{Aut}(\Omega)^n$. Theorem 1 is then an immediate consequence of the following more precise statement.

Proposition 1. Let Ω be a bounded symmetric domain of complex dimension m. Let $n \geq 2$ and $\Gamma \subset \text{Aut}(\Omega)^n$ be a discrete group of holomorphic isometries acting without fixed points. Write X for Ω^n/Γ. Let $S \hookrightarrow X$ be an n-dimensional compact immersed complex submanifold of X and write $\tilde{S} \hookrightarrow \Omega^n$ for the lifting to universal covering spaces. Let Ω_i denote the i-th direct factor of Ω^n. Suppose for every i, $1 \leq i \leq n$, the i-th canonical projection $\nu_i : \tilde{S} \to \Omega_i$ is unramified. Then, $\tilde{S} \hookrightarrow \Omega^n$ is totally geodesic and is equivalent to the diagonal $\delta(\Omega^n)$ of Ω^n.

Proof: Denote by g_i be the canonical Kähler-Einstein metric on Ω_i with Einstein constant -1. Since $\Gamma \subset \text{Aut}(\Omega)^n$, the restriction of g_i to $\tilde{S}$ is invariant under the action of Γ and defines in general a positive semi-definite symmetric 2-tensor h_i on S. Since in our situation by assumption the canonical projection $\nu_i : \tilde{S} \to \Omega_i$ is unramified, h_i is in fact a Riemannian metric on S. As S is of complex dimension $n = \dim_{\mathbf{C}} \Omega$, the Riemanninan metric h_i on S is Kähler-Einstein. By the uniqueness of Kähler-Einstein metrics as a consequence of the Schwarz lemma (Yau [Y]) we conclude that $h_i = h_j$ for $1 \leq i$, $j \leq n$. Fix a point $\tilde{x} = (\tilde{x}_1, ..., \tilde{x}_n) \in \tilde{S}$. There exists an open neigborhood U of $\tilde{x}$ in $\tilde{S}$ such that the canonical projection map $\nu_i : U \to \Omega_i$ maps U biholomorphically onto an open neigborhood U_i of $\tilde{x}_i$ in Ω_i. Thus, for $1 \leq i$, $j \leq n$ the map $\nu_{ji} := \nu_j \circ \nu_i^{-1} : U_i \to U_j$ is a biholomorphism such that $\nu_{ji}^*(g_i) = g_j$. The holomorphic local isometry $\nu_{ji} : U_i \to U_j$ extends via the exponential map to a global holomorphic isometry $\hat{\nu}_{ji} : \Omega_i \to \Omega_j$. This implies readily that $\tilde{S} \hookrightarrow \Omega^n$ is embedded and that it is the image of Ω_1 under the map $(id, \hat{\nu}_{21}, ..., \hat{\nu}_{n1})$. In particular, $\tilde{S}$ is equivalent to the diagonal $\delta(\Omega^n)$ of Ω^n.

Proof of Theorem 1: Theorem 1 follows readily from Proposition 1. In fact, if $S \hookrightarrow X = \Omega^n/\Gamma$ is modelled on $(\Omega^n, \Omega; \delta)$ and is ϵ-pinched for a sufficiently small constant ϵ the canonical projection map $\nu_i : \tilde{S} \to \Omega_i$ is unramified.

Remarks: Since the Ahlfors-Schwarz lemma of Yau [Y] holds for *complete* Kähler manifolds, Theorem 1 remains valid when the compactness assumption on $S \hookrightarrow X = \Omega^n/\Gamma$ (implicit in the formulation of the gap phenomenon) is replaced by the assumption that $S \hookrightarrow X$ is closed and hence a complete submanifold. In fact, in the notations of the proof of Proposition 1, the Kähler metric $g|_{\tilde{S}} = g_1|_{\tilde{S}} + \cdots + g_n|_{\tilde{S}}$ is complete. If $S \hookrightarrow X$ is modelled on $(\Omega^n, \Omega; \delta)$ and if the pinching constant ϵ in [(1.1), Proposition 1] is sufficiently small, all the $g_i|_{\tilde{S}}$ are equivalent, so that the Kähler-Einstein metrics $g_1|_{\tilde{S}}, \cdots, g_n|_{\tilde{S}}$ are complete since their sum $g|_{\tilde{S}}$ is complete. The arguments of Proposition 1 then applies to show that $\tilde{S} \hookrightarrow \Omega^n$ is totally geodesic.

(2.2) In the special case when Ω is of constant holomorphic sectional curvature, *i.e.*, for $\Omega \cong B^m$ the complex unit ball, Proposition 1 implies the following effective result

Theorem 2. Let $(B^m)^n = B^m \times \cdots \times B^m$ be the n-fold Cartesian product of the m-dimensional complex unit ball and Γ be a discrete group of holomorphic isometries on $(B^m)^n$ acting without fixed points. Normalize the Kähler-Einstein g_i metric on each direct factor B_i^m so that (B_i^m, g_i) has constant holomorphic sectional curvatures equal $to - n$. Let $S \hookrightarrow X := (B^m)^n/\Gamma$ be an immersed m-dimensional compact complex submanifold whose holomorphic sectional curvatures are pinched by

$$-\left(1 + \frac{1}{n-1}\right) \leq \text{holomorphic sectional curvatures} \leq -1.$$

Then, S is totally-gedesic and modelled on $((B^m)^n, B^m; \delta)$.

Proof: Let $\tilde{S} \hookrightarrow (B^m)^n$ be the lifting of $S \hookrightarrow X$ to the universal covering spaces. We calculate holomorphic sectional curvatures at a point $\tilde{x} \in \tilde{S}$. Write $\eta = (\eta_1, ..., \eta_n)$ according to the decomposition of $T_{\tilde{x}}^{1,0}((B^m)^n)$. Write Θ_i for the curvature tensor of (B_i^m, g_i) and Θ for that on the product domain $((B^m)^n, g)$. Write R for the curvature tensor on $\tilde{S}$. Denote by $\|\cdot\|$ norms measured with respect to the metrics g_i and g. From the equation of Gauss holomorphic sectional curvatures satisfy

$$R_{\eta\bar{\eta}\eta\bar{\eta}} \leq \Omega_{\eta\bar{\eta}\eta\bar{\eta}} = \sum \Theta_i(\eta_i, \bar{\eta}_i; \eta_i, \bar{\eta}_i) = -n\left(\sum \|\eta_i\|^4\right).$$

Since $\|\eta\| = 1$ we have $\sum \|\eta_i\|^2 = 1$. The minimum for $\sum \|\eta_i\|^4$ (subject only to the condition $\|\eta\| = 1$) is attained when all $\|\eta_i\|$ are the same and thus equal to $\frac{1}{\sqrt{n}}$. Thus we have always $R_{\eta\bar{\eta}\eta\bar{\eta}} \leq (-n)(n)\left(\frac{1}{\sqrt{n}}\right)^4 = -1$. To prove Theorem 2 it suffices to show that under the curvature assumption for each $\tilde{x} \in \tilde{S}$ and for each $\eta \in T_{\tilde{x}}^{1,0}(\tilde{S})$, we have $\eta_i \neq 0$ for each η_i, so that the projection maps $\tilde{S} \to B_i^m$ are unramified. We argue by contradiction. If one of the η_i, say η_n, is zero the sum $\sum_{1 \leq i < n} \|\eta_i\|^4$ is minimum when all the $\|\eta_i\|$, $1 \leq i \leq n-1$, are identical. In this case we have by the Gauss equation

$$R_{\eta\bar{\eta}\eta\bar{\eta}} \leq \Omega_{\eta\bar{\eta}\eta\bar{\eta}} \leq (-n)(n-1)\left(\frac{1}{\sqrt{n-1}}\right)^4 \leq -\frac{n}{n-1} = -\left(1 + \frac{1}{n-1}\right),$$

violating the curvature hypothesis in Theorem 1. In other words, the curvature hypothesis in Theorem 2 implies that the projection maps $\tilde{S} \to B_i^m$ are unramified, so that we can apply Proposition 1 and conclude that $S \hookrightarrow X$ is totally geodesic and modelled on $((B^m)^n, B^m; \delta)$.

Theorem 2 contains the special case of $(\Delta^n, \Delta; \delta)$ where Δ denotes the unit disc and $\delta : \Delta \to \Delta^n$ denotes the diagonal embedding of Δ into the polydisk Δ^n. In the next section we will verify a gap phenomenon on the Siegel upper half-plane, which is an unbounded realization of the bounded symmetric domain D_n^{III}. The Siegel upper half-plane $\mathcal{H}_n$ contains a copy of the n-fold Cartesian product $\mathcal{H}^n$ of the upper half-plane $\mathcal{H} \subset \mathbf{C}$. $\mathcal{H}^n \subset \mathcal{H}_n$ is totally-geodesic. The gap phenomenon for $(\Delta^n, \Delta; \delta)$ as stated in Theorem 2, is the same as that for $(\mathcal{H}^n, \mathcal{H}; \delta)$. Let $i : \mathcal{H} \hookrightarrow \mathcal{H}_n$ be obtained by composing $\delta : \mathcal{H} \hookrightarrow \mathcal{H}^n$ with the inclusion $\mathcal{H}^n \subset \mathcal{H}_n$. We are going to verify the gap phenomenon for $(\mathcal{H}_n, \mathcal{H}; i)$ using the theory of harmonic forms. This can then be regarded as a strengthening of the gap phenomenon for $(\Delta^n, \Delta; \delta)$, except for the fact that we will have a weaker pinching constant for Gauss curvatures.

We remark that in Theorem 2 the pinching condition on holomorphic sectional curvatures

$$-\left(1 + \frac{1}{n-1}\right) \le \text{holomorphic sectional curvatures} \le -1$$

gives the correct growth order for pinching constants (as a function of n), at least in the case of curves. To see this, let C and C' be two compact Riemann surfaces, with fundamental groups Γ resp. Γ', for which there exists a nontrivial holomorphic map $f : C \to C'$ which is somewhere ramified. For $n \ge 2$ let X be $C^{n-1} \times C'$ which is $\Delta^n/(\Gamma^{n-1} \times \Gamma')$. Define $F : C \to C^{n-2} \times C'$ by $F(x) = (x, ..., x; f(x))$ and let $C_n \hookrightarrow X$ be the graph of F. A straightforward calculation then gives the estimate

$$-\left(1 + \frac{A}{n}\right) \le \text{Gauss curvatures of } C_n \le -1$$

for some positive constant A. C_n is obviously not totally geodesic for any $n \ge 2$.

3. The exponential sequence and a gap phenomenon on the Siegel upper half-plane

(3.1) In this section we are going to verify the gap phenomenon for $(\mathcal{H}_n, \mathcal{H}; i)$ for the Siegel upper half-plane $\mathcal{H}_n$ as a generalization of the gap phenomenon for $(\Delta^n, \Delta; \delta)$ in the case of the polydisk. The Siegel upper half-plane $\mathcal{H}_n$ consists of all complex symmetric n-by-n matrices τ with $\text{Im}(\tau) > 0$. It is a Cayley transform of a bounded symmetric domain. In what follows we will choose the canonical Kähler-Einstein metric g on $\mathcal{H}_n$ such that the minimal holomorphic sectional curvature of $(\mathcal{H}_n, g)$ is -2 and the maximal holomorphic sectional curvature $is - \frac{2}{n}$ (*cf.* (3.2)). This means that the Kähler form is $-i\partial\bar{\partial}\log\det(\mathrm{Im}\tau)$, which is $\frac{i}{4y^2}d\tau \wedge d\bar{\tau}$ in the special case of $n = 1$, $\tau = x + iy$. With this convention we formulate the theorem cited in the introduction in the following equivalent form.

Theorem 1. Normalize the canonical metric on $\mathcal{H}_n$ so that the maximum holomorphic sectional curvature is $-\frac{2}{n}$. Let $C \hookrightarrow X = \mathcal{H}_n/\Gamma$ be an immersed compact holomorphic curve on X verifying the curvature condition

$$-\frac{2}{n}\left(1 + \frac{1}{4n}\right) \leq \text{Gauss curvatures of } C \leq -\frac{2}{n}$$

Then, $C \hookrightarrow X$ is totally geodesic in X.

In Mok [Mok2], Theorem 1 was proved using an *infinitesimal* Gauss-Bonnet formula (*cf.* (3.2) for explanation). Here we will give a variation of the proof by considering instead elliptic differential equations on $C \hookrightarrow X$. Our approach consists of finding, in case C is *non-geodesic*, a non-trivial solution η to a certain elliptic differential equation on C. This part is *topological* and hinges on the Riemann-Roch formula. We then deduce Theorem 1 by showing that such a solution cannot possibly exist when C verifies the pinching conditions in Theorem 1 by proving a Bochner-Kodaira formula for η. Here we give a direct proof by interpreting the Siegel upper half-plane as a parameter space for polarized abelian varieties. By interpreting the Siegel upper half-plane equivalently as the parameter space of polarized variations of Hodge structures of weight 1 (*cf.* Barth-Peters-Van de Ven [BPV, §14, *p*.36 − 39]) the proof can be formulated in terms of Hodge theory and harmonic forms (*cf.* (5.7)). In the event X is an arithmetic quotient and non-compact, the first author [Eys1] showed that Theorem 1 remains valid for holomorphic curves $C \hookrightarrow X$ of finite volume. In this article we will only be concerned with compact holomorphic curves.

(3.2) To make our presentation more coherent we will start with some standard background material on the Siegel upper half-plane and recall the topological argument using Riemann-Roch in Mok [Mok2, §4]. For the Bochner-Kodaira formula we will rely on some calculations in Mok-To [MT]. Our choice of Kähler-Einstein metric is then consistent with [MT].

Let Γ be a discrete group of holomorphic isometries on $\mathcal{H}_n$ acting without fixed points and let $C \hookrightarrow X = \mathcal{H}_n/\Gamma$ be an immersed compact holomorphic curve. A point $\tau \in \mathcal{H}_n$ (τ symmetric with $\mathrm{Im}\,\tau > 0$) corresponds to a principally polarized abelian variety $\mathbf{C}^n/L_\tau$, where the lattice L_τ is generated as an abelian group by the standard basis $\{e_i\}$ of $\mathbf{C}^n$ and by the n column vectors of τ. The variation of the uniformizing complex Euclidean spaces $V_\tau \cong \mathbf{C}^n$ constitutes the universal (seminegative) homogeneous holomorphic vector bundle (V, h) over $\mathcal{H}_n$. We denote by $\mathcal{V}$ the associated locally free sheaf of germs of local holomorphic sections of V. The variation of the abelian groups of lattice points L_τ consititutes a locally constant subsheaf $L \subset \mathcal{V}$. Tensoring over $\mathbf{R}$ we obtain $L_\mathbf{R} = L \otimes_\mathbf{Z} \mathbf{R} \hookrightarrow \mathcal{V}$. A local section of $L_\mathbf{R}$ is then a *real* linear combination of lattice points. We have the following short exact sequence which we will call the exponential sequence.

$$0 \to L_\mathbf{R} \to \mathcal{V} \to \mathcal{V}/L_\mathbf{R} \to 0. \qquad (*)$$

The fundamental group $\Gamma \subset Sp(n, \mathbf{R})$ acts on $L_\mathbf{R}$ (but not necessarily on L) and on V so that the exponential sequence descends to X and by restriction to C. From now on we will use the notations in $(*)$ to denote instead the corresponding exponential sequence on C.

We are interested in the real vector space $H^o(C, \mathcal{V}/L_\mathbf{R})$, the real dimension of which will be denoted by $h(C)$. We give here an interpretation of $H^o(C, \mathcal{V}/L_\mathbf{R})$ in terms of holomorphic functions defined on some immersed holomorphic curves on $\mathcal{H}_n$. Let $\tilde{C} \to C$, $\tilde{C} \hookrightarrow \mathcal{H}_n$ be a regular covering map corresponding to the kernel of the canonical homomorphism $\Phi : \pi_1(C) \to \pi_1(X) = \Gamma$. Then, $\mathrm{Im}(\Phi)$ acts on $\tilde{C}$ as deck transformations. A section of $H^o(C, \mathcal{V}/L_\mathbf{R})$ is equivalently a vector-valued holomorphic function $F : \tilde{C} \to \mathbf{C}^n$ satisfying a system of compatibility relations, as follows. For any $\gamma \in \Gamma$ write $\gamma\tau = (A_\gamma\tau + B_\gamma)(C_\gamma\tau + D_\gamma)^{-1}$. Denote a point on $\tilde{C}$ by $\tilde\tau$ and its image in $\mathcal{H}_n$ by τ. Then, the holomorphic map $F : \tilde{C} \to \mathbf{C}^n$ corresponds to an element of $H^o(C, \mathcal{V}/L_\mathbf{R})$ if and only if for any $\gamma \in \mathrm{Im}(\Phi)$ there exist *real* vectors P_γ, $Q_\gamma \in \mathbf{R}^n$ such that

$$F(\gamma\tilde\tau) = ([C_\gamma\tau + D_\gamma]^t)^{-1} F(\tilde\tau) + [\gamma\tau \quad I_n] \begin{bmatrix} P_\gamma \\ Q_\gamma \end{bmatrix} \qquad (1)$$

We will call the last column a period of F. Two holomorphic maps F, $F' : \tilde{C} \to \mathbf{C}^n$ define the same element in $H^o(C, \mathcal{V}/L_\mathbf{R})$ if and only if

$$(F - F')(\tau) = [\tau \quad I_n] \begin{bmatrix} P \\ Q \end{bmatrix} \qquad (2)$$

for some $P, Q \in \mathbf{R}^n$. If $\Gamma \subset Sp(n, \mathbf{Z})$ then we have a polarized family of abelian varieties $\mathcal{A} = \mathcal{V}/L_\mathbf{Z}$ defined over C. In this case $H^o(C, \mathcal{A})$ corresponds to the subset of those $F : \tilde{C} \to \mathbf{C}^n$ with integral periods, and F, $F' : \tilde{C} \to \mathbf{C}^n$ are identified if and only if (2) holds with P, $Q \in \mathbf{Z}^n$.

(3.3) *A non-vanishing theorem for $h(C)$*. We proceed to prove a non-vanishing theorem for $h(C)$ whenever C satisfies the pinching hypothesis in [(3.1), Theorem 1] but is not totally geodesic. (A simple modification actually shows that $h(C) > 0$ for any compact holomorphic curve $C \hookrightarrow X$ unless C is totally geodesic and of constant curvature $-\frac{2}{n}$.) We consider the long exact sequence associated to the exponential sequence. Under the pinching hypothesis we have $H^{\circ}(C, L_{\mathbf{R}}) = 0$, because the non-vanishing would produce a constant sub-local system which would imply that the universal covering map $\tilde{C} \to \mathcal{H}_n$ factors through a Siegel subdomain $\mathcal{H}_n$ $(r < n)$ and that the Gaussian curvature is less than $-\frac{4}{n}$, while $H^2(C, L_{\mathbf{R}}) = 0$ by duality since the standard symplectic pairing on $L_{\mathbf{R}}$ yields a self-duality $L_{\mathbf{R}} \cong L_{\mathbf{R}}^*$. On the other hand the holomorphic vector bundle V, equipped with a canonical Hermitian-Einstein metric, is of seminegative curvature in the sense of Griffiths. The pinching hypothesis on the Gauss curvature implies that V is of strictly negative curvature over the compact holomorphic curve C, so that $H^{\circ}(C, V) = 0$. We obtain therefore from the long exact sequence associated to the exponential sequence over C the exactness of

$$0 \to H^{\circ}(C, V/L_{\mathbf{R}}) \to H^1(C, L_{\mathbf{R}}) \to H^1(C, V) \to H^1(C, V/L_{\mathbf{R}}) \to 0. \tag{1}$$

By considering the Euler-Poincaré characteristic we have

$$\dim_{\mathbf{R}} H^1(C, L_{\mathbf{R}}) = 4n(g-1) \tag{2}$$

where $g = g(C)$ is the genus of the compact Riemann surface C. On the other hand by Riemann-Roch we obtain

$$\dim_{\mathbf{C}} H^1(C, V) = -\chi(V) = n(g-1) - \deg(V). \tag{3}$$

$(\det(V), \det(h))$ is a Hermitian-Einstein bundle on $\mathcal{H}_n$ with Einstein constant -1 while the maximal holomorphic sectional curvature of $(\mathcal{H}_n, g)$ is $-\frac{2}{n}$. By the Gauss-equation the Gauss curvature of $(C, g|_C)$ is everywhere $\leq -\frac{2}{n}$. By considering first Chern forms and integrating over C we obtain

$$\deg(V) \geq \left(\frac{n}{2}\right)(2 - 2g) = -n(g-1), \tag{4}$$

and that equality holds if and only if C is totally geodesic and is of constant Gaussian curvature $-\frac{2}{n}$. Putting (2), (3) and (4) into (1) we have the inequality

$$\dim_{\mathbf{R}} H^{\circ}(C, V/L_{\mathbf{R}}) \geq 2n(g-1) + 2\deg(V) \geq 0 \tag{5}$$

with strict inequality unless the curve $(C, g|_C)$ is totally geodesic and of constant Gaussian curvature $-\frac{2}{n}$. This yields a non-vanishing theorem for $H^0(C, V/L_{\mathbf{R}})$.

(3.4) *A vanishing theorem for $h(C)$.* For $C \hookrightarrow X$ satisfying the hypothesis of Theorem 1 there is on the other hand a vanishing theorem for

$h(C)$ obtained in [Mok2,§4] from geometric considerations. On the to-
tal space $\pi : \mathcal{H}_n \times \mathbf{C}^n \to \mathcal{H}_n$ there is a family of homogeneous metrics
$\{\nu_t\}$ degenerating to $\pi^*\omega$, ω denoting the canonical Einstein metric on
$\mathcal{H}_n$ (*cf.* [Mok2, §3]). Let $f \in H^o(C, \mathcal{V}/L_{\mathbf{R}})$. By considering the graph
of a lifting of f we obtain a one-parameter family of Hermitian metrics
$\{\xi_t\}$ on X degenerating to the restriction of the canonical metric of X
on C. If f is non-trivial we obtain a contradiction by expanding the
Gauss-Bonnet integral over C with respect to $\{\xi_t\}$ in terms of t. More
precisely there is a tensor $\eta(f) \in \mathcal{C}^\infty(C, V \otimes \Omega_C)$ which vanishes if and
only if $F : \tilde{C} \to \mathbf{C}^n$ (as in (3.2)) is a horizontal section, *i.e.*, a *real* linear
combination of lattice points. From the t^2-term in the expansion of the
Gauss-Bonnet integrand we derive the vanishing of $\eta(f)$ and thus $h(C)$.
This contradiction to $h(C) \neq 0$ obtained above means that a non-geodesic
compact holomorphic curve C verifying the pinching hypothesis of [(3.1),
Theorem 1] cannot possibly exist.

Here we give a proof of Theorem 1 by showing that in the geodesic
case of Gaussian curvature $-\frac{2}{n}$, η satisfies the eigensection equation
$\bar{\partial}^*\bar{\partial}\eta = -\eta$, while in general η satisfies a similar equation with a zero-
order perturbation term arising from the second fundamental form (and
not involving its covariant derivatives).

On the Siegel upper-half plane $\mathcal{H}_n$ we have a canonical isomorphism
of the holomorphic tangent bundle T with $S^2 V$. We will choose the
homogeneous Hermitian metric h on V such that $(T, g) \cong (S^2 V, S^2 h)$
as Hermitian holomorphic vector bundles. This is reflected by the fact
that elements of $\mathcal{H}_n$ are symmetric matrices. The canonical isomorphism
$S^2 V \cong T$ defines in an obvious way a canonical bundle embedding $\Xi :
V^* \to T^* \otimes V$. We identify $\overline{V}$ with V^* by contraction with the canonical
Hermitian metric. There is a canonical bundle homomorphism

$$\Phi : S^2\overline{V} \otimes (S^2 V \otimes \overline{V}) \to S^2\overline{V} \otimes V$$

defined by $\Phi(\bar{\alpha} \otimes \bar{\eta}) = \Xi(\overline{\eta(\alpha)})$. We write $\Phi(\bar{\alpha}, \bar{\eta})$ for $\Phi(\bar{\alpha} \otimes \bar{\eta})$. Then,
any $\eta = \eta(f)$ of a local section $f \in H^o(U, \mathcal{V}/L_{\mathbf{R}})$ over an open subset U
of $\mathcal{H}_n$ satisfies the first-order differential equation (Mok-To [MT, (2.4)])

$$\partial_{\bar{\alpha}}\eta = -i\Phi(\bar{\alpha}, \bar{\eta}) = -i\Theta(\overline{\eta(\alpha)}) \qquad (1)$$

The same equation is valid if f is only defined on a complex submanifold S of U, since for S simply-connected and U Stein one can extend f
holomorphically to U and the restriction of (∗) to S is independent of the
extension. We are going to write down formulas for η using coordinates
on the Siegel upper half-plane $\mathcal{H}_n$. Denote by (τ_{ij}) the symmetric matrix
representing a general point $\tau \in \mathcal{H}_n$. Identify the universal holomorphic

vector bundle V as $\mathcal{H}_n \times \mathbf{C}^n$ in the standard way. The canonical Euclidean basis of $\mathbf{C}^n$ will be denoted by (e_k) and the dual basis by (e^k). Furthermore, we regard the space of symmetric n-by-n matrices $M_s(n)$ as a Euclidean subspace of the space of all n-by-n matrices $M_s(n,n)$, with entries $(\tau_{ij})_{1 \le i,j \le n}$. With this understanding $d\tau^{ij}$ will simply mean the restriction of the 1-form $d\tau^{ij}$ on $M(n,n)$ to the subspace $M_s(n,n) \supset \mathcal{H}_n$. Denote by $h_{k\bar{\ell}}$ Hermitian metric tensor for (V,h). From [MT, (3.3)] we have

$$\Xi(e^k) = \sum_p d\tau^{kp} \otimes e_k \tag{2}$$

$$\partial_{\bar{\alpha}} \eta = -i \sum_{A,\ell,k,p} \overline{\alpha^A \, \eta^\ell_A} \, h_{k\bar{\ell}} d\tau^{kp} \otimes e_p \tag{3}$$

(The formula in [MT, (3.3)] was meant for a point where $h_{k\bar{\ell}} = \delta_{k\bar{\ell}}$, at $\tau_o = \frac{i}{2} I_n \in U$). Write ∇° resp. ∇ for covariant differentiation on $\mathcal{H}_n$ resp. on the complex submanifold $S \subset U$. For α of type $(1,0)$ tangent to S at $x \in S$ we want to compute $\nabla_\alpha \partial_\alpha \eta$. In the computation below we assume implicitly that f and hence that $\eta = \eta(f)$ is defined on U. We have for $x = \tau_o$ on U

$$\nabla^{\circ}_\alpha \partial_\alpha \eta = \Phi(\bar{\alpha}, \overline{\Phi(\bar{\alpha}, \bar{\eta})}) = \sum_{k,p,A,q} \alpha^{\overline{kp}} \alpha^A \eta^k_A d\tau^{pq} \otimes e_q \tag{4}$$

The right-hand side can be interpreted in terms of the curvature form Θ of (V,h), as in [MT, (3,4), eqn. (11)] in the form

$$\nabla^{\circ}_\alpha \partial_\alpha \eta = \sum_q \Theta_{\eta(\alpha)\bar{q}\xi\bar{\alpha}} e_q \tag{5}$$

for any $(1,0)$-vector ξ at $x = \tau_o$. On $S \subset U$ write (z_γ) for local holomorphic coordinates on S and extend them to U. Write $(S^{kp}_{\alpha\beta})$ for the second fundamental form of S in U, with the convention that $S(\alpha,\beta) = \sum S^{kp}_{\alpha\beta} \frac{\partial}{\partial \tau_{kp}}$, $S^{kp}_{\alpha\beta} = S^{pk}_{\alpha\beta}$. Then, over U

$$\nabla_\alpha(d\tau^{kp}|_S) = \nabla^{\circ}_\alpha(d\tau^{kp}) - \sum_\beta S^{kp}_{\alpha\beta} dz^\beta. \tag{6}$$

The right-hand side of (7) should be understood as the restriction of a $(1,0)$-form from U to S. The same interpretation applies to what follows. In the special case where S is a holomorphic curve we have for $x = \tau_o \in S$ and for α denoting a holomorphic vector field on S such that $\alpha(x)$ is of unit length

$$\bar{\partial}^* \bar{\partial} \eta = -\nabla_\alpha \partial_{\bar{\alpha}} \eta$$

$$= -\sum_{k,p,A,q} \alpha^{\overline{kp}} \alpha^A \eta^k_A d\tau^{pq} \otimes e_q + \sum_{k,p} S^{kp}_{\alpha\alpha} (\partial_{\bar{\alpha}} \eta)^p_{kp} dz^\alpha \otimes e_p. \tag{7}$$

In (7) the last term does not only depend on $\bar{\partial}(\eta|_S)$ but rather on $\bar{\partial}\eta$ for an the extended η on U. On the other hand by (3) the first order term in (7) can be replaced by a zero-order term, yielding at $x = \tau_o$

$$\bar{\partial}^*\bar{\partial}\eta = -\nabla_\alpha\partial_{\bar{\alpha}}\eta$$

$$= -\sum_{k,p,A,q} \alpha^{\overline{kp}}\alpha^A\eta_A^k d\tau^{pq}\otimes e_q - i\sum_{k,p} S_{\alpha\alpha}^{kp}\bar{\eta}_\alpha^k dz^\alpha\otimes e_p. \tag{8}$$

which is a well-defined elliptic equation for η on the curve C. Integrating $<\bar{\partial}^*\bar{\partial}\eta,\eta>$ over C and using the interpretation (5) for $\nabla_\alpha^o\bar{\partial}_\alpha\eta$, we obtain

$$\int_C \|\bar{\partial}\eta\|^2 = \int_C \left(\tilde{\Theta}(\eta,\bar{\eta}) + S(\bar{\eta},\bar{\eta})\right), \tag{9}$$

where at $x \in C$, $\tilde{\Theta}(\eta,\bar{\eta}) = \Theta_{\eta(\alpha)\overline{\eta(\alpha)}\alpha\bar{\alpha}}$ for a unit vector $\alpha \in T_x^{1,0}(C)$, $S(\cdot,\cdot)$ is a complex symmetric bilinear form on $\overline{(\Omega_C\otimes V)}_x$ which, when lifted to τ_o can be expressed as

$$S(\bar{\eta},\bar{\eta}) = -i\sum_{k,p} S_{\alpha\alpha}^{kp}\bar{\eta}_\alpha^k\bar{\eta}_\alpha^p. \tag{10}$$

In what follows x will be lifted to τ_o. Put the unit vector in the normal form $\frac{1}{\sqrt{n}}\,\mathrm{diag}(\alpha_1,...,\alpha_n)$ with α_i real, $\alpha_1 \geq \cdots \geq \alpha_n \geq 0$. $\sum|\alpha_i|^2 = n$. We have

$$\Theta_{\eta(\alpha)\overline{\eta(\alpha)}\alpha\bar{\alpha}} \leq -\frac{1}{n}|\alpha_n|^2\|\eta\|^2. \tag{11}$$

Suppose the holomorphic sectional curvature in $(\mathcal{H}_n,g)$ is $-\frac{2}{n}\left(1+\frac{a}{4n}\right)$, $0 \leq a \leq 1$, we have (cf. [(2.2), proof of Theorem 2])

$$|\alpha_n|^4 + (n-1)\left(\frac{n-|\alpha_n|^2}{n-1}\right)^2 \geq n\left(1+\frac{a}{4n}\right), \tag{12}$$

implying (as in [Mok2, (4.3), Proposition 1])

$$0 \leq 1 - |\alpha_n|^2 \leq \frac{1}{2}\sqrt{\frac{(n-1)a}{n}} < \frac{\sqrt{a}}{2}. \tag{13}$$

The pinching hypothesis and the Gauss equation imply that for the second fundamental form $\sigma = \sigma_{C|X}$, we have $\|\sigma\|^2 = \frac{b}{2n^2}$ such that $a+b \leq 1$, so that

$$|S(\bar{\eta},\bar{\eta})| \leq \frac{\sqrt{b}}{\sqrt{2}n}\|\eta\|^2. \tag{14}$$

From (9) we can deduce $\eta \equiv 0$ if $\tilde{\Theta}(\eta,\bar{\eta}) + S(\bar{\eta},\bar{\eta}) < 0$. From (11), (13) and (14) this will be the case if for all a, $b \geq 0$, $0 \leq a+b \leq 1$

$$1 - \frac{\sqrt{a}}{2} \geq \frac{\sqrt{b}}{\sqrt{2}}, \tag{15}$$

which follows from $\sqrt{a} + \sqrt{b} \geq 2$. With this we have proved that for any $f \in H^\circ(C, \mathcal{V}/L_{\mathbf{R}})$, $\eta(f) \equiv 0$, contradicting with the non-vanishing theorem $h(C) \neq 0$ obtained in (3.2) by the Riemann-Roch Theorem. With this we have completed the proof of [(3.1), Theorem 1] and verified the gap phenomenon for $(\mathcal{H}_n, \mathcal{H}; i)$ on the Siegel upper half-plane $\mathcal{H}_n$.

In the case of arithmetic quotients $X = \mathcal{H}_n/\Gamma$ Eyssidieux [Eys1] showed that Theorem 1 continues to hold for $C \hookrightarrow X$ of finite volume. The proof makes use of Mumford compactifications and the SL(2)-orbit Theorem of Schmidt.

4. An optimal Arakelov inequality on Chern numbers for compact complex surfaces of quotients of $B^2 \times B^2$

(4.1) Starting from the gap phenomenon on the Siegel upper half-plane $\mathcal{H}_n$ obtained in §3, our perspective is that the same line of thought, formulated in terms of variations of Hodge structures, will lead to the gap phenomenon for compact complex surfaces on quotients of bounded symmetric domains or of period domains. In this section as an illustration we consider complex surfaces on quotients of $B^2 \times B^2$. While the results of §2 already implies the gap phenomenon for $(B^2 \times B^2, B^2; \delta)$ we will derive in this section and the next a new proof, giving a method which can probably be generalized. We will obtain along the way Chern-number inequalities on $B^2 \times B^2$ characterizing certain holomorphic geodesic cycles, a result which is of independent interest.

Recall that in §3 for curves C on quotients of $\mathcal{H}_n$ we obtain a non-vanishing theorem $h(C) \neq 0$ from the Riemann-Roch formula and from the Gauss equation ([(3.2), eqn. 4), which yielded $\deg(V|_C) \geq -n(g(C) - 1)$, an inequality due to Arakelov. Our point of departure is to obtain new inequalites on Chern numbers of complex submanifolds S of quotients X of bounded symmetric domain by interpreting the latter as parameter spaces for polarized Hodge structures, following an idea due to Deligne. We refer the reader back to the Introduction of this article for an informal definition of what we call an Arakelov inequality. We are interested in those which are *optimal* in the sense that equality holds for certain holomorphic geodesic cycles. For the purpose of applying such inequalities to verify the gap phenomenon in the same way as in §3 for the case of $(\mathcal{H}_n, \mathcal{H}; i)$ we will need to show that such inequalities are *local* in the sense that they are obtained from integrals of nonnegative (m, m)-forms ν_S over S, $m = \dim_{\mathbf{C}} S$, such that the vanishing of ν_S over an open set implies that S is totally geodesic.

In this section we will obtain an optimal Arakelov inequality on $B^2 \times B^2$ by considering adjoint representations of $SU(2,1) \times SU(2,1)$ on its Lie algebra. Let Γ be a torsion-free cocompact arithmetic group of automorphisms of $B^2 \times B^2$ and define $X = (B^2 \times B^2)/\Gamma$. We will sometimes denote by B_i^2; $i = 1, 2$; the i–th direct factor of $B^2 \times B^2$. Denote by $\pi : B^2 \times B^2 \to X$ the canonical projection. Write T_X for the holomorphic tangent bundle on X. We define two rank-2 holomorphic vector subbundles T_1 resp. $T_2 \subset T_X$ consisting of vectors tangent to $\pi(B^2 \times \{x_2\})$ resp. $\pi(\{x_1\} \times B^2)$, x_i being arbitrary points on B_i^2. In what follows we will prove a special case of Arakelov inequalities obtained in Eyssidieux [Eys1]. Furthermore, we will verify that the inequality is optimal and local in this special case, leading to a characterization theorem for certain totally geodesic holomorphic cycles.

Theorem 1. Let $S \hookrightarrow X$ be an immersed compact complex surface. Then, on S we have

$$c_2(S) \geq \frac{1}{6}\left(c_1^2(T_1|_S) + c_1^2(T_2|_S)\right)$$

Furthermore, equality is attained in the special case when S is totally geodesic and modelled on $(B^2 \times B^2, B^2; \delta)$, where $\delta : B^2 \to B^2 \times B^2$ denotes the diagonal embedding.

Here and henceforth we will identify an element of $H^4(S, \mathbf{R})$ with its evaluation on the fundamental class $[S]$ of S, equipped with the orientation defined by its complex structure. We remark that the inequality remains valid when we require only that there is a generically finite holomorphic map $f : S \to X$ and $T_i|_S$ is replaced by f^*T_i. From now on we will only consider the case of immersed compact complex surfaces $S \hookrightarrow X$ and denote $T_i|_S$ simply by T_i.

(4.2) *A Chern-number inequality of Gromov's on immersed compact complex surfaces.* As an introduction we start by recalling a special case of some result of Gromov's ([Gro]) giving Chern-number inequalities on immersed compact complex surfaces $i : S \hookrightarrow X$. These inequalities arise from ordinary de Rham cohomology. The inequalites are also valid when $X = (B^2 \times B^2)/\Gamma$ is replaced by any quotient of a bounded symmetric domain by a torsion-free cocompact lattice.

Proposition 1. (from Gromov [Gro, (1.3)]) Let $i : S \hookrightarrow X$ be an immersed compact complex surface. Then, we have

(i) $c_2(S) \geq 0$, and

(ii) $(-1)^r ch(\Omega_S^r) \cdot \mathrm{Todd}(T_S) \geq 0$, $r = 0, 1, 2$;

where Ω_S^0 is the trivial line bundle, Ω_S^1 is the holomorphic cotangent bundle of S, $\Omega_S^2 \equiv K_S$ is the canonical line bundle, $ch(\cdot)$ denotes the Chern character and $\mathrm{Todd}(\cdot)$ denotes the Todd class.

Proof (outline): The crux of Gromov's argument was to show that there are no L^2 harmonic p-forms for $p \neq 2$ on some covering space $\tilde{S} \to S$. Let $\Sigma \subset \pi_1(S)$ be the kernel of the canonical map $\pi_1(S) \to \pi_1(X)$ and denote by $\tilde{S} \to S$ the regular covering space corresponding to the normal subgroup $\Sigma \subset \pi_1(S)$. (Write $\pi : \tilde{X} \to X$ for the canonical projection for the universal covering $\tilde{X} = B^2 \times B^2$. If $S \subset X$ is embedded, then $\tilde{S}$ can be identified with a connected component of $\pi^{-1}(S)$.) By the assumption that $\pi_1(X) = \Gamma$ is arithmetic we know that there is a tower of unramified coverings $X_n \to X$ corresponding to subgroups $\Gamma_n \subset \Gamma$ of finite index such that $\cap \Gamma_n = \{id\}$. We have correspondingly a tower of unramified coverings $S_n \to S$ with $\cap \pi_1(S_n) = \Sigma$. X is Kähler-hyperbolic in the sense of Gromov ([Gro, para. (0.3), $p.265$]). More precisely, the Kähler form $\tilde{\omega}$ on $\tilde{X}$ corresponding to the Kähler-Einstein metric is of the form $d\alpha$, where α is a smooth 1-form on $\tilde{X}$ which is bounded with respect to the Kähler-Einstein metric. Denote also by $i : \tilde{S} \hookrightarrow \tilde{X}$ the immersion arising form $i : S \hookrightarrow X$. By restriction we have $i^*\tilde{\omega} = d(i^*\alpha)$ on $\tilde{S}$, so that the argument of Gromov ([Gro, Theorem $(1.4.A)$, $p.274$]) yields the vanishing of the space of L^2 harmonic p-forms on $\tilde{S}$, provided that $p \neq 2$. Equivalently this means that

$$\lim_{n \to \infty} \frac{\dim_{\mathbf{C}} H^2(S_n, \mathbf{C})}{[\pi_1(S) : \pi_1(S_n)]} = 0 \quad \text{for} \quad p \neq 2,$$

where $[\pi_1(S) : \pi_1(S_n)]$ stands for the index of $\pi_1(S_n)$ in $\pi_1(S)$. As a consequence, the cohomology is asymptotically concentrated in real dimension 2, *i.e.*,

$$c_2(S) = \lim_{n \to \infty} \frac{\dim_{\mathbf{C}} H^2(S_n, \mathbf{C})}{[\pi_1(S) : \pi_1(S_n)]} \geq 0.$$

From Hodge decomposition harmonic forms split into sums of harmonic forms of different bidegrees. The inequalities (ii) then follow from the Riemann-Roch formula.

(4.3) *Locally constant vector bundles and harmonic forms.* The proof of Proposition 1 can be generalized to the situation of variations of polarized Hodge structures. In our situation we are going to interpret $B^2 \times B^2$ as the parameter space for polarized Hodge structures of weight 2 arising from the adjoint representation of $G = SU(2,1) \times SU(2,1)$. Write $\rho : G \to GL(g)$ for the adjoint representation of G on its Lie algebra g. This gives rise to a representation of $\Gamma = \pi_1(X)$ on g and consequently a locally constant sheaf to be denoted by V_{ad}. Denote by $V_{\mathbf{C}}$ the complexification of V_{ad}. We will make no notational distiction between the locally constant sheaf $V_{\mathbf{C}}$ and the underlying locally constant complex vector bundle. We will denote the corresponding objects on the universal covering space $B^2 \times B^2$ by the same symbols. By a generalization of Gromov [Gro,Theorem (1.4.A), p.274] to the context of Hodge theory one can prove the analogue of [(4.2), Proposition 1] for the locally constant sheaf $i^* V^{\mathbf{C}}$ on S. The analogue of [(4.2), Proposition 1, statement (ii)] will then yield inequalities on Chern numbers which turn out to be sharp.

We proceed to formulate this analogy. The Killing form on g descends to a symmetric non-degenerate bilinear form B on V_{ad} which is indefinite. At each point x in $B^2 \times B^2$ we have the Cartan decomposition $g = k_x \oplus m_x$ into an orthogonal direct sum, where k_x is the Lie algebra of the isotropy subgroup at x and m_x is identified with the real tangent space at x. We have $B_x|_{k_x} < 0$ and $B_x|_{m_x} > 0$. In what follows the superscript $\mathbf{C}$ will denote the complexification of a real vector space. Write $g^{\mathbf{C}} = k_x^{\mathbf{C}} \oplus m_x^+ \oplus m_x^-$, where $m_x^{\mathbf{C}} = m^+ \oplus m^-$ is the decomposition of the complexified tangent space $m_x^{\mathbf{C}}$ into subspaces of type (1,0) and type (0,1). At each point $x \in B^2 \times B^2$ let C_x be the Weil operator on $g^{\mathbf{C}}$ which is id on $k_x^{\mathbf{C}}$ and $-id$ on $m_x^{\mathbf{C}}$. Define $b_x(v) = -B(Cv, \bar{v})$. Then, b_x is a Hermitian bilinear form on $g^{\mathbf{C}}$, identified with $V_{\mathbf{C},x}$, the fiber of $V_{\mathbf{C}} \to B^2 \times B^2$ over x. The family of Hermitian bilinear forms b_x then defines a Hermitian metric h on $V_{\mathbf{C}}$ over $B^2 \times B^2$. Furthermore it is invariant under the action of G on $V_{\mathbf{C}}$ over $B^2 \times B^2$ defined by the adjoint representation ρ. This means that h descends to a Hermitian metric, to be denoted by the same symbol, on the locally constant complex vector bundle $V_{\mathbf{C}}$ on X. Since $V_{\mathbf{C}}$ is locally constant over X there is a canonical flat connection d. We note that the Hermitian metric h is not parallel with respect to d. Nonetheless we still have

Proposition 2. Let $i : S \hookrightarrow X$ be an immersed compact complex surface in X and let $\tilde{S} \to S$ be the covering space corresponding to $\Sigma \subset \pi_1(S)$, where Σ is the kernel of the canonical map $\pi_1(S) \to \pi_1(X)$. Denote also by $V_{\mathbf{C}}$ the induced locally constant bundle over $\tilde{S}$. Then, every L^2 harmonic p-form on $\tilde{S}$ with values in $V_{\mathbf{C}}$ is trivial provided that $p \neq 2$.

Here L^2 norms are measured with respect to the Kähler-Einstein metric on $B^2 \times B^2$ and the Hermitian metric h on $V_{\mathbf{C}}$ over $B^2 \times B^2$.

(4.4) *A brief description of the underlying variation of Hodge structures on $V_{\mathbf{C}}$.* The proof of Proposition 2 follows easily from the proof of Gromov ([Gro, Theorem (1.4.A), p.274ff.]). In what follows we will give a brief description of the variation of Hodge structures underlying $V_{\mathbf{C}}$ and state the necessary relevant and standard properties which make the generalization possible. This description of the Hodge structures will also be used in (4.5) to obtain Chern-number inequalities. The following applies to any bounded symmetric domain $X_o = G/K$ in place of $B^2 \times B^2$. Here we adopt notations in conformity with the usual practice in Hodge theory and note that such notations are not necessarily consistent with those used in §3. The description that follows is informal and presupposes some rudimentary knowledge about bounded symmetric domains (*cf.* for instance Mok [Mok1, Chap. 3-5]).

The complexified Cartan decomposition

$$V_{\mathbf{C},x} = m_x^- \oplus k_x \oplus m_x^+ := H_x^{20} \oplus H_x^{11} \oplus H_x^{02}$$

gives rise to a smooth decomposition of $V_{\mathbf{C}} = H^{20} \oplus H^{11} \oplus H^{02}$ over $B^2 \times B^2$ into smooth complex vector bundles. $H^{20} \cong \overline{T}$ can be endowed the holomorphic structure of the holomorphic cotangent bundle by contracting with the Kähler metric and as such it is holomorphically embedded in $V_{\mathbf{C}}$. We can see this as follows. Let X_c be the compact dual of X_o and $X_o \subset\subset X_c$ be the Borel embedding. Then the local system $V_{\mathbf{C}}$ and the complex Cartan decomposition $V_{\mathbf{C},x} \cong g_{\mathbf{C}} = m_x^- \oplus k_x \oplus m_x^+$ can be defined over X_c. Write $V_{\mathbf{C}} = H^{20} \oplus H^{11} \oplus H^{02}$ for the corresponding smooth global decomposition over X_c. To say that $H^{20} \subset V_{\mathbf{C}}$ is a holomorphic subbundle it is equivalent to show that H^{20} is invariant under the action of the complex Lie group $G^{\mathbf{C}}$ of biholomorphisms of X_c. Write $X_c = G^{\mathbf{C}}/P_x$ as a complex homogeneous space where P_x is the Borel subgroup corresponding to the complex Lie subalgebra $p_x = m_x^- \oplus k_x^{\mathbf{C}}$. The infinitesimal action of P_x on $V_{\mathbf{C}}$ at x is given by the adjoint action, *i.e.*, by Lie brackets. But from the theory of bounded symmetric domains we have

$$[m_x^-, m_x^-] = 0; \ [k_x^{\mathbf{C}}, m_x^-] \subset m_x^-; \text{ and } [m_x^+, m_x^-] \subset k_x^{\mathbf{C}} \qquad (*)$$

(*cf.* for instance Mok[Mok1, (3.1), pp.51-52]). The first two relations then imply that $H^{20} \subset V_{\mathbf{C}}$ is a holomorphic subbundle.

Recall that d is the canonical flat connection on $V_{\mathbf{C}}$. In what follows we explain why $H^{20} \subset V_{\mathbf{C}}$ is not a d-parallel subbundle and how the discrepancy from being d-parallel is measured. Denote by $\Omega^{p,q}$ the bundle of (p,q)-covectors. The obstruction on H^{20} to being d-parallel with respect to d is measured by the second fundamental form σ of H^{20} in $V_{\mathbf{C}}$. $\sigma : T \otimes H^{20} \to H^{11} \oplus H^{02}$ is a smooth bundle map. σ is nontrivial since in fact $[m_x^+, m_x^-] = k_x^{\mathbf{C}}$ The last relation in $(*)$ implies that the image of σ lies in H^{11}. The second fundamental form σ corresponds equivalently to a smooth bundle map $\nabla' : H^{20} \to H^{11} \otimes \Omega^{1,0}$, which is called a Gauss-Manin connection. ∇' is in fact a parallel bundle map between homogeneous holmorphic vector bundles which happens in this case to be injective. In the same vein, $(*)$ together with the relationship $[k_x^{\mathbf{C}}, k_x^{\mathbf{C}}] \subset k_x^{\mathbf{C}}$ means that $H^{20} \oplus H^{11} \subset V_{\mathbf{C}}$ is a holomorphic subbundle and that the obstruction to its being d-parallel is given by a Gauss-Manin connection $H^{11} \to H^{02} \otimes \Omega^{1,0}$. The description thus far can be recaptured by describing the locally constant vector bundle $V_{\mathbf{C}}$ over Ω as arising from a variation of Hodge structures, as follows (*cf.* Zucker [Zu]).

The $(1,0)$ part of the flat connection d on $V_{\mathbf{C}}$, to be denoted by ∇, decomposes according to $\nabla = d' + \nabla'$, where

$$d'(\mathcal{C}^{\infty}(H^{pq})) \subset \mathcal{C}^{\infty}(H^{pq} \otimes \Omega^{1,0});$$

$$\nabla'(\mathcal{C}^{\infty}(H^{pq})) \subset \mathcal{C}^{\infty}(H^{p-1,q+1} \otimes \Omega^{1,0}).$$

Here $\mathcal{C}^{\infty}(\cdot)$ denotes the sheaf of germs of smooth sections of the bundles concerned. The $(0,1)$ part of the flat connection d on $V_{\mathbf{C}}$, to be denoted by $\overline{\nabla}$, decomposes according to $\overline{\nabla} = d'' + \nabla''$, where

$$d''(\mathcal{C}^{\infty}(H^{pq})) \subset \mathcal{C}^{\infty}(H^{pq} \otimes \Omega^{0,1});$$

$$\nabla''(\mathcal{C}^{\infty}(H^{pq})) \subset \mathcal{C}^{\infty}(H^{p+1,q-1} \otimes \Omega^{0,1}).$$

The d''-operator endows H^{pq} with holomorphic structures. Furthermore, the smooth complex vector bundle morphism $\nabla' : H^{pq} \to H^{p-1,q+1} \otimes \Omega^{1,0}$ is holomorphic. This morphism is called the Gauss-Manin connection. Putting $D'' = d'' + \nabla'$ and $D' = d' + \nabla''$ and denoting by the same symbols the corresponding exterior differential operators on $V_{\mathbf{C}}$-valued differential forms, the de Rham complex $(V_{\mathbf{C}}, d)$ is bifiltered by

$$\Omega^n(V_{\mathbf{C}}) = \bigoplus_{P+Q=n+2} \Omega^n(V_{\mathbf{C}})^{P,Q};$$

$$\Omega^n(V_{\mathbf{C}})^{P,Q} = \bigoplus_{p+r=P, q+s=Q} H^{pq} \otimes \Omega^{r,s}$$

for $0 \leq n \leq 2\dim_{\mathbf{C}} X_o$ such that

$$D'' \left(\mathcal{C}^{\infty}(\Omega^n(V_{\mathbf{C}})^{P,Q})\right) \subset \mathcal{C}^{\infty}(\Omega^{n+1}(V_{\mathbf{C}})^{P,Q+1});$$

$$D' \left(\mathcal{C}^{\infty}(\Omega^n(V_{\mathbf{C}})^{P,Q})\right) \subset \mathcal{C}^{\infty}(\Omega^{n+1}(V_{\mathbf{C}})^{P+1,Q}). \qquad (*)$$

Denote also by d the exterior de Rham operators and decompose accordingly $d = D' + D''$. Then, from $d^2 = 0$ and $(*)$ it follows readily $D'^2 = D''^2 = 0$. With this formalism the classical Hodge-theoretic identities can be proved and the simple Lefschetz-type argument of Gromov's can be adapted to yield ([4.3], Proposition 1), *i.e*, the non-existence of L^2 harmonic p-forms on $\tilde{S}$ with values in $V_{\mathbf{C}}$ for $p \neq 2$.

(4.5) *An index formula.* Put $\mathcal{E}_S^{P,Q} = \mathcal{C}^{\infty}(S, \Omega^n(V_{\mathbf{C}})^{P,Q}$. As an analogue to [(4.2), Proposition 1] we consider the complex

$$(\mathcal{E}_S^{P,*}; D'') = \left(\cdots \to \mathcal{E}_S^{P,Q} \xrightarrow{D''} \mathcal{E}_S^{P,Q+1} \to \cdots\right)$$

arising from the de Rham complex $(V_{\mathbf{C}}, d)$. We have the following in-equality.

Proposition 3. For the complex $(\mathcal{E}_S^{P,*}; D'')$ the inequality

$$(-1)^p \chi(\mathcal{E}_S^{P,*}; D'') \geq 0$$

holds for the Euler-Poincaré characteristic χ.

Proof: The Kähler identities imply that $\Delta_{D''} = \frac{1}{2}\Delta_d$ for the Laplacians $\Delta_{D''} = D''D''^* + D''^*D''$ and $\Delta_d = dd^* + d^*d$. As a consequence harmonic forms with values in $V_{\mathbf{C}}$ over S decompose according to the (P, Q)-bigrading. Let $S_n \to S$ be a tower of unramified coverings as in [(5.4), Proposition 2]. The vanishing theorem in [(4.2), proof of Proposition 1] then yields

$$(-1)^p \chi(\mathcal{E}_S^{P,*}; D'') = (-1)^p \lim_{n\to\infty} \frac{\chi(\mathcal{E}_{S_n}^{P,*}; D'')}{[\pi_1(S) : \pi_1(S_n)]}$$

$$= \lim_{n\to\infty} \frac{\dim_{\mathbf{C}} H^2(S_n, V_{\mathbf{C}})^{P,4-P}}{[\pi_1(S) : \pi_1(S_n)]} \geq 0.$$

(4.6) *Proof of Theorem 1.* Consider for $p = 3$ the complex $(\mathcal{E}_S^{3,*}; D'')$ as defined in (4.5). There is another complex

$$(\mathcal{E}_S^{3,*}; d'') = \left(0 \to \mathcal{E}_S^{3,0} \xrightarrow{d''} \mathcal{E}_S^{3,1} \xrightarrow{d''} \mathcal{E}_S^{3,2} \xrightarrow{d''} \mathcal{E}_S^{3,3} \to 0\right) \qquad (1)$$

corresponding to

$$0 \to H^{20} \otimes \Omega^{1,0} \xrightarrow{\bar{\partial}} H^{20} \otimes \Omega^{1,1} \xrightarrow{\bar{\partial}} H^{20} \otimes \Omega^{1,2}$$

$$\oplus \qquad\qquad\qquad \oplus$$

$$H^{11} \otimes \Omega^{2,0} \xrightarrow{\bar{\partial}} H^{11} \otimes \Omega^{2,1} \xrightarrow{\bar{\partial}} H^{11} \otimes \Omega^{2,2} \to 0. \qquad (2)$$

The complex $(\mathcal{E}_S^{3,*}; d'')$ has the same index as that of $(\mathcal{E}_S^{3,*}; D'')$ since the Gauss-Manin connection ∇' is of order zero. We have thus

$$0 \leq -\chi(\mathcal{E}_S^{3,*}; D'') = -\chi(\mathcal{E}_S^{3,*}; d'')$$

$$= \chi(S, H^{11} \otimes \Omega^2) - \chi(S, H^{20} \otimes \Omega^1) \qquad (3)$$

where $\Omega^1 = \Omega_S$ denotes the cotangent bundle and $\Omega^2 = K_S$ denotes the canonical line bundle on S. Denote by T the holomorphic tangent bundle on X. Then, $T = T_1 \oplus T_2$ and $H^{20} \cong T^* = T_1^* \oplus T_2^*$, $H^{11} \cong End(T_1) \oplus End(T_2) \cong (T_1^* \otimes T_1) \oplus (T_2^* \otimes T_2)$. We have thus over S

$$H^{20} \otimes \Omega_S = (T_1^* \otimes \Omega_S) \oplus (T_2^* \otimes \Omega_S)$$

$$H^{11} \otimes \Omega^2 = (T_1^* \otimes T_1 \otimes K_S) \oplus (T_2^* \otimes T_2 \otimes K_S) \qquad (4)$$

Suppose $S = S_o$ is totally geodesic and modelled on $(B^2 \times B^2, B^2; \delta)$. Then, $T_1|_{S_o} \cong T_2|_{S_o} \cong T_S$, the holomorphic tangent bundle over S. Then for $i = 1, 2$, we have $T_S \otimes K_S \cong T_S^* = \Omega^1$, so that the two holomorphic vector bundles are isomorphic and we obtain

$$\chi(\mathcal{E}_{S_o}^{3,*}; D'') = \chi(S_o, H^{11} \otimes \Omega^2) - \chi(S_o, H^{20} \otimes \Omega^1) = 0. \qquad (5)$$

In general, by Serre duality we have

$$\chi(S, H^{11} \otimes \Omega^2) = \chi(S, (H^{11})^*) = \chi(S, T_1^* \otimes T_1) + \chi(S, T_2^* \otimes T_2). \qquad (6)$$

For a cohomology class η in $H^*(S, \mathbf{R})$ we denote by η^4 the component in $H^4(S, \mathbf{R})$. We identify furthermore $H^4(S, \mathbf{R})$ with $\mathbf{R}$ by evaluating on the fundamental class $[S]$ determined by the orientation defined by the complex structure. We have

$$\chi(S, H^{20} \otimes \Omega^1) - \chi(S, H^{11} \otimes \Omega^2)$$
$$= [((ch(T_1^* \otimes \Omega_S) - ch(T_1^* \otimes T_1))$$
$$+ (ch(T_2^* \otimes \Omega_S) - ch(T_2^* \otimes T_2))) \cdot Todd(S)]^4. \qquad (7)$$

For $i = 1, 2$ we have

$$ch(T_i^* \otimes T_i) = 4 + [c_1^2(T_i) - 4c_2(T_i)]; \qquad (8)$$
$$ch(T_i^* \otimes \Omega_S) = 4 - 2c_1(T_i) - 2c_1(S) + [c_1^2(T_i) - 2c_2(T_i)] +$$
$$+ [c_1^2(S) - 2c_2(S)] + c_1(T_i) \cdot c_1(S); \qquad (9)$$
$$Todd(S) = 1 + \frac{1}{2}c_1(S) + \frac{1}{12}[c_1^2(S) + c_2(S)]. \qquad (10)$$

Substituting (8), (9) and (10) into (7) and using the fact that $c_1^2(T_i) = 3c_2(T_i)$ we obtain

$$-\frac{1}{3}\left(c_1^2(T_1) + c_1^2(T_2)\right)[c_1(T_1) + c_1(T_2) + 2c_1(S)] \cdot c_1(S)$$
$$-\frac{1}{3}\left(c_1^2(T_1) + c_1^2(T_2)\right) - 2[c_1^2(S) - 2c_2(S)] - c_1(T_1) \cdot c_1(S) - c_1(T_2) \cdot c_1(S) \geq 0,$$

which simplifies to

$$4c_2(S) \geq \frac{2}{3}\left(c_1^2(T_1) + c_1^2(T_2)\right),$$

proving [(4.1), Theorem 1], as desired.

5. Characterization of compact holomorphic geodesic cycles on quotients of $B^2 \times B^2$ modelled on $(B^2 \times B^2, B^2; \delta)$

(5.1) In this section we will show that the Arakelov inequality on Chern numbers obtained in §4 for $(B^2 \times B^2, B^2; \delta)$ is local. First we set up some notations. Let g_i, $i = 1$, 2, be Kähler-Einstein metrics on B_i^2 with the same Einstein constant. Denote by $g = g_1 + g_2$ the Kähler-Einstein metric on $B^2 \times B^2$ and by h_i the Hermitian metric on the canonical line bundle K_i on B_i^2 thus obtained. We will normalize g_i so that (B_i^2, g_i) is of constant holomorphic sectional curvature -2. We also consider (K_i, h_i) as a Hermitian holomorphic line bundle on $B^2 \times B^2$ in the obvious way. We have

Theorem 1. Let $U \subset B^2 \times B^2$ be an open subset and $S \subset U$ be a 2-dimensional complex submanifold on U. Then, in terms of Chern forms arising from the Kähler metric $g|_S$ on S and the Hermitian metrics h_i on K_i, we have the pointwise inequality

$$c_2(S, g|_S) \geq \frac{1}{6} \left(c_1^2(K_1, h_1)|_S + c_1^2(K_2, h_2)|_S \right)$$

between (2,2)-forms on S. Furthermore, equality holds everywhere on U if and only if S is totally geodesic and (U, S) is modelled on $(B^2 \times B^2, \delta(B^2))$ for the diagonal embedding δ.

In this section we will only be considering characteristic forms on $S \subset B^2 \times B^2$. The notation g will stand for $g|_S$ while $c_1(K_i, h_i)$ will mean the restriction of first Chern forms on S. The inequality in Theorem 1 is an inequality between real differential 4-forms on the underlying real 4-manifold S. The Arakelov inequality on Chern numbers in §4 for compact complex surfaces on quotients of $B^2 \times B^2$ is then a consequence of Theorem 1 obtained by integrating real 4-forms. In other words, Theorem 1 asserts that the Arakelov inequality in §4 for $(B^2 \times B^2, B^2; \delta)$ is local in nature. While the argument in §4 via harmonic forms is the origin of Theorem 1, the proof of the latter is completely independent of Hodge-theoretic considerations. It consists of a straight-forward verification using the Chern-Weil theory of characteristic forms.

(5.2) *Preliminaries on the second Chern form.* Let (M, g) be a Kähler manifold with Kähler form ω. From the Chern-Weil theory of characteristic forms the second Chern class $c_2(M)$ is represented by the (2,2) form $c_2(M, g)$

$$\frac{1}{8\pi^2} \sum_{\alpha, \beta, \gamma, \delta} T_{\alpha \bar{\beta} \gamma \bar{\delta}} dz^\alpha \wedge d\bar{z}^\beta \wedge dz^\gamma \wedge d\bar{z}^\delta,$$

where

$$T_{\alpha\bar\beta\gamma\bar\delta} = \sum_{\lambda,\mu,\sigma,\tau} g^{\lambda\bar\sigma} g^{\mu\bar\tau} R_{\lambda\bar\tau\alpha\bar\beta} R_{\mu\bar\sigma\gamma\bar\delta} - \sum_{\lambda,\mu,\sigma,\tau} g^{\lambda\bar\sigma} g^{\mu\bar\tau} R_{\lambda\bar\sigma\alpha\beta} R_{\mu\bar\tau\gamma\bar\delta}. \quad (1)$$

If at a point $x \in U$, we choose coordinates such that $g_{i\bar j}(x) = \delta_{ij}$, then

$$T_{\alpha\bar\beta\gamma\bar\delta} = \sum_{\lambda,\mu} R_{\lambda\bar\mu\alpha\bar\beta} R_{\mu\bar\lambda\gamma\bar\delta} - \sum_{\lambda,\mu} R_{\lambda\bar\lambda\alpha\bar\beta} R_{\mu\bar\mu\gamma\bar\delta}$$

$$= \sum_{\lambda,\mu} R_{\lambda\bar\mu\alpha\bar\beta} R_{\mu\bar\lambda\gamma\bar\delta} - R_{\alpha\bar\beta} R_{\gamma\bar\delta}. \quad (2)$$

Let ω be the Kähler form. Then, $c_2(S,g) \wedge \frac{\omega^{n-2}}{(n-2)!} = \frac{E}{8\pi^2}\frac{\omega^n}{n!}$, where

$$E = \sum_{\lambda,\mu;\alpha\neq\beta} R_{\lambda\bar\mu\alpha\bar\beta} R_{\mu\bar\lambda\beta\bar\alpha} - \sum_{\alpha\neq\beta} R_{\alpha\bar\beta} R_{\beta\bar\alpha}$$

$$- \sum_{\lambda,\mu;alpha\neq\beta} R_{\lambda\bar\mu\alpha\bar\alpha} R_{\mu\bar\lambda\beta\bar\beta} - \sum_{\alpha\neq\beta} R_{\alpha\bar\alpha} R_{\beta\bar\beta} \quad (3)$$

There is a change in signs because $dz^\alpha \wedge d\bar z^\alpha \wedge dz^\beta \wedge d\bar z^\beta = -dz^\alpha \wedge d\bar z^\alpha \wedge dz^\beta \wedge d\bar z^\beta$. In this summation we can remove the restriction that $\alpha \neq \beta$ because the case of $\alpha = \beta$ corresponds to 2 pairs of terms with opposite signs. Using the extended Einstein summation convention (with $g_{ij} = \delta_{ij}$ at a point) we have

$$E = R_{\lambda\bar\mu\alpha\bar\beta} R_{\mu\bar\lambda\beta\bar\alpha} - R_{\lambda\bar\mu\alpha\bar\alpha} R_{\mu\bar\lambda\beta\bar\beta} - R_{\alpha\bar\beta} R_{\beta\bar\alpha} + R_{\alpha\bar\alpha} R_{\beta\bar\beta}$$

$$= \|R\|^2 - 2\|Ric\|^2 + K^2, \quad (4)$$

where K denotes the scalar curvature $K = \sum R_{\alpha\bar\alpha}$. In the special case of a Kähler-Einstein surface, $R_{1\bar1} = R_{2\bar2} = \chi$, $|Ric|^2 = 2\chi^2$, $K^2 = 4\chi^2$, so that $2|Ric|^2 \equiv K^2$, and $E = |R|^2 \geq 0$.

We now return to our situation where $S \subset U \subset B^2 \times B^2$ and g is the restriction of the canonical Kähler-Einstein metric on $B^2 \times B^2$. Recall that (B_i^2, g_i) is of constant holomorphic sectional curvature -2. Consider the submanifold $S_o \subset B^2 \times B^2$ which the image of B^2 under the diagonal embedding $\delta : B^2 \to B^2 \times B^2$ given by $\delta(z) = (z, z)$. Since (B_i^2, g_i) is of constant holomorphic sectional curvature -2 the diagonal (S_o, g) of $(B^2 \times B^2, g)$ is of constant holomorphic sectional curvature -1. For $x \in S_o$ write $\{\alpha, \beta\}$ for an arbitrary orthonormal basis at x. We have

$$R_{\alpha\bar\alpha\alpha\bar\alpha} = R_{\beta\bar\beta\beta\bar\beta} = -1$$

$$R_{\alpha\bar\alpha\beta\bar\beta} = R_{\beta\bar\beta\alpha\bar\alpha} = R_{\alpha\bar\beta\beta\bar\alpha} = R_{\beta\bar\alpha\alpha\bar\beta} = -1, \quad (5)$$

while all other terms are 0. In this case $\|R\|^2 = 1 + 1 + 4 \times \frac{1}{4} = 3$ so that $c_2(S_o, g) = \left(\frac{1}{16\pi^2} \times 3\omega^2\right) = \frac{3\omega^2}{16\pi^2}$, where ω denotes the Kähler form of (S_o, g). On the other hand on (B_i^2, g_i), the Ricci form is given by $Ric = -3\omega_i$, where ω_i denotes the Kähler form corresponding to g_i. We have $\omega = 2\omega_1 = 2\omega_2$ on S_o since S_o is the diagonal in $B^2 \times B^2$. Thus on (S_o, g) we have

$$c_1^2(K_i, h_i) = \frac{1}{4\pi^2} \times (-3\omega_i)^2 = \frac{9\omega_i^2}{4\pi^2} = \frac{9\omega^2}{16\pi^2},$$

so that

$$c_2(S_o, g) = \frac{3\omega^2}{16\pi^2} = \frac{1}{6}\left(\frac{9\omega^2}{16\pi^2} + \frac{9\omega^2}{16\pi^2}\right)$$

$$= \frac{1}{6}\left(c_1^2(K_1, h_1) + c_1^2(K_1, h_1)\right). \tag{6}$$

This shows that for the diagonal $S_o = \delta(B^2)$ we have equality in Theorem 1. To prove Theorem 1 we have to show that for an *arbitrary* complex surface $S \subset B^2 \times B^2$ the inequality on Chern forms is valid and that furthermore pointwise equality on an open set implies that S is an open subset of $\delta(B^2)$ up to a biholomorphism of $B^2 \times B^2$.

(5.3) *Calculation of $c_2(S, g)$ for a complex surface S in $B^2 \times B^2$ at a point with vanishing second fundamental form.* Fix a point x on S which we may consider to be the origin in $B^2 \times B^2$. Denote by Ω the curvature tensor on $B^2 \times B^2$, Ω_i the curvature of the direct summand B_i^2; and R the curvature tensor on the submanifold S and write $(R_{i\bar{j}})$ for its Ricci curvature, K for its scalar curvature. We choose an orthonormal basis for $T_x^{1,0}(S)$ consisting of 2 orthogonal unit vectors $\alpha = (\alpha_1, \alpha_2)$ and $\beta = (\beta_1, \beta_2)$ such that the Ricci tensor is diagonalized with respect to the basis (α, β), i.e., $R_{\alpha\bar{\beta}} = 0$. We assume for the time being that the second fundamental form σ vanishes at x. By (5.2) to calculate the second Chern form $c_2(S, g)$ we have to compute $E = \|R\|^2 - 2\|Ric\|^2 + K^2$. We have

$$R_{\alpha\bar{\alpha}\alpha\bar{\alpha}} = \Omega_1(\alpha_1, \bar{\alpha}_1; \alpha_1, \bar{\alpha}_1) + \Omega_2(\alpha_2, \bar{\alpha}_2; \alpha_2, \bar{\alpha}_2)$$

$$= -2\|\alpha_1\|^4 - 2\|\alpha_2\|^4; \tag{1}$$

$$R_{\beta\bar{\beta}\beta\bar{\beta}} = -2\|\beta_1\|^4 - 2\|\beta_2\|^4; \tag{2}$$

$$R_{\alpha\bar{\alpha}\beta\bar{\beta}} = R_{\beta\bar{\beta}\alpha\bar{\alpha}} = R_{\alpha\bar{\beta}\beta\bar{\alpha}} = R_{\beta\bar{\alpha}\alpha\bar{\beta}}$$

$$= \Omega_1(\alpha_1, \bar{\alpha}_1; \beta_1, \bar{\beta}_1) + \Omega_2(\alpha_2, \bar{\alpha}_2; \beta_2, \bar{\beta}_2)$$

$$= -\|\alpha_1\|^2\|\beta_1\|^2 - \|\alpha_2\|^2\|\beta_2\|^2 - |<\alpha_1,\beta_1>|^2 - |<\alpha_2,\beta_2>|^2$$

$$\leq -\left(\|\alpha_1\|^2\|\beta_1\|^2 + \|\alpha_2\|^2\|\beta_2\|^2\right) \tag{3}$$

Ignoring the terms $R_{\alpha\bar{\beta}\alpha\bar{\alpha}}, R_{\beta\bar{\alpha}\beta\bar{\beta}}, R_{\alpha\bar{\beta}\alpha\bar{\beta}}$ and similar terms we have

$$\|R\|^2 \geq 4\left(\|\alpha_1\|^4 + \|\alpha_2\|^4\right)^2 + 4\left(\|\beta_1\|^4 + \|\beta_2\|^4\right)^2 +$$

$$4\left(\|\alpha_1\|^2\|\beta_1\|^2 + \|\alpha_2\|^2\|\beta_2\|^2\right)^2. \tag{4}$$

Writing without loss of generality $\|\alpha_1\|^2 = 0.5 + \lambda$, $0 \leq \lambda \leq 0.5$, $\|\beta_1\|^2 = 0.5 \pm \mu$; $0 \leq \mu \leq 0.5$, we have

$$\|\alpha_1\|^4 + \|\alpha_2\|^4 = \left(\|\alpha_1\|^2 + \|\alpha_2\|^2\right)^2 - 2\|\alpha_1^2\| \cdot \|\alpha_2\|^2$$

$$= 1 - 2(0.5 + \lambda)(0.5 - \lambda) = 1 - 2(0.25 - \lambda^2) = 0.5 + 2\lambda^2; \tag{5}$$

$$\|\beta_1\|^4 + \|\beta_2\|^4 = 0.5 + 2\mu^2; \tag{6}$$

$$\|\alpha_1\|^2\|\beta_1\|^2 + \|\alpha_2\|^2\|\beta_2\|^2$$

$$= (0.5 + \lambda)(0.5 \pm \mu) + (0.5 - \lambda)(0.5 \mp \mu) = 0.5 \pm 2\lambda\mu. \tag{7}$$

Thus,

$$\frac{1}{4}\|R\|^2 \geq (0.5 + 2\lambda^2)^2 + (0.5 + 2\mu^2)^2 + (0.5 \pm 2\lambda\mu)^2$$

$$\geq 0.75 + 2\lambda^2 + 4\lambda^2 + 2\mu^2 + 4\mu^4 \pm 2\lambda\mu + 4\lambda^2\mu^2. \tag{8}$$

By our choice of coordinates the Ricci tensor is diagonalized, *i.e.*, $R_{\alpha\bar{\beta}} = 0$. Write $R_{\alpha\bar{\alpha}} = A$, $R_{\beta\bar{\beta}} = B$. We have

$$-2\|Ric\|^2 + K^2 = -2(A^2 + B^2) + (A + B)^2$$

$$= -(A^2 + B^2) + 2AB = -(A - B)^2. \tag{9}$$

Since $A - B = (R_{\alpha\bar{\alpha}\alpha\bar{\alpha}} + R_{\alpha\bar{\alpha}\beta\bar{\beta}}) - (R_{\beta\bar{\beta}\beta\bar{\beta}} + R_{\beta\bar{\beta}\alpha\bar{\alpha}}) = R_{\alpha\bar{\alpha}\alpha\bar{\alpha}} - R_{\beta\bar{\beta}\beta\bar{\beta}}$, we have by (8)

$$-2\|Ric\|^2 + K^2 = -\left(2(\|\alpha_1\|^4 + \|\alpha_2\|^4) - 2(\|\beta_1\|^4 + \|\beta_2\|^4)\right)^2$$

$$= -4\left((0.5 + 2\lambda^2) - (0.5 + 2\mu^2)\right)^2$$

$$= -16(\lambda^2 - \mu^2)^2 = -16(\lambda^4 - 2\lambda^2\mu^2 + \mu^4). \tag{10}$$

We conclude thus from (7) and (9) that

$$\frac{1}{4}\left(\|R\|^2 - 2\|Ric\|^2 + K^2\right)$$

$$= (0.75 + 2\lambda^2 + 4\lambda^4 + 2\mu^2 + 4\mu^4 \pm 2\lambda\mu + 4\lambda^2\mu^2) - 4(\lambda^4 - 2\lambda^2\mu^2 + \mu^4)$$

$$= (0.75 + 2\lambda^2 + 2\mu^2 + 12\lambda^2\mu^2 \pm 2\lambda\mu). \tag{11}$$

We summarize the results obtained so far as the following lemma.

Lemma 1. Let $U \subset B^2 \times B^2$ be an open subset and $S \subset U$ be a 2-dimensional complex submanifold. Let $x \in S$ and (α, β) be an orthonormal basis at x such that $R_{\alpha\bar{\beta}} = 0$. Without loss of generality suppose $\alpha = (\alpha_1, \alpha_2)$, $\beta = (\beta_1, \beta_2)$ with $\|\alpha_1\|^2 = 0.5 + \lambda$, $\|\beta_1\|^2 = 0.5 \pm \mu$, $0 \leq \lambda$, $\mu \leq 0.5$. Then, we have the inequality

$$c_2(S,g) \geq \frac{\omega^2}{4\pi^2} \left(0.75 + 2\lambda^2 + 2\mu^2 + 12\lambda^2\mu^2 \pm 2\lambda\mu \right) .$$

(5.4) *Verification of the inequality at a point with vanishing second fundamental form.* Denote by Θ^i the curvature $(1,1)$ form of (K_i, h_i). Identifying Θ^i with the corresponding Hermitian matrix we have

$$\Theta^i = -3 \left[\begin{array}{cc} \| \alpha_i \|^2 & < \alpha_i, \beta_i > \\ \hline < \alpha_i, \beta_i > & \| \beta_i \|^2 \end{array} \right] \tag{1}$$

so that

$$\det(\Theta^i) = \Theta^i_{\alpha\bar{\alpha}} \Theta^i_{\beta\bar{\beta}} - |\Theta^i_{\alpha\bar{\beta}}|^2$$

$$= 9 \left(\|\alpha_i\|^2 \cdot \|\beta_i\|^2 - |< \alpha_i, \beta_i >|^2 \right) \leq 9\|\alpha_i\|^2 \cdot \|\beta_i\|^2. \tag{2}$$

Thus,

$$c_1^2(K_1, h_1) + c_1^2(K_1, h_1)$$

$$= \frac{\omega^2}{4\pi^2} \left(\det(\Theta^1) + \det(\Theta^2) \right) \leq \frac{9\omega^2}{4\pi^2} \left(\|\alpha_1\|^2 \cdot \|\beta_1\|^2 + \|\alpha_2\|^2 \cdot \|\beta_2\|^2 \right)$$

$$= \frac{9\omega^2}{4\pi^2} \left((0.5 + \lambda)(0.5 \pm \mu) + (0.5 - \lambda)(0.5 \mp \mu) \right) = \frac{9\omega^2}{4\pi^2} (0.5 \pm 2\lambda\mu). \tag{3}$$

At a point $x \in S$ where the second fundamental form vanishes we have thus

$$\nu_2(S,g) = c_2(S,g) - \frac{1}{6} \left(c_1^2(K_1, h_1) + c_1^2(K_1, h_1) \right)$$

$$\geq \frac{\omega^2}{4\pi^2} \left(0.75 + 2\lambda^2 + 2\mu^2 + 12\lambda^2\mu^2 \pm 2\lambda\mu - 1.5(0.5 \pm 2\lambda\mu) \right) \omega^2$$

$$\geq \frac{\omega^2}{4\pi^2} \left(2\lambda^2 + 2\mu^2 + 12\lambda^2\mu^2 \mp \lambda\mu \right). \tag{4}$$

If the sign attached to $\lambda\mu$ is $+$, we have obviously $\nu_2(S,g) \geq 0$. If the sign is $-$, then the inequality $\nu_2(S,g) \geq 0$ is a consequence of

$$2\lambda^2 + 2\mu^2 + 12\lambda^2\mu^2 \geq \lambda\mu, \tag{5}$$

which follows readily from $\lambda^2 + \mu^2 \geq 2\lambda\mu$. We have thus proved the inequality at a point $x \in S$ under the assumption that the second fundamental form vanishes.

Under the same assumption $\sigma(x) = 0$ we examine when we have equality $\nu_2(S, g)(x) = 0$. First of all by (5) and (6) we conclude that $\lambda = \mu = 0$. In other words, we have an orthonormal basis $\{\alpha, \beta\}$ of $T_x^{1,0}(S)$ with $\alpha = (\alpha_1, \alpha_2)$, $\beta = (\beta_1, \beta_2)$ such that $\|\alpha_1\|^2 = \|\alpha_2\|^2 = \|\beta_1\|^2 = \|\beta_1\|^2 = \frac{1}{2}$. Furthermore from [(5.3),eqn.(3)] we conclude that $\nu_2(S, g)(x) = 0$ implies $< \alpha_1, \beta_1 >=< \alpha_2, \beta_2 >= 0$. The description of $\{\alpha, \beta\}$ implies readily that, taking x to be the origin, $T_o^{1,0}(S)$ consists of $(\eta, \Phi(\eta))$ where Φ is a unitary transformation. Thus, up to a holomorphic isometry of $B^2 \times B^2$ we have proved that $\sigma(x) = 0$ and $\nu_2(S, g)(x) = 0$ implies that S is tangential to the diagonal $\delta(B^2)$.

(5.5) *Contribution of the second fundamental form.* To complete the proof of the inequality (5.4) we have to consider the contribution of the second fundamental form σ to $c_2(S, g)$. Fix an arbitrary point $x \in S$ and choose an orthonormal basis $\{\alpha, \beta\}$ of $T_x^{1,0}(S)$ such that $R_{\alpha\bar{\beta}} = 0$. We have to prove the semipositivity of the (2,2)-form

$$\nu_2(S, g) = c_2(S, g) - \frac{1}{6}\left(c_1^2(K_1, h_1) + c_1^2(K_1, h_1)\right)$$

$$= \frac{\omega^2}{16\pi^2}\left(\|R\|^2 - 2\|Ric\|^2 + K^2\right)$$

$$-\frac{\omega^2}{24\pi^2}\left((\Theta_{\alpha\bar{\alpha}}^1\Theta_{\beta\bar{\beta}}^1 - |\Theta_{\alpha\bar{\beta}}^1|^2) + (\Theta_{\alpha\bar{\alpha}}^2\Theta_{\beta\bar{\beta}}^2 - |\Theta_{\alpha\bar{\beta}}^2|^2).\right) \qquad (1)$$

In (5.4) we showed that $\nu_2(S, g) \geq 0$ at a point x where the second fundamental form σ vanishes. We now drop the assumption $\sigma(x) = 0$. Recall that Ω stands for the curvature tensor of $(B^2 \times B^2, g)$. From the Gauss equation we have

$$R_{\gamma\bar{\gamma}\delta\bar{\delta}} = \Omega_{\gamma\bar{\gamma}\delta\bar{\delta}} - \|\sigma_{\gamma\delta}\|^2 . \qquad (2)$$

for any pair (γ, δ) of tangent vectors of type (1,0) at x. We have

$$c_2(S, g) \geq \frac{\omega^2}{16\pi^2}\left(R_{\alpha\bar{\alpha}\alpha\bar{\alpha}} + R_{\beta\bar{\beta}\beta\bar{\beta}} + 4R_{\alpha\bar{\alpha}\beta\bar{\beta}}\right)^2 + \frac{\omega^2}{16\pi^2}\left(R_{\alpha\bar{\alpha}\alpha\bar{\alpha}} - R_{\beta\bar{\beta}\beta\bar{\beta}}\right)^2$$

$$= \frac{\omega^2}{16\pi^2}\left(\Omega_{\alpha\bar{\alpha}\alpha\bar{\alpha}} + \|\sigma_{\alpha\alpha}\|^2\right)^2 + \frac{\omega^2}{16\pi^2}\left(\Omega_{\beta\bar{\beta}\beta\bar{\beta}} + \|\sigma_{\beta\beta}\|^2\right)^2 +$$

$$+\frac{\omega^2}{4\pi^2}\left(\Omega_{\alpha\bar{\alpha}\beta\bar{\beta}} + \|\sigma_{\alpha\beta}\|^2\right)^2$$

$$-\frac{\omega^2}{16\pi^2}\left(\Omega_{\alpha\bar{\alpha}\alpha\bar{\alpha}} - \Omega_{\beta\bar{\beta}\beta\bar{\beta}} + \|\sigma_{\alpha\alpha}\|^2 - \|\sigma_{\beta\beta}\|^2 \cdot\right)^2 \qquad (3)$$

We write the inequality (3) in the form $c_2(S,g) \geq \tau(S,g)$. We have proved in (5.4) that $\nu_2(S,g) \geq \tau(S,g) - \frac{1}{6}\left(c_1^2(K_1,h_1) + c_1^2(K_1,h_1)\right) \geq 0$. Let I be the sum of terms in $\tau(S,g)$ which are quadratic in σ and II be those which are quartic in σ. The same proof as in (5.4) shows that in general

$$\nu_2(S,g) \geq I + II. \qquad (4)$$

We are going to verify that in fact I, $II \geq 0$ as (2,2)-forms. We have first of all

$$II = \frac{\omega^2}{16\pi^2}\left(\|\sigma_{\alpha\alpha}\|^4 + \|\sigma_{\beta\beta}\|^4 + 4\|\sigma_{\alpha\beta}\|^4 - \left(\|\sigma_{\alpha\alpha}\|^2 - \|\sigma_{\beta\beta}\|^2\right)^2\right)$$

$$= \frac{\omega^2}{16\pi^2}\left(4\|\sigma_{\alpha\beta}\|^4 + 2\|\sigma_{\alpha\alpha}\|^2\|\sigma_{\beta\beta}\|^2\right) \geq 0. \qquad (5)$$

Furthermore

$$I = \frac{\omega^2}{16\pi^2}[2\Omega_{\alpha\bar{\alpha}\alpha\bar{\alpha}}\|\sigma_{\alpha\alpha}\|^2 + 2\Omega_{\beta\bar{\beta}\beta\bar{\beta}}\|\sigma_{\beta\beta}\|^2 + 8\Omega_{\alpha\bar{\alpha}\beta\bar{\beta}}\|\sigma_{\alpha\beta}\|^2$$

$$-2(\Omega_{\alpha\bar{\alpha}\alpha\bar{\alpha}} - \Omega_{\beta\bar{\beta}\beta\bar{\beta}})(\|\sigma_{\alpha\alpha}\|^2 - \|\sigma_{\beta\beta}\|^2)^2]$$

$$= \frac{\omega^2}{8\pi^2}\left(\Omega_{\alpha\bar{\alpha}\alpha\bar{\alpha}}\|\sigma_{\beta\beta}\|^2 + \Omega_{\beta\bar{\beta}\beta\bar{\beta}}\|\sigma_{\alpha\alpha}\|^2 + 4\Omega_{\alpha\bar{\alpha}\beta\bar{\beta}}\|\sigma_{\alpha\beta}\|^2\right) \geq 0. \qquad (6)$$

We have thus proved that $\nu_2(S,g) \geq I + II \geq 0$ in general.

We examine now the situation when we have strict equality everywhere on S. Since $(B^2 \times B^2, g)$ is of negative holomorphic bisectional curvature the vanishing of II implies by (6) that the second fundamental form σ must vanish identically on S, so that $S \subset U \subset B^2 \times B^2$ is totally geodesic. On the other hand from $\sigma \equiv 0$ it follows from [(2.4), last paragraph] that for each $x \in S$ there exists an automorphism Ψ of $B^2 \times B^2$ such that S is tangent to $\Psi(\delta(B^2))$. This together with the total-geodesy of S implies readily that (U,S) is modelled on $(B^2 \times B^2, B^2; \delta)$. The proof of Theorem 1 is complete.

(5.6) *The gap phenomenon for* $(B^2 \times B^2, B^2; \delta)$ *using the optimal Arakelov inequality.* In this section we are going to give another verification of the gap phenomenon for $(B^2 \times B^2, B^2; \delta)$ based on the optimal Arakelov inequality in a way analogous to the situation in §3 in the case of compact curves on quotients of the Siegel upper half-plane modelled on $(\mathcal{H}_n, \mathcal{H}; i)$. The optimal Arakelov inequality [(4.1),Proposition 1] is a special case of such inequalities given in Eyssidieux [Eys1] in the context of variation of Hodge structures. One can then in a similar way formulate the gap phenomenon for $(\mathcal{D}, B^2; i)$ by replacing $B^2 \times B^2$ by some period domain of Griffiths $\mathcal{D}$. In this situation we are then interested in compact immersed surfaces $S \hookrightarrow \mathcal{D}/\Gamma$ arising from a classifying map. The merit of the new derivation here for the gap phenomenon on $(B^2 \times B^2, B^2; \delta)$ is that the set-up applies to the situation of $(\mathcal{D}, B^2; i)$, except that it remains to be proved that the compact complex surface S is totally-geodesic when the corresponding Arakelov inequality is an equality.

From the local property of the Arakelov inequality in [(5.1), Theorem 1] we conclude that $-\chi(\mathcal{E}_S^{3,*}, D'') > 0$ unless S is modelled on $(B^2 \times B^2, B^2; \delta)$. Take a tower of unramified finite coverings $S_n \to S$ as in [(4.2), proof of Proposition 1]. By [(4.3), Proposition 2] and $-\chi(\mathcal{E}_S^{3,*}, D'') > 0$ we conclude that there exist D''-harmonic forms with value in $\mathcal{E}_{S_n}^{31}$ for n sufficiently large. We will denote the space of such harmonic forms by $H^2_{harm}(S_n, V_\mathbf{C})^{3,1}$ and denote the corresponding subspace in $H^2(S_n, V_\mathbf{C})$ by $H^2(S_n, V_\mathbf{C})^{3,1}$. On the other hand we are going to show

Proposition 1. There exists $\epsilon > 0$ such that whenever $S \hookrightarrow X = (B^2 \times B^2)/\Gamma$ is ϵ-pinched and modelled on $(B^2 \times B^2, B^2; \delta)$, we have the vanishing theorem $H^2(S_n, V_\mathbf{C})^{3,1} = 0$.

Proof: Suppose $\xi \in H^2_{harm}(S_n, V_\mathbf{C})^{3,1}$. By definition $\Delta_{D''}\xi = 0$. Write $D'' = d'' + \nabla'$ where ∇' denotes the Gauss-Manin connection. We have

$$\Delta_{D''} = \Delta_{d''} + (\nabla'\nabla'^* + \nabla'^*\nabla') + (d''\nabla'^* + \nabla'^*d'') + (d''^*\nabla' + \nabla'd''^*). \quad (1)$$

In the special case where $S = S_o$ is totally geodesic and is modelled on $(B^2 \times B^2, B^2; \delta)$ we have $(d''\nabla'^* + \nabla'^*d'')\xi = (d''^*\nabla' + \nabla'd''^*)\xi = 0$ and we conclude that $d''\xi = \nabla'\eta = \nabla'^*\xi = 0$. If S is modelled on $(B^2 \times B^2, B^2; \delta)$, a simple calculation shows that the zero-order operator $\nabla'\nabla'^* + \nabla'^*\nabla'$ is strictly positive. In fact in this case the Gauss-Manin connection $\nabla' : H^{20} \otimes \Omega^{1,1} \to H^{11} \otimes \Omega^{2,1}$ is an isomorphism. We have thus

$$< (\nabla'\nabla'^* + \nabla'^*\nabla')\xi; \xi > (x) \geq C_1 \|\xi(x)\|^2 \quad (2)$$

for some $C_1 > 0$. As a consequence, any D''-harmonic form ξ in

$$H^2_{\text{harm}}(S_n, V_\mathbf{C})^{3,1}$$

must vanish identically. In general one can show that in terms of the second fundamental form σ we have at each $x \in S$ the inequality

$$< (d''\nabla'^* + \nabla'^*d'') + (d''^*\nabla' + \nabla'd''^*)\xi; \xi > (x) \geq -C_2 \|\sigma(x)\| \|\xi(x)\|^2 \quad (3)$$

for some absolute constant C_2. On the other hand, if holomorphic sectional curvatures on S are sufficiently close to that of the diagonal $\delta(B^2)$, we have

$$< (\nabla'\nabla'^* + \nabla'^*\nabla')\xi; \xi > (x) \geq \frac{1}{2}C_1 \|\xi(x)\|^2, \quad (4)$$

say. Integrating $< \Delta_{D''}\xi, \xi >$ over S and applying (2) and (3) we see that $\xi \equiv 0$ provided that $\|\sigma(x)\| < \frac{C_2}{2C_1}$ everywhere over S.

Remarks: There is another approach to verifying the gap phenomenon for $(B^2 \times B^2, B^2; \delta)$ by using the optimal Arakelov inequality which recovers $[(2.1),$ Proposition 1] in this case, except for the fact that we assume that $i : S \hookrightarrow X = (B^2 \times B^2)/\Gamma$ with X compact and Γ arithmetic. The point of departure is that by $[(5.1),$ Theorem 1] we have equality if and only if $S \hookrightarrow X$ is totally geodesic and modelled on $(B^2 \times B^2, B^2; \delta)$. On the other hand, we have

$$0 \leq 4c_2(S) - \frac{2}{3}\left(c_1^2(T_1) + c_1^2(T_2)\right) = -\chi(\mathcal{E}_S^{3,1}, D'')$$

$$= [(ch(H^{20} \otimes \Omega^1) - ch(H^{11} \otimes \Omega^2)) \cdot \text{Todd}(S)]^4,$$

so that equality holds whenever $H^{20} \otimes \Omega^1$ and $H^{11} \otimes \Omega^2$ are isomorphic holomorphic vector bundles. This is the case under the assumption that both canonical projections $\tilde{S} \to B_i^2$ are unramified, as in $[(4.6),$ equation (4)]. In fact, the isomorphism given there is the same as that given by the restriction of the Gauss-Manin connection $\nabla' : H^{20} \otimes \Omega^1 \to H^{11} \otimes \Omega^2$ to S. Thus the hypothesis that $\tilde{S} \to B_i^2$ is unramified implies that S is totally geodesic and modelled on $(B^2 \times B^2, B^2; \delta)$ by $[(5.1),$ Theorem 1]. This proof uses however explicitly the product structure of $B^2 \times B^2$ and does not appear applicable to the situation of irreducible bounded symmetric domains.

(5.7) *Comparison with the gap phenomenon for* $(\mathcal{H}_n, \mathcal{H}; \delta)$. Results of §3 on the gap phenomenon for $(\mathcal{H}_n, \mathcal{H}; \delta)$ can be formulated in terms of variations of Hodge structures. The Siegel upper half-plane $\mathcal{H}_n$ is equivalently the parameter space for variations of polarized Hodge structure of weight 1. There is an underlying locally constant vector bundle $V_{\mathbf{C}}$ which is the direct sum $V \oplus V^*$ as smooth complex vector bundles, where V represents the universal nonpositive bundle over $\mathcal{H}_n$. Here $V^* = H^{10} \hookrightarrow V_{\mathbf{C}}$ is a holomorphic subbundle and $V = H^{01}$. The Gauss-Manin connection is given by $\nabla' : H^{10} \to H^{01} \otimes \Omega^1$, which is the same as the parallel bundle map $\Xi : V^* \to V \otimes \Omega^1$ in (3.2). We have the same formalism as in (4.4) except that in the bigrading we take $P + Q = n + 1$ (corresponding to the fact that the variation of Hodge structure is of weight 1). Equation (1) there corresponds to asserting that $\eta + \bar{\eta}$ is a D''-harmonic form and equation (8) involving the second fundamental form σ gives the analogue of $[(5.6),$ equation (3) and inequality (4)] for $H^1_{harm}(C, V_{\mathbf{C}})^{1,1}$ on an immersed curve $C \hookrightarrow X = \mathcal{H}_n/\Gamma$. The inequality $[(3.1),$ inequality (4)] $\deg(V) \geq -n(g-1)$, due to Arakelov, is then the simplest optimal Arakelov inequality. The local property of this inequality is a trivial consequence of the Gauss equation for submanifolds which shows that the

Gauss curvatures on C are less than or equal to corresponding sectional curvatures on the ambient manifold X. In general, it is far from clear that optimal Arakelov inequalities are local, and as shown in the particular example on $B^2 \times B^2$ given here in [(5.1), Theorem 1], when such inequalities are local the verification is far from being immediate.

References

[BPV] Barth, W., Peters, C. and Van de Ven, A.: *Compact Complex Surfaces*, Ergebnisse der Mathematik und ihrer Grenzgebiete, 3. Folge, Band 4, Springer Verlag, Berlin-Heidelberg-New York-Tokyo, 1984.

[Eys1] Eyssidieux, P.: Variations de structure de Hodge: In/'egalit/'es d'Ara-kelov locales et globales, Thèse, Université de Paris-Sud, Centre d'Orsay, 1994.

[Gro1] Gromov, M.: Almost flat manifolds, J. Diff. Geom. **13** (1978), 231-242.

[Gro2] Gromov, M.: Kähler hyperbolicity and L^2-Hodge theory, J. Diff. Geom. **33** (1991), 263-292.

[Har] Hartman, P.: On homotopic harmonic maps, Canad. J. Math. **19** (1967), 637-687.

[Iha] Ihara, S.: Holomorphic imbeddings of symmetric domains, J. Math. Soc. Japan **19** (1967), 261-302; Supplement, ibid., 534-544.

[Kli] Klingenberg, W.: *Riemannian Geometry*, de Gruyter Studies in Mathematics, Vol. 1, Verlag Walter de Gruyter, Berlin-New York, 1982.

[Mok1] Mok, N.: *Metric Rigidity Theorems on Hermitian Locally Symmetric Manifolds*, Ser. Pure Math. **6**, Singapore-New Jersey-London-Hong Kong 1989.

[Mok2] Mok, N.: Aspects of Kähler geometry on arithmetic varieties, Am. Math. Soc. Proc. Symp. Pure Math. **52** (1991), 335-396.

[MR] Min-Oo and Ruh, E. A.: Vanishing theorems and almost symmetric spaces of noncompact type, Math. Ann. **257** (1981), 419-433.

[MT] Mok. N and To, W.-K.: Eigensections on Kuga families of Abelian varieties and finiteness of their Mordell-Weil groups, J. reine angew. Math. **444** (1993), 29-78.

[Ros] Ros, A.: A characterization of seven compact Kähler submanifolds by holomorphic pinching, Ann. of Math. **121** (1985), 377-382.

[Sa] Satake, I.: *Algebraic Structures of Symmetric Spaces*, Iwanami Shoten & Princeton Univ. Press, Princeton, 1980.

[Yau] Yau, S.-T. A general Schwarz lemma for Kähler manifolds, Amer. J. Math. **100** (1978), 197-203.

[Zu] Zucker, S. W. Locally homogeneous variations of Hodge structures, Enseig. Math. (2) **27** (1981), no. 3-4, 243-276.

Philippe Eyssidieux: Université de Paris-Sud, Centre d'Orsay, Mathématique, Bât. 425, F-91405 Orsay Cedex, France

Ngaiming Mok: Department of Mathematics, The University of Hong Kong, Pokfulam, Hong Kong

ON KOHN'S MICROLOCALIZATION
OF $\overline{\partial}$ PROBLEMS

CHARLES FEFFERMAN

The purpose of this expository paper is to explain an important idea of Kohn that reduces certain $\overline{\partial}$ problems to the study of a certain pseudodifferential operator A. After a brief, historical review of $\overline{\partial}$, I'll explain what A is, why $\overline{\partial}$-problems reduce to it, and how one analyzes it.

To begin the historical review, let me recall the basic $\overline{\partial}$-problems. Let $D \subset \mathbf{C}^n$ be a domain with smooth boundary. We want to construct analytic functions with prescribed singularities at the boundary ∂D. To do so, we need to solve the inhomogeneous Cauchy-Riemann equations $\overline{\partial} u = \alpha$, where u is an unknown function on D, and $\alpha = \sum_{k=1}^{n} g_k d\overline{z}_k$ is a (0,1)-form. Here, $\overline{\partial} u = \sum_k \frac{\partial u}{\partial \overline{z}_k} d\overline{z}_k$, and α is assume to satisfy the obvious consistency condition $\overline{\partial}\alpha = \sum_{j<k}\left(\frac{\partial g_k}{\partial \overline{z}_j} - \frac{\partial g_j}{\partial \overline{z}_k}\right)d\overline{z}_j \wedge d\overline{z}_k = 0$. The solution u, if it exists, is clearly non-unique, since we can add any analytic function to u without changing $\overline{\partial} u$. To remedy the non-uniqueness, we demand that u be orthogonal to analytic functions in $L^2(D)$. Thus we obtain the first of our $\overline{\partial}$-problems:

(1) $\overline{\partial} u = \alpha$, $u \perp$ nullspace $(\overline{\partial})$, given that $\overline{\partial}\alpha = 0$.

There is an analogous problem on the boundary ∂D, namely

(2) $\overline{\partial}_b u = \alpha$, $u \perp$ nullspace $(\overline{\partial}_b)$, given that $\overline{\partial}_b\alpha = 0$.

Here u is an unknown function on ∂D, and the inner product is taken in $L^2(\partial D)$.

To write down $\overline{\partial}_b$ explicitly, let P_0 be a point in ∂D. In a small neighborhood of P_0 in ∂D we can pick smooth vector fields $L_1, \cdots, L_{n-1}$ that form a basis for the vectors of type (1,0) tangent to ∂D. Then we can regard α in (2) as a vector of functions $\alpha = (g_1, \cdots, g_{n-1})$, and $\overline{\partial}_b u = \alpha$ means that $\overline{L}_j u = g_j (1 \leq j \leq n-1)$. The consistency condition $\overline{\partial}_b\alpha = 0$ means that $\overline{L}_j g_k - \overline{L}_k g_j = \sum_m \theta_{jk}^m g_m$, where the θ_{jk}^m are defined by the equation

(2a) $$[\overline{L}_j, \overline{L}_k] = \sum_m \theta_{jk}^m \overline{L}_m.$$

There are such θ_{jk}^m because the commutation of two vector fields of type (1,0) is again of type (1,0). Thus we obtain the second $\overline{\partial}$-problem, (2).

Associated to $\bar{\partial}$ and $\bar{\partial}_b$, there are two second-order systems of P.D.E., namely

(3) $\square u = \alpha$ on D with $\bar{\partial}$–Neumann boundary conditions at ∂D,

and

(4) $\square_b u = \alpha$ on ∂D.

These problems are related to (1) and (2) as harmonic functions are related to analytic functions. In particular, from solutions of (3) and (4) we can read off solutions of (1) and (2). In (3), u and α are $(0,1)$-forms with u unknown and α given, while in (4), u and α are forms of the same type as the α in (2). The passage from (1), (2) to (3), (4) is a big step forward, because we have rid ourselves of the consistency conditions $\bar{\partial}\alpha = 0$, $\bar{\partial}_b \alpha = 0$ and the side conditions $u \perp$ nullspace $(\bar{\partial})$, $u \perp$ nullspace $(\bar{\partial}_b)$. Equation (3) was first proposed by Garabedian and Spencer, while (4) is due to Kohn and Rossi [10].

The operator $\square$ is essentially the Laplacian; the difficulty of (3) arises from the boundary conditions. On the other hand, $\square_b$ is a non-elliptic system on a manifold without boundary.

A problem intimately related to (1) $\cdots$ (4) is to understand the Bergman and Szegö kernels $K^B(z,w), K^S(z,w)$ on $D \times D$. Here K^B is defined by the reproducing formula

(5) $$\pi^B f(z) = \int_D K^B(z,w)f(w)d \, vol(w),$$

where π^B is the orthogonal projection from $L^2(D)$ to the subspace of analytic functions. Similarly, $K^S(z,w)$ is defined by

(6) $$\pi^S f(z) = \int_{\partial D} K^S(z,w)f(w)d \, area(w),$$

where π^S is the orthogonal projection from $L^2(\partial D)$ to the subspace of boundary values of analytic functions on D.

Next, let me recall what one wants to know about the $\bar{\partial}$-problems. There is a hierarchy of four levels of understanding. The first task is to prove C^∞ regularity of the problems (1) $\cdots$ (4). This comes from subelliptic estimates for Sobolev norms. The next level of understanding is to derive sharp estimates for Sobolev norms, and estimates in spaces other than Sobolev spaces, such as the Lipschitz spaces $\mathrm{Lip}(\alpha)$. Progressing further, we want to write down the singularities of the integral kernels that solve the $\bar{\partial}$-problems, as well as the Bergman and Szegö kernels. Finally the most complete understanding comes when we derive sharp estimates directly from the singularities of the integral kernels in question.

It is essential to work on the right class of domains. These turn out to be the weakly pseudoconvex domains of finite type, which we now explain.

A basic phenomenon in several complex variables is that of envelopes of holomorphy: For a given domain $D \subset \mathbf{C}^n$, there may be a strictly larger domain $\hat{D} \supset D$ such that every analytic function on D continues analytically to $\hat{D}$. When we think we are studying analytic functions on D, we are really studying them on $\hat{D}$. Hence it is natural to restrict attention to smooth domains D for which $D = \hat{D}$. This amounts to saying that the domain is "pseudoconvex". Pseudoconvexity may be defined in terms of the smooth vector fields $L_1, \cdots, L_{n-1}$ that form a basis for the vectors of type (1,0) tangent to ∂D, as in (2a). The real vector fields $Re(L_j)$, $Im(L_j)(1 \leq j \leq n-1)$ span a codimension 1 subspace of the tangent space $T(\partial D)$ at each boundary point. Let T be a real vector field in $T(\partial D)$, defined in a neighborhood of a given $P_0 \epsilon\ \partial D$, transversal to span $\{Re(L_j), Im(L_j)\}$. (We ignore here the normalization of the sign of T.)

Thus, $Re(L_1), \cdots, Re(L_{n-1}), Im(L_1), \cdots, Im(L_{n-1}), T$ form a basis for $T(\partial D)$ at each point in a neighborhood of P_0. For suitable smooth coefficients $\lambda_{j\overline{k}}$, $b_{j\overline{k}}^m$, $b_{j\overline{k}}^{\overline{m}}$, we have

$$(7) \qquad [L_j, \overline{L}_k] = \frac{1}{i}\lambda_{j\overline{k}}T + \sum_m (b_{j\overline{k}}^m L^m + b_{j\overline{k}}^{\overline{m}} L^{\overline{m}}).$$

The matrix $(\lambda_{j\overline{k}})$ is called the *Levi form* for the domain D. A domain is called *pseudoconvex* if its Levi form is positive semidefinite, $(\lambda_{j\overline{k}}) \geq 0$. If the Levi form is strictly positive definite, then the domain is called *strictly pseudoconvex*. (These ideas generalize from ∂D to a *CR manifold* M, on which we can introduce locally suitable complex vector fields L_j and a real vector field T.)

Strictly pseudoconvex domains are much easier to understand than general pseudoconvex domains. The reason is that we can make a holomorphic change of coordinates in a neighborhood of a given boundary point of a strictly pseudoconvex domain, so that in the new coordinate system, the domain is a small perturbation of the unit ball. Thus, if we can understand the unit ball, then we can start to understand strictly pseudoconvex domains. Weakly pseudoconvex domains (*i.e.* pseudoconvex but not strictly pseudoconvex) cannot be similarly understood in terms of a short list of basic examples.

Weakly pesudoconvex domains can be bad for the $\bar{\partial}$-problems because their boundaries are too flat. For instance, a half-space $D = \{(z_1, \cdots, z_n)\epsilon\ \mathbf{C}^n : \operatorname{Im} z_1 > 0\}$ has the property that the restriction of an analytic function to ∂D must be analytic in $(z_2, \cdots, z_n)$. Clearly, then, there can be no analytic functions on D with a singularity at a single boundary point. The construction of such "peak functions" is a typical application of $\bar{\partial}$-regularity, so we cannot hope for $\bar{\partial}$-regularity when the boundary is too flat.

To make sure our domain D has a boundary that isn't too flat, Kohn introduced the condition in $\mathbf{C}^2$ that ∂D be of *finite type*. His definition involved iterated commutators of vector fields. The higher dimensional generalization evolved from work of Kohn and D'Angelo, and may be stated as follows:

If $P \epsilon\ \partial D \subset \mathbf{C}^n$, and if $V \subset \mathbf{C}^n$ is a complex analytic curve containing P, then for all $Q \epsilon\ V$ sufficiently close to P, we have the estimate

(8) $\qquad \operatorname{dist}(Q,\ \partial D) \geq c\ [\operatorname{dist}(Q,\ P)]^m$ for suitable $c,\ m > 0$.

It is essential to allow singular curves V in this definition. The definition can also be expressed more algebraically (See D'Angelo's article in these proceedings.)

Kohn saw that the weakly pseudoconvex domains of finite type form the correct class of domains for the regularity of the $\bar{\partial}$-problems.

Next, let me sketch what is known about $\bar{\partial}$. The first important results were Kohn's proof of C^∞ regularity of problems $(1)\cdots(4)$ for strictly pseudoconvex domains, using subelliptic estimates [8]. Aside from its own importance, this work led quickly to two historic results, namely the invention of pseudodifferential operators, and the study of commutators in PDE. In fact, the modern theory of pseudodifferential operators was first given by Kohn and Nirenberg, in connection with a paper on $\bar{\partial}$. (Calderon had previously introduced a calculus of singular integral operators, equivalent for many purposes to the Kohn-Nirenberg calculus. The Kohn-Nirenberg calculus allows more facility in computation. Kohn and Nirenberg make a point of acknowledging generously Calderon's contribution.) Regarding commutators, suppose, say, $\bar{\partial}_b u = \alpha$, and let $L_1, \cdots, L_{n-1}, T$ be as in (7). Then one can control $L_j u$ and $\overline{L}_j u$ directly in terms of α, but controlling Tu is a deeper problem. The crucial point is that the missing direction T may be written essentially as a commutator of directions in which one already has control. Kohn saw how to exploit this in order to prove C^∞ regularity for the $\bar{\partial}$ problems for strictly pseudoconvex domains. Soon after Kohn's work, Hörmander [7] gave a basic regularity theorem for sums of squares of vector fields which, together with their repeated commutators, span the tangent space.

The next stage in the study of $\overline{\partial}$ was to understand the strictly pseudoconvex case completely, and not just on the level of C^∞ regularity. This was carried out by Stein, Folland, Greiner, Rothschild and others.

Then came the work of Catlin [1,2], D'Angelo [4] and Kohn [9] on weakly pseudoconvex domains. Kohn formulated the notion of finite type, and proved the basic subelliptic estimates that give C^∞ regularity for $\overline{\partial}$ on weakly pseudoconvex domains of finite type with a real analytic boundary. In doing so he introduced the important new technique of ideals of subelliptic multipliers. (This idea has proved useful in many other problems; see the discussion of Kohn's work in this proceedings for some indication of this.) In the smooth case D'Angelo proved the crucial geometrical result that finite type (8) at $P_0 \epsilon\; \partial D$ implies also finite type at all $P \; \epsilon \; \partial D$ in a neighborhood of P_0, even though the exponent m in (8) may be larger for P than for P_0. This gave strong evidence at an early stage that finite type was the right condition. Catlin proved subellipticity, hence C^∞ regularity, for C^∞ weakly pseudoconvex domains of finite type. He also showed that finite type is necessary for subellipticity. It should be noted that analysis of commutators isn't enough to deal with weakly pseudoconvex domains.

Deeper understanding of general weakly pseudoconvex domains of finite type remains a hard and distant goal, which relates to issues of algebraic geometry with bounds, over the reals. (This is clear, just by glancing at the definition (8).) Such questions, at the border of analysis and algebra, are fascinating but up to now virtually untouched.

In the mid 1980's, weakly pseudoconvex domains of finite type in two complex variables were completely understood. To obtain sharp estimates for $\overline{\partial}$, Kohn introduced a certain pseudodifferential operator A. Christ [1] analyzed a variant of A, and Kohn and I [5] understood A itself. Soon after, many people independently discovered overlapping pieces of the big picture, and the $\overline{\partial}$ problems in two complex variables were completely understood. The contributors to this effort were Christ; Kohn and me; McNeal; Nagel, Rosay, Stein and Wainger; and others.

Kohn, Machedon and I [6] derived "almost sharp" estimates for $\overline{\partial}$ in more than two complex variables using the operator A, but only under a restrictive hypothesis: We assume that the Levi form on ∂D is diagonalizable. That is, we assume that the vector fields $L_1, \cdots, L_{n-1}, T$ in (7) can be picked smoothly so that the Levi form $(\lambda_{j\overline{k}})$ is a diagonal matrix,

$$(9) \qquad\qquad \lambda_{j\overline{k}} = \lambda_j \delta_{jk},\; \lambda_j \epsilon\; C^\infty.$$

To remove this restriction seems very hard.

This concludes the brief historical survey of the $\overline{\partial}$ problems. The rest of this paper explains Kohn's pseudodifferential operator A and its connection with $\overline{\partial}$. We start with the $\overline{\partial}$ problems in two complex variables. The problems here are disjoint from those arising in higher dimensions. Unlike higher dimensions, the key estimates in two complex variables come from commutators. Unlike higher dimensions, $\overline{\partial}_b u = \alpha$ is a scalar equation, essentially $\overline{L}u = f$ for the single vector field $\overline{L}$ in (7). There is no consistency condition $\overline{\partial}_b \alpha = 0$. However, Image($\overline{\partial}_b$) has infinite codimension, because L has an infinite-dimensional nullspace. So the natural $\overline{\partial}$ problems on a 3-dimensional CR-manifold M (such as $M = \partial D \subset \mathbf{C}^2$) are as follows:

(10) $\overline{\partial}_b u = f$ modulo (Image $\overline{\partial}_b$)$^{\perp}$, with $u \,\epsilon$(Nullspace $\overline{\partial}_b$)$^{\perp}$.

(11) $\square_b u = f$ modulo (Image $\square_b$)$^{\perp}$, with $u \,\epsilon$(Nullspace $\square_b$)$^{\perp}$.

(12) $u =$ Szegö Projection of f.

In local coordinates on M, we need a single complex vector field L to span $T^{1,0}M$, the vector fields of type (1,0). The Levi form is a scalar function λ, given by $[L, \overline{L}] = i\,\lambda\,T \mod L, \overline{L}$, where T is a real vector field complementary to the span $\{Re\,L,\,Im\,L\}$. We assume that M is pseudoconvex, *i.e.*, $\lambda \geq 0$. Also, we assume that M is of finite type m, *i.e.*, commutators of up to of m $L's$ and $\overline{L}'s$ span the tangent space of M.

A remarkable example of Rossi shows that Image $\overline{\partial}_b$ needn't be closed on a three-dimensional CR manifold M, even if M is strictly pseudoconvex. However, if $M = \partial D$ with $D \subset \mathbf{C}^2$, then Image $\overline{\partial}_b$ is closed. We make the key additional assumption: Image $\overline{\partial}_b$ is closed.

Next we motivate and introduce Kohn's pseudodifferential operator A. We work microlocally in (x, ξ) space, where we have the following picture.

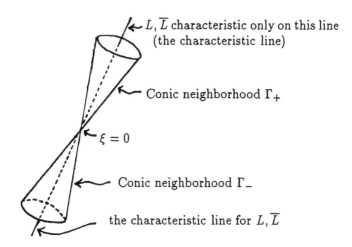

$L, \overline{L}$ characteristic only on this line (the characteristic line)

Conic neighborhood Γ_+

$\xi = 0$

Conic neighborhood Γ_-

the characteristic line for $L, \overline{L}$

In Γ_+, we have $iT \geq 0$, so $L\overline{L} \geq \overline{L}L$.
In Γ_-, we have $iT \leq 0$, so $\overline{L}L \geq L\overline{L}$
Outside $\Gamma_+ \cup \Gamma_-$, L and $\overline{L}$ are elliptic.

The problem that makes $\overline{\partial}$ hard is that neither $L\overline{L}$ nor $\overline{L}L$ is invertible. Kohn's insight is that microlocally, one operator or the other is invertible. Hence, we should form an operator $A \approx L\overline{L} + \overline{L}L + |[L, \overline{L}]| \approx L\overline{L} + \overline{L}L + \lambda|T|$. More precisely, we form a pseudodifferential operator A which looks microlocally like $L\overline{L}$ in Γ_+, like $\overline{L}L$ in Γ_-, and is elliptic outside $\Gamma_+ \cup \Gamma_-$. In Γ_+, $L\overline{L}$ is invertible, but $\overline{L}L$ is not; In Γ_-, $\overline{L}L$ is invertible, but $L\overline{L}$ is not. Hence, A should be invertible, since it is invertible microlocally in every direction. Kohn's idea is to reduce $\overline{\partial}$-problems to A, then understand A.

To see how $\overline{\partial}$-problems reduce to A, let us look at (10), *i.e.*,

(13) $\overline{L}u = f \bmod (\text{Image } \overline{L})^{\perp}$, with $u \perp (\text{Nullspace } \overline{L})$.

We deal with (13) differently in Γ_+, Γ_-, and in the complement of $(\Gamma_+ \cup \Gamma_-)$. Microlocally in Γ_+, we argue as follows. $\overline{L}u = f \bmod (\text{Image } \overline{L})^{\perp}$ means $\overline{L}u = f \bmod (\text{Nullspace } L)$, *i.e.*, $L\overline{L}u = Lf$. Also, microlocally in Γ_+, we have $L\overline{L} = A$, so that

(14) $u = A^{-1}Lf$, microlocally in Γ_+.

Microlocally in Γ_-, we argue as follows. $\overline{L}u = f \bmod (\text{Image } \overline{L})^{\perp}$ will follow if we can solve $\overline{L}u = f$. We want also $u \in (\text{Nullspace}(\overline{L}))^{\perp} = \text{Image } (L)$, since $\overline{L}$ has closed image. Thus, we take $u = Lv$, so that microlocally in Γ_-, we have to solve $\overline{L}Lv = f$. Also, $\overline{L}L = A$ microlocally in Γ_-, so we get $v = A^{-1}f$, hence

(15) $u = L A^{-1}f$, microlocally in Γ_-.

Outside $\Gamma_+ \cup \Gamma_-$, the problem (13) is elliptic, and may therefore be solved microlocally without difficulty.

From (14) and (15) we see that the solution of the $\bar{\partial}$-problem (13) can be given in terms of the inverse of Kohn's operator A. Examples of theorems on $\bar{\partial}_b$ that come from the reduction to A are as follows.

Suppose M is a 3-dimensional weakly pseudoconvex CR manifold of type m, with $\bar{\partial}_b$ having a closed image. If $f \in L^\infty(M)$, the the equation

$$\bar{\partial}_b u = f \mod (\text{range } \bar{\partial}_b)^\perp, \ u \perp (\text{Nullspace } \bar{\partial}_b)$$

has a solution $u \in \text{Lip}(\frac{1}{m})$. If $f \in \text{Lip}(\alpha)(0 < \alpha < 1)$, then the Szegö projection $\pi^S f$ belongs also to $\text{Lip}(\alpha)$. These results are easily seen to be best possible.

The basic result on the Lipschitz regularity of A is as follows (See [5]).

Theorem. Let M be a 3-dimensional CR manifold, weakly pseudoconvex and of type m. Assume $\bar{\partial}_b$ has closed range. Let A be Kohn's pseudodifferential operator. Then

$A^{-1} : \text{Lip}(\alpha) \to \text{Lip}(\alpha + \frac{2}{m})$;
$LA^{-1}, \overline{L}A^{-1}, A^{-1}L, A^{-1}\overline{L} : \text{Lip}(\alpha) \to \text{Lip}(\alpha + \frac{2}{m})$;
$LA^{-1}\overline{L}, \overline{L}A^{-1}L, L\overline{L}A^{-1}, \overline{L}LA^{-1}, A^{-1}L\overline{L}, A^{-1}\overline{L}L : \text{Lip}(\alpha) \to \text{Lip}(\alpha)$;
$LA^{-1}\overline{L}, L^2A^{-1}, A^{-1}L^2, \overline{L}A^{-1}\overline{L}, \overline{L}^2A^{-1}, A^{-1}\overline{L}^2 : \text{Lip}(\alpha) \to \text{Lip}(\alpha - \epsilon)$,
any $\epsilon > 0$.

These statements must be modified in case α, $\alpha + \frac{1}{m}$ or $\alpha + \frac{2}{m}$ is an integer. We omit the details.

Let us put off until later the idea of the proof of the regularity of A^{-1}. Now, we explain how to apply A to study $\bar{\partial}$ in more than two complex variables, when the Levi form is diagonalizable. We will be discussing joint work of Kohn, Machedon, and me [6].

Our setting is a CR-manifold M. In local coordinates, the complex tangent space is spanned by $L_1, \cdots, L_{n-1}, \overline{L}_1, \cdots, \overline{L}_{n-1}, T$. $\bar{\partial}_b u$ means $(\overline{L}_1 u, \cdots, \overline{L}_{n-1} u)$ in local coordinates. The Levi form is $(\lambda_{j\overline{k}})$, given by $[L_j, \overline{L}_k] = \sqrt{-1}\lambda_{j\overline{k}}T \mod L$'s and $\overline{L}$'s.

Diagonalizable Levi form means we can pick the L_j to make $(\lambda_{j\overline{k}})$ diagonal,

$$(\lambda_{j\overline{k}}) = \begin{pmatrix} \lambda_1 & & \\ & \ddots & \\ & & \lambda_{n-1} \end{pmatrix}.$$

Pseudoconvexity means that each of the $\lambda_j \geq 0$.

Finite type m means that for each j, there is a commutator consisting of at most m $L'_j s$ and $\overline{L}'_j s$, which is not in the span of $L_1, \cdots, L_{n-1}, \overline{L}_1, \cdots$

$\cdot, \overline{L}_{n-1}$, and thus captures the T-direction. Note that the idea of finite type is formulated in terms of commutators, because the Levi form is diagonalizable.

In this setting, our main results on $\overline{\partial}_b$ are as follows.

Theorem. Let M be as above, with (real) dimension at least 5. Then $\square_b u = f$, $f \in L^\infty$ imply $u \in \text{Lip}(\frac{2}{m} - \epsilon)$, any $\epsilon > 0$; $\overline{\partial}_b u = f$, $f \in L^\infty$, $u \perp \text{Nullspace}(\overline{\partial}_b)$ imply $u \in \text{Lip}(\frac{1}{m} - \epsilon)$, any $\epsilon > 0$; $f \in \text{Lip}(\alpha)$ implies that the Szegö projection $\pi^S f$ belongs to $\text{Lip}(\alpha - \epsilon)$, any $\epsilon > 0$.

We explain how this theorem reduces to the study of Kohn's operator A when $\dim M = 5$ (*e.g.* $M = \partial D \subset \mathbf{C}^3$). Dimension 5 still doesn't explain all the ideas needed for the general case. Our explanation of the 5-dimensional case contains "little white lies."

In 5 dimensions, we have

$$\square_b \approx \begin{pmatrix} L_1 \overline{L}_1 + \overline{L}_2 L_2 & 0 \\ 0 & \overline{L}_1 L_1 + L_2 \overline{L}_2 \end{pmatrix} .$$

Thus, although $\square_b u = f$ is a system, it decouples into two scalar equations. Let us pick out one of the scalar equations.

$$(16) \qquad (L_1 \overline{L}_1 + \overline{L}_2 L_2)u = f.$$

Say $f \in \text{Lip}(\alpha)$, and we want $u \in \text{Lip}(\alpha + \frac{2}{m} - \epsilon)$. Again, we work microlocally in the cones Γ_+, Γ_-. In Γ_+, we have $L_j \overline{L}_j \geq \overline{L}_j L_j$. In Γ_-, we have $\overline{L}_j L_j \geq L_j \overline{L}_j$. Outside $\Gamma_+ \cup \Gamma_-$, $\square_b$ is elliptic, hence trivial to invert.

The analysis of Γ_- is analogous to that of Γ_+. Here, we will just work microlocally in Γ_+.

In our equation (16), the term $L_1 \overline{L}_1$ is good, while the term $\overline{L}_2 L_2$ is bad. The first main idea is to apply L_2 to both sides of (16). We can commute L_2 past $L_1 \overline{L}_1$ without grave harm, because the Levi form is diagonal, and hence $[L_2, \overline{L}_1] \epsilon$ span $\{L_1, L_2, \overline{L}_1, \overline{L}_2\}$. Thus (16) yields $L_2(L_1 \overline{L}_1 + \overline{L}_2 L_2)u = L_2 f$, and hence

$$(17) \qquad L_2(L_1 \overline{L}_1 + L_2 \overline{L}_2)u = L_2 f + (\text{harmless junk}).$$

We ignore the harmless junk.

As in our work on two complex variables, we introduce a pseudodifferential operator A, equal to $L_1 \overline{L}_1 + L_2 \overline{L}_2$ microlocally in Γ_+, equal to $\overline{L}_1 L_1 + \overline{L}_2 L_2$ microlocally in Γ_-, and elliptic outside $(\Gamma_+ \cup \Gamma_-)$. Since are working microlocally in Γ_+, (17) becomes $A(L_2 u) = L_2 f$, so that $L_2 u = A^{-1} L_2 f$, and

$$(18) \qquad \overline{L}_2 L_2 u = \overline{L}_2 A^{-1} L_2 f.$$

Lipschitz regularity holds for A^{-1}. In particular, $f \in \text{Lip}(\alpha)$ implies $\overline{L}_2 A^{-1} L_2 f \in \text{Lip}(\alpha - \epsilon)$, any $\epsilon > 0$. Therefore, (16) and (18) show that the two terms $L_1 \overline{L}_1 u$ and $\overline{L}_2 L_2 u$ in (16) both belong separately to $\text{Lip}(\alpha - \epsilon)$.

The second main idea of the reduction is to recombine $L_1 \overline{L}_1 u$ and $\overline{L}_2 L_2 u$ using a weight factor, by noting that $(L_1 \overline{L}_1 + c\, \lambda_1 \overline{L}_2 L_2) u \equiv g \in \text{Lip}(\alpha - \epsilon)$, where λ_1 is a component of the Levi form, and c is a small, positive constant. Microlocally in Γ_+, $\tilde{A} = L_1 \overline{L}_1 + c\, \lambda_1 \overline{L}_2 L_2$ is an operator of the same type as A, because the bad commutator $c\, \lambda_1 [\overline{L}_2, L_2] = -c\, \lambda_1 \lambda_2 (iT + junk)$ may be absorbed into the good commutator $[L_1, \overline{L}_1] = \lambda_1 (iT + junk)$, $iT \geq 0$. Hence $\tilde{A} u = g \in \text{Lip}(\alpha - \epsilon)$, with $\tilde{A}$ one of Kohn's pseudodifferential operators of type m. It follows that $u \in \text{Lip}(\alpha - \epsilon + \frac{2}{m})$, as in the above Theorem. Thus, if we can understand operators like A, $\tilde{A}$, then we can control $\square_b^{-1}$ in (real) dimension 5.

In higher dimensions, the reduction of $\square_b^{-1}$ to the operator A is more elaborate. In general, we have to invert

$$(19) \qquad (L_1 \overline{L}_1 + \cdots + L_{n-2} \overline{L}_{n-2}) + \overline{L}_{n-1} L_{n-1}.$$

As before we work with cones Γ_+ and Γ_-. Microlocally in Γ_+, all but one of the terms in (19) are good, and we can proceed as in the 5-dimensional case, with L_{n-1} playing the role of L_2 there. However, microlocally in Γ_-, only one term in (19) is good; the rest are all bad. We use induction on the number k of "bad" terms, to invert the operator

$$(20) \quad \square_b^{(k)} = (L_1 \overline{L}_1 + \cdots + L_k \overline{L}_k) + (\overline{L}_{k+1} L_{k+1} + \cdots + \overline{L}_{n-1} L_{n-1}), \quad k \leq n-2.$$

The idea is to proceed as in 5 dimensions, to reduce the study of $\square_b^{(k)}$ to that of $\square_b^{(k-1)}$. The induction is quite tricky, and I won't try to explain it. However, let me mention one point. For the induction, we need to control the heat operators $e^{-t\square_b^{(k)}}$, not just $(\square_b^{(k)})^{-1}$. The reason is that if one can bound the operators $\overline{L}_\mu e^{-t\square_b^{(k)}}$ and $e^{-t\square_b^{(k)}} L_\nu$ for $0 < t << 1$, then one can automatically bound $\overline{L}_\mu e^{-t\square_b^{(k)}} L_\nu$, simply by writing $\overline{L}_\mu e^{-t\square_b^{(k)}} L_\nu = (\overline{L}_\mu e^{-\frac{1}{2}\square_b^{(k)}})(e^{-\frac{1}{2}\square_b^{(k)}} L_\nu)$. This turns out to be essential for the induction.

Now that we see (more or less) how to reduce the $\overline{\partial}$-problems to the study of Kohn's operator A, it is time to understand A. We work in the following setting, which isn't quite general enough for applications, but is enough to see the main ideas. Let $X_1, \cdots, X_N$ be smooth, real vector fields on a compact manifold M. Assume that the X_j and their repeated commutators containing up to m factors span the tangent space. We will study the pseudodifferential operator

$$(21) \qquad A = -\sum_{j=1}^{N} X_j^2 + S,$$

where S is a first-order operator, whose symbol $S(x,\xi)$ is nonnegative and satisfies

$$(22) \qquad S(x,\xi) \le \sum_{j,k} |\{\sigma X_j,\, \sigma X_k\}| + \sum_{j} |\sigma X_j|.$$

In (22), σX_j denotes the symbol of the vector field X_j, and the curly brackets $\{\}$ denote the Poisson bracket. Under these assumptions, our main result is as follows.

Lipschitz Regularity Theorem. $A^{-1} : \mathrm{Lip}(\alpha) \to \mathrm{Lip}(\alpha + \frac{2}{m})$;
$X_j A^{-1},\ A^{-1} X_j : \mathrm{Lip}(\alpha) \to \mathrm{Lip}(\alpha + \frac{1}{m})$;
$X_j X_k A^{-1},\ X_j A^{-1} X_k,\ A^{-1} X_j X_k : \mathrm{Lip}(\alpha) \to \mathrm{Lip}(\alpha - \epsilon)$, any $\epsilon > 0$.

Before explaining the proof, let me give a conjecture on the fundamental solution of A. This conjecture hasn't been proven, but it's probably true. Kohn and I are grateful to Eli Stein for a very useful conversation on the analogous operator $-\Delta_x + |\frac{\partial}{\partial t}|$, from which the conjecture arose.

Recall how the fundamental solution for $L = \sum_j X_j^2$ behaves. The basic geometric idea is to define a non-Euclidean ball $B_L(x,\rho)$, which consists of all the points we can reach from x in time $\le \rho$ by travelling on broken paths, on each segment of which we move along an integral curve of one of the X_j. Stein pioneered the study of Hörmander operators and $\overline{\partial}$-problems in terms of such non-Euclidean balls. On one hand, we understand how the $B_L(x,\rho)$ looks; on the other hand, we know how the fundamental solution of L behaves in terms of the $B_L(x,\rho)$. (See Rothschild-Stein [14] and Nagel-Stein-Wainger [13].) In fact:

(23) After a controlled change of coordinates, $B_L(x,\rho)$ is essentially a box $I_{\delta_1} \times I_{\delta_2} \times \cdots \times I_{\delta_n}$, $I_\delta \equiv [-\delta,\delta]$, in the sense that
$$I_{c\delta_1} \times I_{c\delta_2} \times \cdots \times I_{c\delta_n} \subset B_L(x,\rho) \subset I_{C\delta_1} \times I_{C\delta_2} \times \cdots \times I_{C\delta_n}.$$

(24) $\qquad -\sum_{j} X_j^2 u = f$ is solved (in 3 or more dimensions) by

$$u(x) = \int_M K(x,y) f(y)\, dy, \text{ where } |K(x,y)| \le \frac{C\,\delta^2(x,y)}{Vol(x,y)},$$
$$\delta(x,y) = \inf\{\rho > 0 : y \in B_L(x,\rho)\},$$
$$Vol(x,y) = \text{Volume of } B_L(x,\rho) \text{ with } \rho = \delta(x,y).$$

There are analogous estimates for the derivates of $K(x, y)$.

An elementary example of (24) is the fundamental solution $K(x, y) = c|x - y|^{n-2}$ of the Laplacian on $\mathbf{R}^n$, for which $\delta(x, y) \sim |x - y|$.

Now we can explain how the above discussion of $-\sum_j X_j^2$ applies with modifications to the operator A. Let us try to solve $Au = f$ microlocally in $\{(x, \xi) \epsilon\ T^*M : |\xi| \sim R\}$. According to the uncertainty principle, we can only resolve $x \epsilon M$ to an accuracy $\sim R^{-1}$. Hence, it is natural to set

$$B_L^*(x, \rho; R) = \{y \epsilon M : |y - y'| < R^{-1} \text{ for some } y' \epsilon B_L(x, \rho)\}.$$

Then $B_L^*(x, \rho; R)$ should play the role of $B_L(x, \rho)$ for A. Moreover, not all ρ are relevant: If all the δ_k in (23) are $>> R^{-1}$, then $B_L(x, \rho)$ really pertains to microlocal analysis of $\sum_j X_j^2$ on $\{(x, \xi) : |\xi| << R\}$, so ρ is too big to matter here.

If all the δ_k in (23) are $<< R^{-1}$, then $B_L(x, \rho)$ really pertains to microlocal analysis of $\sum_j X_j^2$ on $\{(x, \xi) : |\xi| >> R\}$, so ρ is too small to matter here.

Hence we call ρ *relevant* if $c\ \min_k \delta_k \leq R^{-1} \leq C\ \max_k \delta_k$. One conjecture on the microlocalized fundamental solution of A is as follows: Microlocally in $\{(x, \xi) \epsilon\ T^*M : |\xi| \sim R\}$, the equation $Au = f$ is solved by $u(x) = \int_M K_R(x, y) f(y)\ dy$, with

$$(25)\quad |K_R(x, y)| \leq C \sum_{\rho=2^{-k},\ \rho \text{ relevant}} \frac{\rho^2}{\text{Vol } B_L^*(x, \rho; R)} (\chi_{B_L^*(x, \rho; R)})(y),$$

and with analogous estimates on the derivatives of K_R. In (25), χ denotes the characteristic function. The full fundamental solution of $Au = f$ is obtained by adding together $K_R(x, y)$ for $R = 1, 2, 4, 8, \cdots$.

We end this article with a quick sketch of the proof of the Lipschitz Regularity Theorem for A. The first thing to note is that $f \epsilon \text{Lip}(\alpha)$ if and only if $\|T_R f\|_{L^\infty} \leq C\ R^{-\alpha}$, where T_R microlocalizes f to $\{(x, \xi) : R < |\xi| < 2R\}$, $R >> 1$. Therefore, Lipschitz regularity of A^{-1} amounts to proving L^∞ estimates microlocally in $\{(x, \xi) : R < |\xi| < 2R\}$. The basic idea of our proof is rescaling. We illustrate the use of rescaling in the simplest case: Once we know that $\Delta^{-1} f(x) = \int_{\mathbf{R}^n} K(x, y) f(y)\ dy$ with $|K(x, y)| \leq C$ for $|x - y| \sim 1$, then rescaling gives $|K(x, y)| \leq C\rho^{2-n}$ for $|x - y| \sim \rho$. A typical proof by rescaling starts with a standard subelliptic estimate, e.g.

$$(26)\qquad\qquad \langle Au, u \rangle \geq c\|u\|_{(\epsilon)}^2 - C\|u\|^2.$$

After appropriate rescaling, we get sharp estimates on u in terms of Au. The enemy in a proof by rescaling is the "junk" term $C\|u\|^2$ in (26), which rescales to be of the same order of magnitude as the "main" term $c\|u\|^2_{(\epsilon)}$. The main effort in the proof is to overcome this. Here is how we do so.

We prove L^∞ estimates in two steps. The first step involves a mixture between the L^2 and L^∞ norms. The second step is to pass from this mixed norm to L^∞. For the first step, we work microlocally in $\{(x,\xi)\epsilon\ T^*M : DR \leq |\xi| \leq 2DR\}$. Here, R is to be taken arbitrarily large, while D is a fixed, large constant. For each $x\ \epsilon\ M$, let $B(x) = B_L(x,\rho)$ with $\rho = \rho(x)$ picked so that $\min_k \delta_k \sim R^{-1}$ in (23). Then introduce the mixed norm

$$|||u|||^2_s = \sup_{x\epsilon M}\{\frac{(\rho(x))^{-2s}}{vol\ B(x)}\int_{B(x)}|u|^2\}.$$

Clearly $|||u|||_0 \leq \|u\|_{L^\infty}$. We control $|||u|||^2_{s+2}$ in terms of $|||f|||^2_s$ microlocally in $\{|\xi| \sim DR\}$ when $Au = f$. To do so, for fixed $B(x)$, we apply (23) to pass to a box $I_{\delta_1} \times I_{\delta_2} \times \cdots \times I_{\delta_n}$, and then we rescale to the unit cube. After rescaling, we can apply the subelliptic estimate (26). At this stage, the "enemy" appears: the "junk" terms have the same order of magnitude as the "main" term in (26) after rescaling. However, the large parameter D saves the day. The "junk" term contains a lower power of D than the "main" term. Hence, by picking D once and for all (and sufficiently large), we can assume that the "junk" term is at most half of the main term in the rescaled (26). Thus the "junk" term may be absorbed in the "main" term, and the enemy is defeated. We obtain the desired mixed norm estimate, controlling $|||u|||_{s+2}$ by $|||f|||_s$.

It remains to pass from mixed norms to L^∞ norms. We introduce the L^∞ norms

$$\|u\|_{\infty,s} = \sup_{x\epsilon M}(\rho(x))^{-s}|u(x)|, \text{ with } \rho(x) \text{ as before.}$$

Recall that $Au = f$, i.e.,

$$(27) \qquad\qquad -\sum_j X_j^2 u = g$$

$$(28) \qquad\qquad g = f - Su.$$

We start with a standard estimate for (27) and rescale. The standard estimate we use is

$$(29) \quad \|u\|_{L^\infty(B)} \leq c_1\|g\|_{L^\infty(B^*)} + C_1\|g\|_{L^2(B^*)} + C_1\|u\|_{L^2(B^*)}.$$

Here, B is the unit ball and B^* its double. We can take c_1 as small as we please, by taking C_1 large enough. Rescaling (29) from the unit ball to $B(x)$, we find that

$$(30) \qquad \|u\|_{\infty,s+2} \le c_1 \|g\|_{\infty,s} + C_1 \|\|g\|\|_s + C_1 \|\|u\|\|_{s+2}.$$

From (28) and the fact that we work microlocally in $\{|\xi| \sim DR\}$, we deduce that

$$(31) \qquad \|g\|_{\infty,s} \le \|f\|_{\infty,s} + C_0 \|u\|_{\infty,s+2},$$

where C_0 is a large constant independent of c_1, C_1 in (29), (30). Similarly

$$(32) \qquad \|\|g\|\|_{\infty,s} \le \|\|f\|\|_{\infty,s} + C_0 \|\|u\|\|_{\infty,s+2}.$$

Putting (31) and (32) into (30), we see that

$$(33) \quad \|u\|_{\infty,s+2} \le C_0 c_1 \|u\|_{\infty,s+2} + C_0 c_1 \|f\|_{\infty,s} + C' \|\|u\|\|_{s+2} + C' \|\|f\|\|_s.$$

Here, C_0 is fixed, c_1 may be taken arbitrary small, and C' depends on c_1 (via C_1). In (33) we see the usual enemy – one of the terms on the right in (33) is of the same order of magnitude as the left-hand side. However, we may simply pick c_1 so small that $C_0 c_1 < \frac{1}{2}$, so that the first term on the right in (33) may be absorbed into the left-hand side. Thus we obtain

$$(34) \qquad \|u\|_{\infty,s+2} \le C \|f\|_{\infty,s} + C' \|\|u\|\|_{s+2} + C' \|\|f\|\|_s.$$

The triple norm $\|\|u\|\|_{s+2}$ is already under control, thanks to our previous discussion of mixed norms. Thus, (34) provides sharp, microlocal L^∞ estimates for u. From these L^∞ estimates, we can read off the conclusions of the Lipschitz Regularity Theorem, since $\rho(x) \le C \, R^{-1/m}$.

References

[1] D. Catlin, Subelliptic estimates for the $\bar{\partial}$-Neumann problem on pseudoconex domains, Ann. of Math. 126 (1987), pp. 131-191.

[2] D. Catlin, Necessary conditions for subellipticity of the $\bar{\partial}$-Neumann problem, Ann. of Math. 117 (1983), pp. 147-171.

[3] M. Christ, Regularity properties of the $\bar{\partial}_b$-equation on weakly pseudoconvex CR manifolds of dimension three.

[4] J. P. D'Angelo, Real hypersurfaces, orders of contact, and applications, Ann. of Math. 115 (1982), pp. 615-637.

[5] C. Fefferman and J.J. Kohn, Hölder estimates on domains of complex dimension two and on three-dimensional CR manifolds, Advances in Math. 69, no.2 (1988), pp. 223-303.

[6] C. Fefferman, J.J. Kohn and M. Machedon, Hölder estimates on CR manifolds with a diagonalizable Levi form, Advances in Math. 84, no.1 (1990), pp. 1-90.

[7] L. Hörmander, Hypoelliptic second-order differential equations, Acta Math. 119 (1967), pp. 147-171.

[8] J. J. Kohn, Harmonic integrals on strongly pseudoconvex domains I & II, Ann. of Math. 78 (1963), pp. 112-148, and 79 (1964), pp. 450-472.

[9] J. J. Kohn, Subellipticity of the $\overline{\partial}$-Neumann problem on pseudoconvex domains: sufficient conditions, Acta Math. 142 (1979), pp. 79-122.

[10] J. J. Kohn and H. Rossi, On the extension of holomorphic functions from the boundary of a complex manifold, Ann. of Math. 81 (1965), pp. 451-472.

[11] J. McNeal, Boundary behavior of the Bergman kernel function in $\mathbf{C}^2$, Duke Math J. 58(1989), pp. 499-512.

[12] A. Nagel, J.-P. Rosay, E. M. Stein, and S. Wainger, Estimates for the Bergman and Szegö kernels in $\mathbf{C}^2$, Ann. of Math. 129 (1989), pp. 113-149.

[13] A. Nagel, E. M. Stein, and S. Wainger, Balls and metrics defined by vector fields I: basic properties, Acta Math. 155 (1985), pp. 103-147.

[14] L.P. Rothschild and E. M. Stein, Hypoelliptic differential operators and nilpotent groups, Acta Math. 137 (1976), pp. 247-320.

COMPLEX DYNAMICS IN HIGHER DIMENSION. II

JOHN ERIK FORNAESS AND NESSIM SIBONY

Introduction

In this paper we continue the investigations started in [FS1] in order to construct a Fatou-Julia theory for holomorphic (respectively mero-morphic) self maps in $\mathbb{P}^2$. We start by considering maps arising in two dimensions from Newton's method. This leads to the study of iteration of meromorphic maps in $\mathbb{P}^2$. More precisely if we consider the problem in $\mathbb{C}^2$ to find the zeroes of a complex polynomial map, we are led to study iteration of maps on $\mathbb{P}^2$. The results can afterwards be interpreted back down in $\mathbb{C}^2$.

It turns out that generically, when one applies Newton's method, the map one has to study in $\mathbb{P}^2$ is not holomorphic but just meromorphic. This is why we are also interested in the dynamics of meromorphic maps.

One of the main tools in iteration theory in one complex variable is the Montel Theorem, i.e. the fact that $\mathbb{P}^1 \backslash \{0, 1, \infty\}$ is Kobayashi hyperbolic. This approach was explored in [FS1].

It is however possible to prove many results of the Fatou-Julia theory in one variable using potential theory. This was started by Brolin [Br] and continued in [Si], [T]. In paragraph 6 we show how to work out this approach in the context of rational maps in $\mathbb{P}^1$ in order to obtain recent results due to Lyubich [Ly] and Lopez-Freire-Mane [FLM], see also Hubbard-Papadopol [HP]. In the above mentioned articles the use of the Koebe distortion theorem is crucial in order to construct a measure of maximal entropy for a rational map in $\mathbb{P}^1$ and to prove convergence results. Such a distortion theorem is not valid in several variables.

After generalities on meromorphic maps in $\mathbb{P}^2$, we consider the Green function associated to a "generic meromorphic" map in $\mathbb{P}^2$. Such a function was studied in the context of Hénon maps by Hubbard, from the topological point of view, and it was extended to the case of holomorphic maps by Hubbard and Papadopol [HP].

To the Green function we associate an invariant closed positive $(1, 1)$ current T in $\mathbb{P}^2$. Such currents and their wedge products where considered in the context of Hénon maps by Sibony. Bedford and Sibony established their first properties. Some of the results they obtained appeared in § 3 of

Bedford-Smillie [BS1], see also the introduction of [BLS]. The structure of these Green-currents was studied extensively by Bedford-Lyubich-Smillie [BS1], [BS2], [BLS] and Fornaess-Sibony [FS]. The notion was adapted in the context of holomorphic maps in $\mathbb{P}^k$ by Hubbard-Papadopol [HP].

Let f be a holomorphic self map in $\mathbb{P}^2$ of degree $d \geq 2$. Let J_0 denote the Julia set of f, i.e. $p \in J_0$ if the family (f^n) is not normal in a neighborhood of p. We show that there exists a closed positive $(1,1)$ current T satisfying the functional equation $f^*T = dT$ and whose support is exactly J_0, (this is done in the context of "generic meromorphic normal" maps). Moreover T is extremal among the currents satisfying the previous functional equation. We then show that the Julia set J_0 is always connected and the Fatou set i.e. $\mathbb{P}^2 \backslash J_0$ is a domain of holomorphy. Hence the critical set of f always intersects the Julia set J_0.

In § 6 we study the probability measure $\mu := T \wedge T$ and we show that it is an invariant ergodic measure of maximal entropy.

The analysis of a simple example such as $f[z : w : t] = [z^d : w^d : t^d]$ shows that the Julia set is not indecomposable as in one variable. We introduce a first order Julia set J_1, see Definition 5.8, and we show that $J_1' := \text{supp}\,\mu \subset J_1$. If J_1' has nonempty interior then $J_1' = \mathbb{P}^2$. If K is a compact invariant set, hyperbolic and of unstable dimension 2, then necessarily $K \subset J_1$. We also show that no f can be hyperbolic on the whole Julia set J_0. Finally if the nonwandering set of a holomorphic map of degree $d \geq 2$ is hyperbolic, then the Fatou components are preperiodic to finitely many periodic basins.

It will be natural to consider similar questions in $\mathbb{P}^k$, $k \geq 2$. For simplicity we have written this article, restricting ourselves to the case of $\mathbb{P}^2$. We will continue our study in forthcoming articles, see [FS4].

1. Newton's method and meromorphic maps in $\mathbb{P}^2$.

Given two complex polynomials P, Q in two variables z, w, Newton's method provides a way to approximate the roots (z_0, w_0) of the equations $P(z, w) = Q(z, w) = (0, 0)$ by starting with an initial guess (z_1, w_1) and inductively define $(z_{n+1}, w_{n+1}) := (z_n, w_n) - F'^{-1}(z_n, w_n)F(z_n, w_n)$ where $F : \mathbf{C}^2 \to \mathbf{C}^2, F := (P, Q)$ and the $2 * 2$ matrix F' denotes the derivative of F.

Hence Newton's method consists of iteration of the map

$$R(z, w) := (z, w) - F'^{-1}F.$$

The natural way to consider this map is as a map on $\mathbb{P}^2$ rather than $\mathbb{C}^2$ since it depends on quotients of polynomials. To write this as a map on $\mathbb{P}^2$, introduce first the determinant D of F', and the two determinants $D_w := Q_w \cdot P - P_w \cdot Q$ and $D_z := P_z \cdot Q - Q_z \cdot P$ where we have used subscripts to denote partial derivatives. Hence we write R as a map on $\mathbb{P}^2$ in homogeneous coordinates as $R : [z : w : 1] \to [zD - D_w : wD - D_z : D]$.

We will assume that the map F has maximal rank 2, otherwise $F = 0$ would in general have no root. Also we can assume that the map F is at least of degree two. Otherwise Newton's method immediately gives the root after one step. This amounts to the same thing as requiring that R is non constant.

Observe that at points where $D_z = D_w = 0$, which means that $\begin{pmatrix} P_z \\ Q_z \end{pmatrix}, \begin{pmatrix} P \\ Q \end{pmatrix}$ are linearly dependent and $\begin{pmatrix} P_w \\ Q_w \end{pmatrix}, \begin{pmatrix} P \\ Q \end{pmatrix}$ are linearly dependent we have generically that $D = 0$, i.e. $\begin{pmatrix} P_z \\ Q_z \end{pmatrix}, \begin{pmatrix} P_w \\ Q_w \end{pmatrix}$ are linearly dependent. So in general R has poles.

If the map R has maximal rank 1, then it is easy to show that the image X of R is a $\mathbb{P}^1$ and that R restricted to X is a rational map. We consider, for the purpose of this paper, this as a known case. So we will assume from now on that R has maximal rank 2. For the one variable theory we refer to [B], [Ca] or [Mi].

Notice that if $R = [A : B : C]$ are homogeneous polynomials of degree d, we may assume they have at most finitely many common zeroes (lines of common zeroes in $\mathbb{C}^3$). If not, they have a common factor which can be divided out. The remaining points p if any in $\mathbb{P}^2$ are called points of indeterminacy.

In the case where R is linear, the dynamics is rather simple. Hence we will in the rest of the paper restrict ourselves to the case when R has at least degree 2.

We will confine ourselves to giving one situation in which a nontrivial polynomial equation gives rise to a linear Newton's map - in complete analogy with an important case in one variable.

Proposition 1.1. *If $(P,Q)(z,w)$ are homogeneous of the same degree $n > 1$, and $F = (P,Q)$ has maximal rank 2, then $R(z,w) = ((1 - 1/n)z, (1 - 1/n)w)$. If (P,Q) are two polynomials of degree $n > 1$ such that the highest degree terms (P_n, Q_n) have maximal rank 2, then R is the identity map at infinity $(t = 0)$.*

Proof: Immediate. $\square$

The situation can be quite different for homogeneous polynomials of different degree. Let $F = (P, Q) = (z^3 + w^3, z^2 + w^2)$. Then Newton's map $R = [4z^3 w - 3z^2 w^2 + w^4 : -z^4 + 3z^2 w^2 - 4zw^3 : 6zwt(z - w)]$. Then the point $(0, 0)$ is a root of $F = 0$. However Newton's method applied to points $[\epsilon : \epsilon : 1]$ arbitrarily close to the root, $\epsilon \neq 0$, are all mapped to the fixed point $[1 : -1 : 0]$ at infinity, so Newton's method diverges arbitrarily close to the root ! The problem is that at such points $D = 0$ but D_w, $D_z \neq 0$. The root $[0 : 0 : 1]$ is a point of indeterminacy for R while $(z = w)$ is mapped to a different fixed point.

Schröder ([1871]) was the first to study Newton's method in one complex variable. He observed that there are infinitely many variations of Newton's method $R = z - (f'(z))^{-1} f(z)$. For example, one can replace $f'(z)$ by the derivative at some fixed point close to the root. Doing this in two dimensions increases the class of maps obtained from Newton's method significantly. Any polynomial map $R = [A : B : t^d]$ can be written in the form, $(t = 1)$,
$(A, B) = (z, w) - M^{-1}(P, Q)$ where M is an invertible constant matrix and (P, Q) is polynomial. There is also the relaxed Newton's method, $R = z - \epsilon F'^{-1} \cdot F$ for some constant $0 < \epsilon < 1$ (which approximates the Newton Flow $X = -F'^{-1} \cdot F$). Based on these remarks, we will from now on study the class of all meromorphic maps R on $\mathbb{P}^2$ with maximal rank 2 and degree at least two. (The Newton Flow would lead rather to the study of complex foliations of $\mathbb{P}^2$).

With the degree d of any homogeneous map $R = [A : B : C]$ we mean the degrees of A, B or C, which are equal, after cancellation of all common irreducible factors. So we assume that $d \geq 2$. Let $I = I(R) = \{q_k\}$ denote the (finite) indeterminacy set consisting of the points q_k in $\mathbb{P}^2$ where $A = B = C = 0$. Also let $V = \cup V_j$ denote the finite union of irreducible compact complex curves V_j on each of which R has a constant value (at least outside I). Say $R(V_j) = p_j$. We call such curves V_j, R-*constant*.

Proposition 1.2. *If V is nonempty, then also I is nonempty. In fact each irreducible branch of V contains at least one point of indeterminacy. It can happen that V is empty while I is nonempty.*

Proof: Suppose that V is nonempty. Then we may after a rotation assume that some irreducible component $V_j = \{h = 0\}$ for some irreducible homogeneous polynomial h and that $R(V_j) = [0 : 0 : 1]$. Hence h divides both A and B. But then the set $C = 0$ and $h = 0$ must intersect and such an intersection point is a point of indeterminacy.

For the converse, consider the example $R = [zw : z^2 + wt : t^2]$. For this example, there is one point of indeterminacy, $[0{:}1{:}0]$, while the map is not constant on any curve. $\Box$

Let R be a meromorphic map $\mathbb{P}^2 \to \mathbb{P}^2$. Let I be the indeterminacy set. Given $a \in \mathbb{P}^2$ we want to discuss the number of preimages of a. Recall that Bezout's Theorem asserts that if $(P_1, \cdots, P_k)$ are k homogeneous polynomials in $\mathbb{P}^k$ with discrete set of zeroes then the number of zeroes counted with multiplicities is equal to the product of the degrees.

Proposition 1.3. *Let* $R : \mathbb{P}^2 \to \mathbb{P}^2$ *be a meromorphic map of degree* d. *Assume* $I \neq \emptyset$.
Assume R *is of rank 2. Then for any* a *which is not one of the finitely many points which is the image of an* $R-$ *constant curve,* $R^{-1}(a) = d' < d^2$. *Here we count the number of points with multiplicity.*

Proof: Consider the map $\mathbb{C}^4 \to \mathbb{C}^3$, $(z_0, z_1, z_2, t) = (z, t) \to R(z) - at^d$. We have 3 polynomials in $\mathbb{P}^3$. Assume there is no 1 dimensional variety in $\mathbb{P}^2$ such that $R(V) = a$. Then the number of zeroes of $R(z) - at^d$ is finite in $\mathbb{P}^3$. So it is d^3 counting multiplicity. For $p \in I$, $[p : 0]$ is a zero of multiplicity at least d. Hence number of zeroes in $t = 1$ is $< d^3$. Since rotation of t by a dth root of unity produces an equivalent solution in $\mathbb{P}^2$ we get that $d' < d^2$. $\Box$

Remark 1.4. *If a point* $a \notin R(\mathbb{P}^2 \backslash I)$ *then* $\sum_{p \in I}$ *multiplicity of* $[p : 0]$ *for* $R(z) - at^d = 0$ *is* d^3.

Consider the forward orbit of the points p_j. The variety V_j is called *degree lowering* if for some (smallest) $n = n_j \geq 0$, $R^n(p_j) \in I$.

We will next discuss the growth of the degrees of the iterates of maps R.

When there is a degree lowering variety, all the components of the iterates of R^{n_j+1} vanish on $V_j = \{h = 0\}$. Hence one need to factor out a power of h in order to describe the map properly. Hence the degree of the iterate will drop below d^{n_j+1}.

We will not study the class of maps with degree lowering varieties in this paper. If a map R has an $R-$ constant variety V with $R(V) = p$ which is not degree lowering, most likely the complement of the preimages of this variety is Kobayashi hyperbolic. So on these varieties, the map eventually lands on the R constant variety after which the orbit reduces to the orbit of a point, and in the complement of these varieties, the iterates is a normal family, even on any subvariety disjoint from $\bigcup_{n \geq 0} R^{-n}(V)$.

If the variety is degree lowering, the iterations can be much more complicated and worthy of further study. However, the method we will be pursuing in this paper, pluripotential theory, is more difficult to carry out for such maps. So we will pursue these in a separate paper with other methods.

2. Green Function

In this section we will study *generic meromorphic maps* on $\mathbb{P}^2$, i.e. meromorphic maps of maximal rank 2, which have degree at least 2 and which have no degree lowering curves.

We have first to define Fatou sets and Julia sets of $R : \mathbb{P}^2 \to \mathbb{P}^2$.

Since we would like to have a notion of Fatou and Julia set which is invariant under R, we need to be precise about what one means with the preimage of a point. We say that for a given p, a point q is a preimage of p if R is defined at q and $R(q) = p$. (If $q \in I(R)$ and $p \in W$, the blow up of q, so (q, p) is in the closure of the graph of R, then q is with this convention *not* in the preimage of p). Here I denote as usual the indeterminacy set of the meromorphic map R.

Definition 2.1. *An orbit* $\{p_n\}_{n=-k}^{0}$ *is called complete if*
  (i) $R(p_n) = p_{n+1}$,
  (ii) $p_0 \in I$,
  (iii) $p_n \notin I$, $n < 0$,
  (iv) *If k is finite, $p_{-k} \notin R(\mathbb{P}^2 \backslash I)$.*

We call $k + 1$ the length of the orbit.

Lemma 2.2. *A point $p \in \mathbb{P}^2$ is a point of indeterminacy for R^n if and only if $\{p, R(p), \cdots, R^{k-1}(p)\}$ is a right tail of some complete orbit for some $1 \le k \le n$.*

Proof: Immediate. $\square$

Corollary 2.3. *If I_n denotes the indeterminacy set of R^n, then $I_n \subset I_m$ $\forall m > n$.*
Proof: Immediate. $\square$

The set $I(R)$ should belong naturally to the "Julia set". So does $\cup I_n$. Hence the closure $E := \overline{\cup I_n}$, *the extended indeterminacy set*, belongs naturally to the Julia set as well.

Proposition 2.4. *If $p \in \mathbb{P}^2 \backslash E$ and $R^n(p) \in E$, $n \geq 1$, then p is on an R^n-constant curve.*

Proof: If p is not on an R^n- constant curve there are arbitrarily small neighborhoods $U(R^n(p))$, $V(p)$ so that $R^n \mid V : V \to U$ is a finite, proper, surjective holomorphic map. Moreover we may assume that $V \cap E = \emptyset$. Since every such open set U contains a point from some I_k, it follows that V contains a point from some I_{n+k}. Hence $p \in E$, a contradiction. $\square$

Definition 2.5. *Let $R : \mathbb{P}^2 \to \mathbb{P}^2$ be a generic meromorphic map. A point $p \in \mathbb{P}^2$ is in the Fatou set if and only if there exists for every $\epsilon > 0$ some neighborhood $U(p)$ such that $\operatorname{diam} R^n(U \backslash I_n) < \epsilon$ for all n.*

Note that this implies that p cannot belong to the extended indeterminacy set. We say that the Julia set is the complement of the Fatou set. By a normal family argument it follows that the Fatou set is an open set and that the Julia set is closed. Also we conclude that the extended indeterminacy set belongs to the Julia set. We denote the Julia set of R by $J(R)$ or J_0, since we introduce also higher order Julia sets.

We have complete invariance of the Julia set :

Proposition 2.6. *Suppose that $p \in \mathbb{P}^2 \backslash I$ and that $R(p) \in J(R)$, the Julia set of R. Then p belongs to the Julia set also. On the other hand, if $p \in J(R)$, and $p \notin I$, then $R(p) \in J(R)$.*

Lemma 2.7. *Suppose that $p \in \mathbb{P}^2 \backslash I$ and that $R(p) \in F(R)$, the Fatou set of R. Then $p \in F(R)$ also.*

Proof: This is obvious from the definition since R is continuous at p.

Lemma 2.8. *Suppose that $p \in F(R)$. So in particular, $p \in \mathbb{P}^2 \backslash I$. Then $R(p)$ is also in $F(R)$.*

Proof: Suppose first that p does not belong to any $R-$ constant curve. Then R is locally finite to one and proper near p. Hence arbitrarily small neighborhoods of p are mapped properly onto arbitrarily small neighborhoods of $R(p)$. The conclusion follows.

Suppose next that p belongs to an $R-$ constant curve X. We can select a small complex disc Δ centered at p such that $\Delta \backslash \{p\}$ does not intersect X nor the critical set of R. Considering a small neighborhood of $\Delta\backslash\{p\}$, and taking the image of it, we obtain a piece of a complex curve Y through $R(p)$ and a neighborhood V of $Y\backslash R(p)$ on which the iterates of R is an equicontinuous family. Note that $R(p)$ cannot belong to $\cup I_n$ since this would imply that p belongs to this set also, contradicting that p belongs to the Fatou set. Pushing discs, we can conclude that any iterate R^n is holomorphic in a fixed neighborhood of $R(p)$ so $R(p) \notin E$. But then it follows that equicontinuity extends to a neighborhood of $R(p)$, so $R(p)$ is in the Fatou set.

So the proposition follows from the two lemmas. $\square$

Definition 2.9. *A point $p \in \mathbb{P}^2$ has a nice orbit if there is an open neighborhood $U(p)$ and an open neighborhood $V(I)$ so that $R^n(U)\cap V = \emptyset$ for all $n \geq 0$.*

So if p has a nice orbit, R^n is well defined for all n on some fixed neighborhood of p. The set of nice points is an open subset of $\mathbb{P}^2\backslash E$.

Definition 2.10. *A generic meromorphic map is said to be normal if N, the set of nice points equals $\mathbb{P}^2\backslash\overline{\cup I_n}$.*

Let R be a generic meromorphic map in $\mathbb{P}^2$. With an abuse of notation we will also denote by R a lifting of R to $\mathbb{C}^3$. If $\|\ \|$ is a norm on $\mathbb{C}^3$ we define the nth Green function G_n on $\mathbb{C}^3$ by the formula $G_n := \frac{1}{d^n}\text{Log}\ \|\ R^n\ \|$. Here d is the common degree of the components of R. Observe that if R is meromorphic, G_n has other poles in $\mathbb{C}^3$ than just the origin. Let π denote the canonical map $\mathbb{C}^3\mid\{0\} \to \mathbb{P}^2$.

Proposition 2.11. *The functions G_n converge u. c. c. to a function G on the set $\pi^{-1}(N)$ of points with nice orbits.*

Proof: If $p \in N$ there exists $U(p)$ and $c > 0$ such that on $\pi^{-1}(U(p))$

$$\|\ R^{n+1}(z)\ \|\geq c\ \|\ R^n(z)\ \|^d .$$

On the other hand the reverse inequality

$$\|\ R^{n+1}(z)\ \|\leq C\ \|\ R^n(z)\ \|^d$$

holds always. Hence the sequence

$$\frac{\log\|\ R^n\ \|}{d^n}$$

converges u. c. c. on $\pi^{-1}(U(p))$. $\square$

Because of the second inequality, the limit G always exists and is a plurisubharmonic function on $\mathbf{C}^3$, possibly $\equiv -\infty$, although we don't believe this can happen. (You just need one periodic orbit to show that the limit is not identically $-\infty$.)

Obviously $N \subset \bigcup_n \mathrm{int}\{G > -n\}$.

The other inclusion does not hold in general, as the following example shows.

Example 2.1. *Let $R = [z^d : w^d : t^{d-1}w]$, $d \geq 3$. This map has one point of indeterminacy, $[0 : 0 : 1]$, which has no preimage. Hence $E = [0 : 0 : 1]$. Also, the only $R-$ constant curve is $(w = 0)$ whose image is the fixed point $[1 : 0 : 0]$. Hence the map is generic. We get that $R^n = [z^{d^n} : w^{d^n} : t^{(d-1)^n}w^{d^n-(d-1)^n}]$. Computing the Green's function G,*

$$G_n = 1/d^n \log \| (z^{d^n}, w^{d^n}, t^{(d-1)^n}w^{d^n-(d-1)^n}) \|$$

and we obtain $G = \max\{\log | z |, \log | w |\}$. Letting $\Omega := \{| z | < | w | < | t |\}$, G is pluriharmonic, on $\pi^{-1}(\Omega)$, but $\Omega \not\subset N$ since $R^n \to [0 : 0 : 1]$. Furthermore this is pluriharmonic when $| z | < | w |, | w |$ is close to 1 and $t = 1$. But then $R^n = [(z/w)^{d^n} : 1 : (1 : w)^{(d-1)^n}]$. Notice that when $| w | < 1$ the limit becomes $[0 : 0 : 1]$ while if $| w | > 1$ the limit becomes $[0 : 1 : 0]$ so in particular the points where $| z | < | w | = | t |$ are in the Julia set even though G is pluriharmonic there. In the region $t = 1$, $| z | > | w |$, $R^n = [1 : (w/z)^{d^n} : \left(\frac{w}{zw^{(d-1)^n/d^n}}\right)^{d^n}]$ so $R^n \to [1 : 0 : 0]$, so the Julia set also contains $\{| z | = | w |\}$. There are three Fatou components : $\Omega_1 = \{| z | < | w | < | t |\}$ on which $R^n \to [0 : 0 : 1]$, $\Omega_2 = \{| w | < | z |\}$, on which $R^n \to [1 : 0 : 0]$, $\Omega_3 = \{| z | < | w |, | t | < | w |\}$ on which $R^n \to [0 : 1 : 0]$. The blow up of $[0 : 0 : 1]$ is the z-axis (which happens to coincide with the $R-$ constant curve).

For the behaviour on the Julia set : If $| z | = | w | < 1 = t$, then $R^n \to [0 : 0 : 1]$. If $| z | = | w | > 1 = t$, then $R^n \to [1 : 1 : 0]$. If $| z | < | w | = 1 = t$, then R^n converges to the invariant circle $z = 0$, $| w | = | t |$. Notice that these points are not in a nice component. Finally the set $| z | = | w | = | t |$ is an invariant torus.

We just recall that plurisubharmonic (p.s.h. for short) functions on a complex manifold are upper semicontinuous functions that are subharmonic on one dimensional analytic discs, see [Le] or [Kl].

Theorem 2.12. *Let R be a generic meromorphic map on $\mathbb{P}^2$. Then $G(z) = \lim \frac{1}{d^n} \log \parallel R^n \parallel = \lim \searrow H_n$ where $H_n := G_n + \sum_{k=n+1}^{\infty} \frac{\log M}{d^k}$, M is some constant.*

(i)   *The function G is plurisubharmonic in $\mathbf{C}^3$ (or $\equiv -\infty$).*
(ii) *G is pluriharmonic on $\pi^{-1}(\Omega)$ if Ω is a Fatou component.*
(iii) *If N is the set of nice points of R then G is continuous on $\pi^{-1}(N)$ and if G is pluriharmonic on $\pi^{-1}(\omega)$ where ω is an open subset of N, then ω is contained in a Fatou component.*

Proof: Let $M := \sup\{\parallel R(z) \parallel; \parallel z \parallel = 1\}$. Then

$$\parallel R^{n+1}(z) \parallel \leq M \parallel R^n(z) \parallel^d$$

by homogeneity. Hence

$$G_{n+1}(z) \leq \frac{\log M}{d^{n+1}} + G_n(z).$$

Replacing G_n by $H_n := G_n + \sum_{k=n+1}^{\infty} \frac{\log M}{d^k}$ we get

$$H_{n+1} = G_{n+1} + \sum_{k=n+2}^{\infty} \frac{\log M}{d^k} \leq G_n + \sum_{k=n+1}^{\infty} \frac{\log M}{d^k} = H_n.$$

The function G is a decreasing limit of p.s.h. functions, hence it is p.s.h. or $\equiv -\infty$.

Next we prove ii).

Suppose that $p \in U$ is a point in the Fatou set, and U is a small neighborhood inside the Fatou set. Choose a subsequence R^{n_k} which converges uniformly to a holomorphic map R^∞ on U. Shrinking U if necessary, taking a thinner subsequence and renaming the coordinates, we may assume that $R^{n_k}(U) \subset \{|z|, |w| < 2, t = 1\}$. We can then write $R^{n_k} = R_3^{n_k}(A_k, B_k, 1)$ for uniformly bounded holomorphic functions A_k, B_k over U in $\mathbf{C}^3$. Hence $G_{n_k} = 1/d^{n_k} * \log | R_3^{n_k} | + 1/d^{n_k} * \log \parallel (A_k, B_k, 1) \parallel$. Since the last term converges uniformly to 0 and the first term is always pluriharmonic, the result follows.

We prove iii). On a compact subset of N, we have

$$| G_{n+k} - G_n | \leq C/d^n$$

since $| R^{n+1}(z) | \geq c_1 | R^n(z) |^d$, c_1 independent of n, k. So $| G - G_n | \leq C/d^n$ and if $G = \log | h |$, h is a nonvanishing holomorphic function, we get $| 1/d^n \log \frac{\| R^n \|}{|h^{d^n}|} | \leq C/d^n$ i.e.

$$ e^{-C} \leq \frac{\| R^n \|}{| h^{d^n} |} \leq e^C. \quad \square $$

Remark 2.13. The same theorem holds for holomorphic maps on $\mathbb{P}^k \geq 1$.

Corollary 2.14. *Let $p \in E$. The Hausdorff dimension of J near p is at least 2. If R is normal, then the Hausdorff dimension near any point of J is at least 2.*

Proof: Let $q \in \mathbb{C}^3 \backslash \{0\}$ such that $\pi(q) = p$. Assume the 4 dimensional Hausdorff measure $\Lambda^4(\pi^{-1}(J) \cap U(q)) = 0$ for some neighborhood $U(q)$ of q. Then using a theorem of Harvey-Polking [Ha.P] G would extend as a pluriharmonic function in $U(q)$. But $G = -\infty$ on $\pi^{-1}(\cup I_n) \cap U(q)$, a contradiction. In the normal case we apply the same extension result and Theorem 2.12.

Proposition 2.15. *In the generic meromorphic case, the Green function G satisfies the functional equation*

$$ G(R(z)) = dG(z). $$

Moreover for $\lambda \in \mathbb{C}$, $G(\lambda z) = \log | \lambda | + G(z)$.

Proof: Direct computation gives

$$ G(R(z)) = \lim \frac{\log \| R^n(R(z)) \|}{d^n} = d \lim \frac{\log \| R^{n+1}(z) \|}{d^{n+1}} = dG(z). $$

The proof of the second assertion is clear. $\square$

We give an example showing that the pole set of G is not just $\pi^{-1}(\cup I_n)$. The example has also interesting dynamics.

Example 2.2. $R = [w^4 : w^2(w - 2z)^2 : t^4]$.

The indeterminacy set I consists of one point, $[1 : 0 : 0]$. The only $R-$ constant variety is $(w = 0)$ which is mapped to the fixed point $[0 : 0 : 1]$, so the map is generic. The map R is a polynomial map, sending the hyperplane at infinity, $(t = 0)$, to itself. At infinity, the map is $w \to \left(\frac{w-2}{w}\right)^2$. This map is a critically finite map on $\mathbb{P}^1$, and all critical points are preperiodic. Hence the Julia set of this map on $\mathbb{P}^1$ is all of $\mathbb{P}^1$, [Ca]. In particular, $\cup I_n$ is dense in $(t = 0)$, so $E = (t = 0)$. Since for example $[1 : 1 : 0]$ is a fixed point for R, $G \mid (t = 0) \not\equiv -\infty$. However, $G \equiv -\infty$ on $\bigcup_n I_n$. Since $(G = -\infty)$ is a G_δ dense set in $(t = 0)$, $(G = -\infty)$ is uncountable. Hence $(G = -\infty) \neq \bigcup_n I_n$.

The point $[0 : 0 : 1]$ is a superattractive fixed point and the punctured line $(w = 0)\backslash[1 : 0 : 0] =: L$ is mapped to it. Hence L is contained in the attractive basin Ω of $[0 : 0 : 1]$. But R maps lines through $[0 : 0 : 1]$ to lines through $[0 : 0 : 1]$. Since the map $R \mid (t = 0)$ is chaotic, it follows that Ω contains a dense set of lines through $[0 : 0 : 1]$, punctured at $(t = 0)$.

On the fixed line $(w = 2iz)$ the map is $z \to 16z^4$. Hence the Julia set contains the disc $\{[z : 2iz]; \mid z \mid \geq 2^{4/3}\}$ centered at $(t = 0)$. It follows that for a dense set of points p in $(t = 0)$, the straight line through $[0 : 0 : 1]$ and p contains a disc centered at p which is contained in $J(R)$.

3. Special Generic Meromorphic Maps

If one wants to study various well known phenomena from complex dynamics in the context of generic meromorphic maps one sees that many such phenomena are often no longer always true, but are true for large subclasses. The subclass depends often on which phenomena one wishes to study. We will illustrate this here by discussing some such phenomena, showing by examples that they don't generally hold, and give conditions which make them hold.

There are two particular cases of generic meromorphic maps which have been previously studied. First there are the complex Hénon maps. These are invertible polynomial maps which have one point of indeterminacy at infinity. Also the hyperplane at infinity is the unique $R-$ constant variety and its image is a fixed point at infinity, different from the point of indeterminacy. This point of view is developped in [FS2]. Also there is the class of holomorphic maps on $\mathbb{P}^2$, maps without points of indeterminacy.

We will define various subclasses of the meromorphic maps. These classes will be usually large enough to contain all holomorphic maps and all Hénon maps, i.e. maps of the form $R[z : w : 1] = [p(z) + aw : z : 1]$ where p is a one variable polynomial of degree $d \geq 2$.

Definition 3.1. *A generic meromorphic map is said to belong to the class of indeterminacy repellors - IR - if there exist arbitrarily small neighborhoods $U \subset\subset V$ of the indeterminacy set for which $R(\mathbb{P}^2 \backslash U) \subset \mathbb{P}^2 \backslash V$.*

Both Hénon maps and holomorphic maps belong to IR. Nevertheless the definition is rather strong. It implies that every point in $\mathbb{P}^2 \backslash I$ has a nice orbit, which also implies that the points in I have no preimages (recall that points in I are not considered as preimages).

Definition 3.2. *We say that a generic meromorphic map belongs to the class with no $R-$ constant blow ups, NRB, if there is no point of indeterminacy q for which the blow up is R^n- constant for some $n \geq 1$. The complement of this class is the set RB.*

Hénon maps are in RB. The map $R = [zw : z^2 + wt : t^2]$ is in NRB since it has no R-constant variety.

Definition 3.3. *We say that a generic meromorphic map R is a meromorphic Hénon map, MH, if there exists a generic meromorphic map S such that $R \circ S = Id = S \circ R$ in the complement of some hypersurface. We say that S is the inverse of R.*

Example 3.1. *Example of a meromorphic Hénon map : $R[z : w : t]$ $= [t^2 : zt : z^2 + ct^2 + awt]$, $a \neq 0$. $I = \cup I_n = [0 : 1 : 0]$ $(t = 0) \to [0 : 0 : 1]$ $\to [1 : 0 : c] \to [c^2 : c : 1 + c^3] \to \cdots$ Note that $[0 : 1 : 0]$ has no preimage so $(t = 0)$ cannot be degree lowering. It follows that the map is generic. If $c = 0$ we have a cycle of period 2. Let $S = \left[wz : \frac{tz - w^2 - cz^2}{a} : z^2 \right]$. Then S is also generic and $RS = Id$ out of $z = 0$ and $SR = Id$ out of $t = 0$. The map R belongs also to the class IR if $|a| < 1$.*

Note that a shear S, $S[z : w : t] = [zt : wt + z^2 : t^2]$ has an inverse which is also a shear, but they are not generic, so are not in the class MH.

Proposition 3.4. *Suppose that R and S are meromorphic maps on $\mathbb{P}^2$ such that $R \circ S = Id$ outside a hypersurface. Suppose that R is a generic meromorphic map. Then they are both meromorphic Hénon maps.*

Proof: It suffices to show that S is a generic meromorphic map. Clearly S has maximal rank 2. Since R is not linear, S cannot be linear either. Suppose that S is not generic. Then there must exist an irreducible compact curve V which is $S-$ constant and which is degree lowering. Let $q = q_0$ denote the image $S(V)$. Let $q_{n+1} := S(q_n)$ be the inductively defined orbit for q_0 up to q_m which is a point of indeterminacy for S. Note that this means that there exists a curve W such that $R(W) = q_m$. Hence W is $R-$ constant. Moreover $R^{m+1}(W) = q_0$ and q_0 is a point of indeterminacy for R. (Unless some q_j already is a point of indeterminay for R, which is also fine.) Hence W is degree lowering, a contradiction. $\square$

Proposition 3.5. *If a generic meromorphic map R belongs to NRB or is normal, then $F(R) = F(R^n)$ for all $n \geq 1$.*

Proof: Clearly $F(R) \subset F(R^n)$. Suppose next that $p \in F(R^n)\backslash F(R)$. Then there exists a subsequence $\{R_m^{j+nkm}\}$, $0 < j < n$, such that R^{nkm} converges uniformly on some neighborhood $U(p)$ while R^{j+nkm} diverges on all neighborhoods of p. This implies that $R^{nkm}(p) \to I_j$. The diameter of the images under R^{j+nkm} must remain large. Since R belongs to NRB, this must remain true for R^{n+nkm}. But this contradicts that R^n is a normal family. If R is normal, this is a consequence of Theorem 2. 12.

4. The invariant current T.

Let R be a generic meromorphic map in $\mathbb{P}^2$. Let G be the Green function in $\mathbf{C}^3$ associated to R. We study, in this section, the properties of the current T defined by the relation $\pi^* T = dd^c G$.

We refer to de Rham [de Rh] for general properties of currents. For results concerning positive currents on complex analytic varieties, see [Le] or [Kl]. We recall here a few facts. For simplicity we restrict to domains Ω in $\mathbf{C}^n$, but since the definitions are invariant under holomorphic change of coordinates, they make sense on any complex manifold.

Let $\Omega \subset \mathbf{C}^n$ be an open set, let $\mathcal{D}^{p,q}$ denote the smooth compactly supported (p,q) forms $\varphi = \sum \varphi_{IJ} dz^I \wedge d\bar{z}^J$, $|I| = p$, $|J| = q$, with its usual topology [de Rh]. The dual $\mathcal{D}_{p,q}$ of $\mathcal{D}^{p,q}$ is the space of currents of bidimension (p,q). So a current is just a differential form with distributions as coefficients.

Let $f : M \to N$ be a proper holomorphic map from the complex manifold M to the manifold N. If S is a current on M, the direct image of S under f, which we denote $f_* S$ is defined by

(1) $< f_* S, \varphi > := < S, f^* \varphi >$.

Observe that the definition make sense if for every compact $K \subset N$, $f^{-1}(K) \cap \text{Support}(S)$ is compact. If S has locally integrable coefficients, the form $f_* S$ is obtained by integrating S on the fibres of f.

When f is a submersion and S is a current on N we define f^*S as the current acting on test forms as follows, see [Sc],

$$(2) \qquad\qquad < f^*S, \varphi >:=< S, f_*\varphi > .$$

The operation f^*, f_* have the same functorial properties as when they are applied to smooth forms.

A current S of bidegree $(n-p, n-p)$ on a complex analytic manifold M is positive if for any given forms $\varphi_1, \cdots, \varphi_p$ of bidegree $(0,1)$, smooth with compact support the distribution

$$S \wedge i\varphi_1 \wedge \bar{\varphi}_1 \wedge \cdots \wedge i\varphi_p \wedge \bar{\varphi}_p$$

is positive. When S is of bidegree $(1,1)$ and is written in coordinates as

$$S = i \sum_{j,k} S_{jk} dz_j \wedge d\bar{z}_k,$$

the condition of positivity is equivalent to

$$\sum_{j,k} S_{jk} \lambda_j \bar{\lambda}_k \geq 0 \ \text{ for all } \lambda_j \in \mathbf{C}.$$

The 0-currents S_{jk} are then measures.

Recall also that $d = \partial + \bar{\partial}$ and $d^c = i(\bar{\partial} - \partial)/(2\pi)$. An upper semi-continuous function V with values in $[-\infty, \infty($ is p.s.h. iff $dd^c V \geq 0$. Recall that $E \subset \Omega$ is pluripolar if $E \subset \{u = -\infty\}$ where u is p.s.h. and $u \not\equiv -\infty$.

If S is a positive current and ω is a smooth $(1,1)$ positive form, then the current $S \wedge \omega$ is positive. If S is positive and f^*S, f_*S are well defined, then they are positive.

The mass norm of a current S is given by

$$M(S) = \sup_{|\varphi| \leq 1} |< S, \varphi >| .$$

The norm on the test form φ, is just the supremum over all coefficients after fixing an atlas, [de Rh]. When S is a positive distribution then $M(S)$ is comparable to the total variation of the positive measure $\sum_{i=1}^n S_{ii}$. The mass of a positive current on a compact set K is given by

$$M_K(S) = \sup_{|\varphi| \leq 1} |< \chi_K S, \varphi >| = M(\chi_K S)$$

here χ_K denotes the characteristic function of K. The multiplication of S by χ_K makes sense since S has measure coefficients.

We now describe the current on $\mathbb{P}^2$ associated to a p.s.h. function on $\mathbf{C}^3$, with the right homogeneity properties.

Note that G has the following homogeneity of a plurisubharmonic function H

$$H(\alpha z, \alpha w, \alpha t) = \log |\alpha| + H(z, w, t)(*).$$

We denote by $[H]$ the class of functions equal to H up to a constant. Let P denote the class of plurisubharmonic function classes $[H]$ on $\mathbf{C}^3$ with $H(\alpha z) = c \log |\alpha| + H(z)$,
$c \geq 0$, and let Q denote the class of closed, positive $(1,1)$ currents T on $\mathbb{P}^2$.

Let $\pi : \mathbf{C}^3 \backslash 0 \to \mathbb{P}^2$ be the natural projection. Consider any local holomorphic inverse $s : U \to \mathbf{C}^3 \backslash 0$ such that $\pi \circ s = Id$. Then we can define $T = T_s$ on U by $T_s = dd^c(H \circ s)$, $H \in P$. The important fact is that T_s is independent of s : If s' is another section of U, then $s' = \varphi s$ for some invertible holomorphic function φ on U. Hence

$$T_{s'} = dd^c(H \circ s') = dd^c(H(\varphi s)) =$$

$$dd^c c \log |\varphi| + dd^c H(s) = T_s.$$

So using this local definition we can define $T = L_1(H)$, $L_1 : P \to Q$ globally on $\mathbb{P}^2$ and write with abuse of notation $T = dd^c H$. Since H is plurisubharmonic, it follows that T is positive [Le], so T is a positive, closed $(1,1)$ current. For the reader's convenience we prove the following result.

Theorem 4.1. *The map L_1 is a bijection between P and Q.*

Proof: We may suppose that we are in the coordinate system where $t \neq 0$. Use the section $s(z, w) = (z, w, 1)$. Using that π^* and dd^c commutes ([deRh]), we get that

$$\pi^* T = \pi^*(dd^c(H \circ s)) = dd^c \pi^*(H \circ s) =$$

$$dd^c(H \circ s \circ \pi) =$$

$$dd^c(H(z/t, w/t, 1)) =$$

$$dd^c(H \circ (1/t)(z, w, t)) =$$

$$dd^c(H(z, w, t) - c \log |t|) =$$

$$dd^c H.$$

Suppose next that $L_1(H) = L_1(H')$. By the first part of the proof, $dd^c H = dd^c H'$. Hence it follows that $H - H'$ is a pluriharmonic function on $\mathbf{C}^3 \backslash 0$. But then $H - H'$ extends through the origin as a pluriharmonic function. Since $H - H'$ also grows at most like $\log \| z \|$ at infinity, it follows that $H - H'$ is constant and hence that $[H] = [H']$ and hence L_1 is $1 \to 1$.

Next we study the inverse. So let T be a positive, closed $(1, 1)$ current on $\mathbf{IP}^2$. Define $\nu = \pi^* T$ on $\mathbf{C}^3 \backslash (0)$. Since π is a submersion on $\mathbf{C}^3 \backslash 0$, ν is a closed, positive $(1, 1)$ current [deRh]. Since the Hausdorff dimension of (0) is zero, it follows by a theorem of [HaP] that the trivial extension of ν is positive and closed. Hence, by Lelong's Theorem [Le], there exists a plurisubharmonic function H on $\mathbf{C}^3$ such that $\nu = dd^c H$. Then H is unique modulo pluriharmonic additions.

Lemma 4.2. There is a unique plurisubharmonic $[H]$, $H = O(\log | z |)$ at infinity, such that $\nu = dd^c H$.

Proof. Pick any H with $dd^c H = \nu$. For $\theta \in \mathbb{R}$, define $H_\theta(z) = H(e^{i\theta} z)$. We show at first that $dd^c H_\theta = \nu$ as well. Let $T_\theta(z) := e^{i\theta} z$ on $\mathbf{C}^3$. Then

$$dd^c H_\theta = dd^c (H \circ T_\theta) = T_\theta^* dd^c H =$$

$$T_\theta^* \nu = T_\theta^* (\pi^* T) = (\pi \circ T_\theta)^* T =$$

$$\pi^* T = \nu.$$

Hence it follows that if we define

$$\tilde{H}(z) = \frac{1}{2\pi} \int_0^{2\pi} H_\theta(z) d\theta,$$

then $dd^c \tilde{H} = \nu$ also. Now $\tilde{H} = \tilde{H}_\theta$ $\forall \theta$. Since $\tilde{H}$ therefore is a radial subharmonic function on any complex line through zero, either $\tilde{H} \equiv -\infty$ (which can happen at most on a set of lines of measure zero) on the line or $\tilde{H} = -\infty$ at most at the center. Pluriharmonic functions like this are constant.

This implies that except for constant additions, $\tilde{H}$ is the only function with $\tilde{H} = \tilde{H}_\theta$, $\forall \theta$ such that $dd^c \tilde{H} = \nu$. However, for any $\alpha \in \mathbf{C}^*$, $\tilde{H}_\alpha(z) := \tilde{H}(\alpha z)$ is also such a solution. Hence

$$\tilde{H}(\alpha z) \equiv C(\alpha) + \tilde{H}(z) (**)$$

for some constant $C(\alpha)$. So

$$\tilde{H}(\alpha^n z) = nC(\alpha) + \tilde{H}(z).$$

Hence $\tilde{H}$ is of sublogarithmic growth at ∞. The Lemma follows, since pluriharmonic functions of sublogarithmic growth at ∞ are constant. $\square$

Lemma 4.3. *There is a constant $c \geq 0$ so that $\tilde{H}(\alpha z) = c \log | \alpha |$ $+ \tilde{H}(z)$, $\tilde{H}$ as in the previous lemma, so we write $[\tilde{H}] = L_2(T)$, $L_2 : Q \to P$.*

Proof: Let $\alpha = 2$ and define $H' = \tilde{H} - \frac{C(2)}{\log 2} \log \| z \|$. Then H' is radial and subharmonic on each punctured complex line through 0. Moreover

$$H'(2z) = \tilde{H}(2z) - \frac{C(2)}{\log 2} \log \| 2z \| =$$

$$C(2) + \tilde{H}(z) - \frac{C(2)}{\log 2} \log 2 - \frac{C(2)}{\log 2} \log \| z \| = H'(z).$$

Hence H' is bounded from above, hence constant on each complex line. Therefore on the complex line through z, we have

$$\tilde{H}(\tau z) = H'(\tau z) + \frac{C(2)}{\log 2} \log \| \tau z \| = H'(z) + \frac{C(2)}{\log 2} \log | \tau | + \frac{C(2)}{\log 2} \log \| z \|$$

$$= \tilde{H}(z) + \frac{C(2)}{\log 2} \log | \tau |. \square$$

Lemma 4.4. *The maps L_1, L_2 are inverses of each other.*

Proof: Suppose $[H] \in P$. We define $T = L_1(H)$. Then $\pi^* T = dd^c H$. Hence, $H = L_2(T)$, so $L_2 \circ L_1 = Id$.

Next, let $T \in Q$ and define $\tilde{H} = L_2(T)$ as above. Then $\tilde{H} \in P$ and $dd^c \tilde{H} = \pi^* T$. Next, consider $T' = L_1(\tilde{H})$. Then T' is a current so that $\pi^* T = \pi^* T'$. Composing with a section of π we see that $T' = T$. Hence $L_1 \circ L_2 = Id$ as well. With these lemmas the theorem follows. $\square$

Proposition 4.5. *Let f be a generic meromorphic normal map in $\mathbb{P}^2$ with $G \not\equiv -\infty$. Then support $T = J$. If (f_λ) is a holomorphic family of holomorphic self maps of $\mathbb{P}^2$ of degree d, then the function $(\lambda, z) \to G_\lambda(z)$ is p.s.h. and the currents (T_λ) vary continuously.*

Proof: That Support $T = J$ is a consequence of Theorem 2.12. Assume λ varies in a complex manifold Δ. For $\lambda \in \Delta$ let F_λ be a lifting of the mapping f_λ, we can assume that $\lambda \to F_\lambda$ is also a holomorphic family. Fix $\lambda_0 \in \Delta$, and define

$$M_\delta = \sup\{\| F_\lambda(z) \| \; ; \; \| z \| = 1, \quad \lambda \in \Delta_\delta \text{ a } \delta \text{ neighborhood of } \lambda_0\}.$$

If δ is small enough M_δ is finite. Then $G_\lambda(z)$ is a limit of an almost decreasing sequence of p.s.h. functions and is p.s.h. with respect to (λ, z), the proof is just as in Theorem 2.12 with parameter.

When (f_λ) is a holomorphic family of holomorphic maps of degree d then, given $\epsilon > 0$ if λ is in a small enough neighborhood of λ_0, Δ_δ, we have

$$\frac{1}{C} \| z \|^d \leq \| F_\lambda(z) \| \leq C \| z \|^d \quad \text{uniformly}.$$

Hence for $\lambda \in \Delta_\delta$ and $n \in \mathbb{N}$

$$\left| \frac{1}{d^{n+1}} \log \| F_\lambda^{n+1}(z) \| - \frac{1}{d^n} \log \| F_\lambda^n(z) \| \right| \leq \frac{\log C}{d^{n+1}}.$$

Hence $G_{n,\lambda}$ converges uniformly to G_λ. Since each $G_{n,\lambda}$ is continuous in (λ, z) it follows that $G_\lambda(z)$ is continuous in (λ, z). As a consequence G_λ varies continuously with λ and hence $\lambda \to T_\lambda$ is a continuous map with values in Q. Here Q carries the weak topology of currents. $\square$

Remark 4.6. This remark is a continuation of example 2.1. Let $R = [z^d : w^d : t^{d-1}w]$. Then $G(z, w, t) = \sup(\log | z |, \log | w |)$ and $J = (| z | = | w |) \cup (| z | \leq | w | = | t |)$. As we observed G is pluriharmonic near in $(| z | < | w | = | t |)$ which is in J. Hence the support of T does not coincide with the Julia set J. In this example there is no positive closed $(1, 1)$ current whose support is equal to J.

Suppose S is such a current. Let $\sigma : (| w | = | z |) \to S^1$, $\sigma(z, w) = w/z$ and let $[c^{-1}(t)]$ denote the current of integration on $\{[z : w : t]\}$; $w/z = t, | w | > | t |$. It follows from a Theorem by Demailly [De] that there exists a measure ν whose support is S^1 such that $S = \int [\sigma^{-1}(t)] d\nu(t)$ on a neighborhood of $\{| w | = | z | > | t |\}$. Let $\tilde{S} = \int [\tilde{\sigma}^{-1}(t)] d\nu(t)$ where $[\tilde{\sigma}^{-1}(t)]$ stands for the current of integration on the whole line, not only for $| w | = | z | > | t |$. Then $\tilde{S}$ is closed and the support of $S - \tilde{S}$ is $K := \{| z | \leq | w | = | t |\} \cup \{| w | = | z | \leq | t |\}$. This is a compact set in $\mathbb{C}^2$. No compact set in $\mathbb{C}^2$ is the support of a nonzero positive closed $(1, 1)$ current. If S_1 is such a current, $< S_1, dd^c \| z \|^2 >= 0$ since S_1 have compact support and on the other hand positivity of S_1 implies that $< S_1, dd^c \| z \|^2 >$ represents the mass of S_1. Here $z = (z_1, z_2)$.

We study now sets of zero mass for the currents T.

Proposition 4.7. *Let f be a generic meromorphic map in $\mathbb{P}^2$, with Green function $G \not\equiv -\infty$. Let X, Y be closed sets in $\mathbb{P}^2$ with $Y \subset X$ and $\Lambda^2(Y) = 0$. Assume that G is locally bounded on $\pi^{-1}(X\backslash Y)$ and that X is locally $\mathbb{R}^4$-polar. Then T has no mass on X.*

Proof: Recall that a positive current is of order zero and hence has measure coefficients in a local chart. To say that T has no mass on a set E, means that all such measures, in the expression of T in a chart, have zero mass on E. For the reader's convenience we prove a well known lemma. Here Δ denotes the Laplacian in $\mathbb{R}^k$. Recall also that a polar set is a set contained in $(v = -\infty)$ where v is a subharmonic function in $\mathbb{R}^k$, not identically $-\infty$. In particular, analytic varieties in $\mathbb{P}^2$ are locally $\mathbb{R}^4$ polar.

Lemma 4.8. *Let v be a bounded subharmonic function on an open set $U \subset \mathbb{R}^k$. Then Δv (which is a locally finite measure), has no mass on $\mathbb{R}^k$ polar sets.*

Proof: Since a polar set is contained in a G_δ polar set and since Δv is a regular measure, it is enough to show that if K is a compact polar set then $(\Delta v)(K) = 0$. Suppose not. Then choose u such that $\Delta u = \chi_K \Delta v$, u is subharmonic, u is harmonic out of K. If u is locally bounded below near K it would have a harmonic extension, which is impossible. Also, let u' be a subharmonic function with $\Delta u' = \Delta v - \chi_K \Delta v$. Then $u' + u$ must differ from v by a harmonic function, a contradiction since v is bounded, it would follow that u is bounded below. $\square$

We now continue the proof of the proposition. Since G is locally bounded near $\pi^{-1}(X\backslash Y)$ it follows that $\pi^*(T)$ has no mass on $\pi^{-1}(X\backslash Y)$. Since $\pi^*(T)$ is a $(1,1)$ positive current in $\mathbf{C}^3$, it has no mass on $\pi^{-1}(Y)$ which is locally of 4-Hausdorff measure 0. It follows that π^*T and hence T has no mass on X. $\square$

We want to show next that, under quite general conditions, the current T satisfies the functional equation $f^*T = dT$ and that it is an extremal current among the currents satisfying this equation. Given a $(1,1)$ closed positive current S on $\mathbb{P}^2$ and f a generic meromorphic map of degree d, we want to define the current f^*S. Observe however that f is not a submersion.

Let F be a lifting for f. Using Theorem 4.1 we can find $u \in L$ such that $\pi^*S = dd^c u$. We define f^*S as the closed positive $(1,1)$ current such that $\pi^*(f^*S) = dd^c(u \circ F)$. If f is a submersion on $\Omega \subset \mathbb{P}^2$, then F is a submersion on $\pi^{-1}(\Omega)$ and hence $dd^c(u \circ F) = F^* dd^c u = F^* \pi^* S$. So if C denotes the critical set of f, we have, on $\Omega = \mathbb{P}^2 \backslash C$, $\pi^* f^* S = F^* \pi^* S$. We also have that on $\mathbb{P}^2 \backslash C$, f^*S defined above coincides with f^*S defined by relation (2) when f is a submersion.

Let Ω be a complex manifold and A a closed set in Ω. Let S be a closed positive (p, p) current on $\Omega \backslash A$, with locally bounded mass near A. We call the trivial extension of S to Ω, the extension of S to Ω giving zero mass to A. A Theorem of Skoda [Sk] asserts that such an extension is closed if A is an analytic variety.

Definition 4.9. *Let $\mathcal{N}$ denote the set of $(1, 1)$ positive currents S on $\mathbb{P}^2$ which do not charge any compact complex curve V. In other words, S agrees with the trivial extension of $S_{|\mathbb{P}^2 \backslash V}$.*

Theorem 4.10. *Let R be a generic meromorphic map in $\mathbb{P}^2$ of degree d, with Green function $G \not\equiv -\infty$.*

(i) The currents T and R^*T belong to $\mathcal{N}$ and satisfy the functional equation $R^*T = dT$.
(ii) If R is normal and E, the extended indeterminacy set, has Lebesgue measure zero, then the current T is on an extremal ray of the cone of positive closed currents satisfying $R^*S = dS$.

Proof: We first show that $T \in \mathcal{N}$. Let V be an irreducible analytic variety in $\mathbb{P}^2$. If T has mass on V, then a Theorem of Siu [Siu] implies that the nonzero current $\chi_V T$ is closed and there exists a constant $C > 0$ such that $\chi_V T = c[V]$, here χ_V denotes the characteristic function of V and $[V]$ denotes the current of integration on V.

Let h be a polynomial of degree ℓ such that $h^{-1}(0) = V$ and $\pi^*[V] = dd^c \log | h |$. Hence $G = c \log | h | + U$, where U is p.s.h. But we have

$$G(R(z)) = dG(z) = cd \log | h(z) | + dU(z)$$

$$= c \log | h(R(z)) | + U(R(z)).$$

Hence

$$G(z) = \frac{c}{d} \log | h(R(z)) | + \frac{1}{d} U(R(z)).$$

So, the current T has also mass $c\ell$ on $(h \circ R = 0)$. Since the mass of T on $\mathbb{P}^2$ is bounded the varieties $(h \circ R^s = 0)$ cannot be all distinct as s varies. Without loss of generality assume $(h \circ R = 0) = (h = 0)$. So $R : V \to V$.

If V is R-constant then $R(V) = p \notin I$, since R is generic. Hence p is fixed and if $\pi(q) = p$ we cannot have $G(q) = -\infty$. So we can assume that R is a non constant self map on V. If the normalisation $\hat{V}$ of V is Kobayashi hyperbolic or a $\mathbb{P}^1$ then R has periodic points and we can conclude as above. If $\hat{V}$ is a torus we also have periodic points except if R

is an irrational translation on $\hat{V}$. In this case the argument is more delicate since we don't have periodic points. Then V has no cusp singularities. But we know that Green's function is identically $-\infty$ on $\pi^{-1}(V)$. Hence there are points of indeterminacy of R on V, otherwise all orbits stay away from I and G is not $-\infty$. For simplicity we write the argument assuming that there is just one point of indeterminacy $p \in V$. We are going to show that the area on V of the set where $\frac{1}{d^n}\log \parallel R^n \parallel < \log \epsilon$ is small if ϵ is small. Given a point $z \in \pi^{-1}(V)$ we can write $\parallel R(z) \parallel = \delta([z]) \parallel z \parallel^d$ where $\delta([z])$ satisfies $\frac{1}{c}d([z], [p])^l \leq \delta([z]) \leq cd([z], [p])$ for some finite $c, l \geq 1$ and d is the distance in $\mathbb{P}^2$. If (z_n) is the orbit of z_0, $\parallel z_0 \parallel = 1$, $z_n = R^n(z_0)$, $\delta_k = \delta([z_k])$ we have

$$\parallel z_1 \parallel = \delta_0 \parallel z_0 \parallel^d \quad , \quad \parallel z_{n+1} \parallel = \delta_n \delta_{n-1}^d \cdots \delta_0^{d^n}.$$

If all $\delta_k \geq \epsilon^{d^k/(k+2)^2}$ then $\mid R(z_n) \mid > \epsilon^{d^n}$. So we measure the area of

$$N_k =: \{z; \delta_k(z) < \epsilon^{d^k/(k+2)^2}\}.$$

Since in this case R is essentially area preserving on V, N_k is contained in a disc of radius $(c' \epsilon^{d^k/(k+2)^2})^{1/r'}$ for some $c' > 0$, $r' \geq 1$ measured in a smooth metric on V, the sum of the area of $\bigcup_k N_k$ is very small. On $\pi^{-1}(V \setminus \bigcup N_k)$, G is not $-\infty$, hence no such V exists and $T \in \mathcal{N}$. The potential for the current $\frac{1}{d}R^*T$ is $\frac{G(R(z))}{d}$, so the same analysis shows that $R^*T \in \mathcal{N}$.

We have

$$\pi^* R^* T = dd^c G(R) = d(dd^c G) = d\pi^* T.$$

Hence

$$R^* T = dT.$$

We prove now, that the ray λT, $\lambda > 0$, is extremal in the cone of positive closed currents S, satisfying the functional equation $R^* S = dS$.

Assume $T = T_1 + T_2$. By positivity $T_j \in \mathcal{N}$. Let G_1, G_2 be plurisubharmonic functions in L such that $\pi^* T_j = b_j dd^c G_j$, $j = 1, 2$, $b_j > 0$. Since $G - (b_1 G_1 + b_2 G_2)$ is pluriharmonic of logarithmic growth, it is constant. We can assume, without loss of generality, that $G = b_1 G_1 + b_2 G_2$. Let $\Omega := \mathbb{P}^2 \setminus E$. Since R is normal, given $z_0 \in \Omega$ there is a neighborhood $B(z_0, r)$ of z_0 such that on $\pi^{-1}(\bigcup_{n=0}^{\infty} R^n(B(z_0, r)))$ we have : $\log \parallel z \parallel - a \leq G(z) \leq \log \parallel z \parallel + a$. The functions G_j are u.s.c. so, there are constants a_j, such that $\log \parallel z \parallel - a_j \leq G_j(z) \leq \log \parallel z \parallel + a_j$ on $\pi^{-1}(\bigcup_{n=0}^{\infty} R^n(B(z_0, r)))$. If we compose by R^n divide by d^n and let $n \to \infty$, we find that $\frac{G_j(R^n(z))}{d^n} \to G(z)$. But since $R^* T_j = dT_j$ we have that $\frac{G_j \circ R}{d} - G_j = c_j$ where c_j is a constant and $b_1 c_1 + b_2 c_2 = 0$. Adding a suitable constant to G_j we can assume $c_j = 0$, so $G_j = G$ on $\pi^{-1}(\Omega)$. If E is of Lebesgue measure zero on $\mathbb{P}^2$, $\pi^{-1}(E)$ is of Lebesgue measure zero in $\mathbf{C}^3$, hence $G_j = G$ and therefore T is extremal. $\square$

We want next to show that, under quite general conditions, given a positive closed $(1,1)$ current S on $\mathbb{P}^2$, $\frac{(f^n)^* S}{d^n} \to cT$ where c is a positive constant.

We need some preliminary results.

Let K be a compact set in $\mathbf{C}^2$, and let $f : U \to \mathbf{C}^2$ be a holomorphic map on a bounded neighborhood U of K. Assume that for every $w \in f(U)$, the fiber $S_w = f^{-1}(w)$ is discrete.

We prove a Lojasiewicz type inequality.

Proposition 4.11. *Let n be the maximum multiplicity of points in S_w, $w \in f(K)$. There is a constant $c > 0$ such that if $w \in f(U)$, $z \in K$, then*

$$\| f(z) - w \| \geq c \operatorname{dist}(z, S_w)^n.$$

Proof: It suffices to prove the proposition locally. Assume that $f(0) = 0$ and that f has multiplicity n at 0. We will suppose z, w are close enough to 0.

The graph of f is a branched covering with multiplicity n of the w-plane. The branches are locally given by $z = \{g_1(w), \cdots, g_n(w)\}$, $g_j(w) = (g_j^1(w), g_j^2(w))$. Hence we can form the symmetric products

$$\prod_{j=1}^{n}(z_1 - g_j^1(w)) = z_1^n + a_{n-1}(w)z_1^{n-1} + \cdots + a_0(w) = P_1(z_1, w)$$

$$\prod_{j=1}^{n}(z_2 - g_j^2(w)) = z_2^n + b_{n-1}(w)z_2^{n-1} + \cdots + b_0(w) = P_2(z_2, w)$$

to obtain two Weierstrass polynomials.

Fix z^0, w^0 close to 0. Then $(z^0, f(z^0))$ belongs to the graph of f. Hence $P_1(z_1^0, f(z^0)) = P_2(z_2^0, f(z^0)) = 0$. We will try to find a point $(\tilde{z}, w^0)$ on the graph with $\tilde{z}$ close to z^0. In that case $\tilde{z} \in S_{w^0}$ and we obtain the proposition by proving a good estimate on $z^0 - \tilde{z}$.

Let C be a constant to be determined below. This constant only depends on the size of the first derivatives of the coefficients $a_j(w)$, $b_i(w)$. Say $w^0 \neq f(z^0)$. If $w^0 = f(z_0)$ we are done.

There exists an integer $2 \leq k \leq 4n$ so that $P_r(t, f(z^0))$ has no root t with

$$(k-1)C\left(\sqrt[n]{\| w^0 - f(z^0) \|}\right) \leq |t - z_1^0|$$

$$\leq (k+1)C \left({}^n\sqrt{\| w^0 - f(z^0) \|} \right), \ r = 1 \text{ or } 2.$$

For any $\theta \in \mathbb{R}$, let $\zeta_r(\theta) = \zeta_r = kC \left({}^n\sqrt{\| w^0 - f(z^0) \|} \right) e^{i\theta} + z_r^0$ and consider the symmetric product

$$\prod_{j=1}^{n} (g_j^r(w) - \zeta_r) = G_\theta^r(w).$$

Then $G_\theta^r(w)$ is uniformly Lipschitz. Moreover,

$$| G_\theta^r(f(z^0)) | = | P_r(\zeta_r, f(z^0)) | \geq C^n \| w^0 - f(z^0) \|,$$

using the choice of k.

Hence G_θ^r has no zeroes in the ball $\{w; \| w - f(z^0) \| \leq 2 \| w^0 - f(z^0) \| \}$, if C is chosen large enough. Hence $(\zeta_1(\theta), \zeta_2(\psi), w)$ is not on the graph for any w in this ball. By continuity this means that when $w = w^0$ there must exist a point (z, w^0) on the graph with $(| z_r - z_r^0 | \leq kC(\| w^0 - f(z^0) \|)^{1/n}, r = 1, 2.$

The proposition now follows immediately. $\square$

Let $\mathcal{H}_d$ denote the space of non degenerate holomorphic self maps of degree d in $\mathbb{P}^2$.

Corollary 4.12. *Let $f \in \mathcal{H}_d$ be holomorphic on $\mathbb{P}^2$, $d \geq 2$. Then there exists a $c > 0$ so that if $z, w \in \mathbb{P}^2$, then*

$$\text{dist}(f(z), w) \geq c \, \text{dist}(z, f^{-1}(w))^{(d^2)}.$$

Proof: The maximum multiplicity possible is d^2. $\square$

Let us return to the notation of the proposition 4.11.

Corollary 4.13. *There are constants $a > 0$, $r_0 > 0$, so that if $z \in K$, $0 < r < r_0$, then*
$$f(\mathbb{B}(z, r)) \supset \mathbb{B}(f(z), ar^n).$$

Proof: The conclusion of proposition 4.11 holds in a neighborhood of K.

Pick r_0 so small that $\overline{\mathbb{B}(z, r_0)}$ is in this neighborhood for all $z \in K$, and such that no point has more than n preimages in any such ball.

Let $z_0 \in K$, $0 < r < r_0$. Then there exists k, $2n \leq k \leq 4n$, so that there is no preimage of $f(z_0)$ in $\{r \cdot \frac{k-1}{4n+1} \leq \| z - z_0 \| \leq r \cdot \frac{k+1}{4n+1}\}$. Hence if

$z \in \partial \mathbb{B}(z_0, r \cdot \frac{k}{4n})$, then $\text{dist}(z, S_{f(z_0)}) \geq \frac{r}{4n+1}$. Hence $\| f(z) - f(z_0) \| \geq ar^n$ for some $a > 0$. But then

$$f(\mathbb{B}(z_0, r)) \supset f(\mathbb{B}(z_0, \frac{rk}{4n})) \supset \mathbb{B}(f(z_0), ar^n). \ \Box$$

Applying this to $f \in \mathcal{H}_d$, we get

Corollary 4.14. *Let $f \in \mathcal{H}_d$. Then there exist constants $c > 0$, $r_0 > 0$ so that for $z \in \mathbb{P}^2$ and $0 < r < r_0$; then $f(\mathbb{B}(z, r)) \supset \mathbb{B}(f(z), cr^{d^2})$.*

Next we discuss the size of the image of a ball $B(z, r)$ under iteration of f.

Theorem 4.15. *Assume $f \in \mathcal{H}_d$ is holomorphic on $\mathbb{P}^2$, $d \geq 3$. Suppose that the local multiplicity of f is at most $(d-1)$ except on a finite set S. We assume that S contains no periodic points. There exists a constant $c > 0$, so that if $\{z_j\}_{j=0}^\infty$ is any orbit of f and $0 < r < 1$, then there exist radii $\{r_j\}_{j=0}^\infty$ with $f(\mathbb{B}(z_j, r_j)) \supset \mathbb{B}(z_{j+1}, r_{j+1})$ for every j. Moreover $r_0 = r$, $r_{j+1} = cr_j^{d_j}$ where $1 \leq d_j \leq d^2$ is an integer and $d_0 d_1 \cdots d_n \leq \frac{1}{c}(d - \frac{1}{2})^n$ for every n.*

Proof: The hypothesis on f implies that if N is sufficiently large the local multiplicity of f^N is at most

$$(d^2)^\ell (d - 1)^{N-\ell} \leq (d - \frac{1}{2})^N$$

where ℓ denotes the number of points in S. But then the result follows easily from Proposition 4.11 and Corollary 4.13.

Theorem 4.16. *Let $R \in \mathcal{H}_d$ be a holomorphic map on $\mathbb{P}^2$. Assume that the local multiplicity of R is at most $(d - 1)$, except possibly on a finite set without periodic points. Let $u \in P$, $u(\lambda z) = \log | \lambda | + u(z)$. Then $u(R^n)/d^n \to G$ in L^1_{loc}. Hence if $\pi^* S = dd^c u$, then $(R^n)^* S/d^n \to T$ in the sense of currents.*

Proof: The sequence $u_n := u(R^n)/d^n$ is uniformly bounded above on $\{\| z \| \leq 1\}$. We show first that no subsequence $u_{n_i} \to -\infty$ uniformly on compact sets. If so, then

$$\frac{1}{d^{n_i}} u \left(\frac{R^{n_i}}{\| R^{n_i} \|} \right) = \frac{1}{d^{n_i}} u(R^{n_i}) - \frac{1}{d^{n_i}} \log \| R^{n_i} \|$$

$$= \frac{1}{d^{n_i}} u(R^{n_i}) - G_{n_i} \to -\infty.$$

Hence the map $\frac{R^{n_i}}{\|R^{n_i}\|}$ on $|z|=1$ cannot be surjective, a contradiction.

Assume that $u_{n_i} \to G_1$ in L^1_{loc}. We want to show that $G_1 = G$. Clearly $G_1 \le G$. Since G_1 is upper semi continuous and G is continuous, $\{G_1 < G\}$ is open. Let $\omega \in \mathbb{P}^2$ be an open set such that $G_1 < G - 2\delta$ on $\pi^{-1}(\omega)$, $\delta > 0$. By Hartogs Lemma it follows that for n_i large enough,

$$\frac{1}{d^{n_i}} u(tR^{n_i}/\|R^{n_i}\|) < -\delta$$

on $\pi^{-1}(\omega)$, $t \in [\frac{1}{2}, 1]$ arbitrary. Hence, $\pi^{-1}(R^{n_i}(\omega)) \cap \{\frac{1}{2} \le \| z \| \le 1\}$ is contained in $X := \{u < -\delta d^{n_i}\}$. Let L be any line on which u is not identically $-\infty$. Since u has sublogarithmic growth at infinity, the logarithmic capacity of $X \cap L$ is at most $e^{-\delta d^{n_i}}$. But a classical estimate, see [Ts], shows that any disc contained in $X \cap L$ has a radius of order of magnitude at most $e^{-\delta d^{n_i}}$. However, by the previous theorem $X \cap \{\frac{1}{2} \le \| z \| \le 1\}$ contains balls of radius of order of magnitude $\epsilon^{(d-1/2)^{n_i}}$ for some $\epsilon > 0$ in the image of $\pi^{-1}(\omega)$. This contradiction completes the proof. $\square$

Proposition 4.17. *The hypothesis of Theorem 4.16 is satisfied in the complement of a countable union of closed, proper, subvarieties of $\mathcal{H}_d$, $d \ge 3$.*

Proof: For $f \in \mathcal{H}_d$, let $(J_f = 0)$ be the equation for the critical set. Then $\sum := \{(f,z) \in \mathcal{H}_d \times \mathbb{P}^2 ; \text{grad } J_f = 0, J_f = 0\}$ is an analytic variety. It follows from the example in the proof of Lemma 5.9 in [FS1] that the projection $\pi(\sum)$ in $\mathcal{H}_d$ of $\sum$ is not all of $\mathcal{H}_d$. Hence $\pi(\sum)$ is a proper subvariety of $\mathcal{H}_d$, and for $f \notin \pi(\sum)$, the local multiplicity is at most 2 except for finitely many points in $(J_f = 0)$.

For ϵ small, consider the map

$$R_\epsilon = [z^d + \epsilon z(w^{d-1} + 2t^{d-1}) : w^d + \epsilon w(z^{d-1} + 2t^{d-1}) :$$

$$t^d + \epsilon t(z^{d-1} + 2w^{d-1})].$$

One sees easily that the periodic orbits are the points where two axes cross, or are in the axes close to the unit circles or close to the torus $|z|=|w|=|t|=1$.

However one easily checks that none of these points are on the critical set for $\epsilon \ne 0$, ϵ small enough.

For every n, consider the analytic set $\sum_n \subset \mathcal{H}_d \times \mathbb{P}^2$ given by

$$\sum_n = \{(f,z) ; f^n(z) = z \text{ and } J_f(z) = 0\}.$$

Then each $\pi(\sum_n)$ is a proper subvariety of $\mathcal{H}_d$.

The hypothesis of the theorem is then satisfied for any

$$f \in \mathcal{H}_d \backslash (\pi(\sum) \cup \bigcup_n \pi(\sum_n)). \quad \square$$

Remark 4.18. *If f is a normal, generic, meromorphic map on $\mathbb{P}^2$, and if the local multiplicity of f is $\leq d-1$ in the set of normality, except on a finite set without points belonging to periodic orbits, then the conclusion of the theorem holds provided E has zero volume.*

Remark 4.19. *A natural question is whether positive closed currents S satisfying $f^*S = dS$ are unique. The answer is no in general. Let $F = (F_1(z,w), F_2(z,w), t^d)$, then $(t = 0)$ is exceptional, i.e. $f^{-1}(t = 0) = (t = 0)$. Define $G_1(z,w,t) = G(z,w,0)$. Then $G_1 \circ F = G(F_1, F_2, 0) = G(F(z,w,0)) = dG(z,w,0) = dG_1(z,w,t)$. It is easy to check in this case that $G(z,w,t) = \sup(G(z,w,0), \log |t|)$. If $F_1(z,w) = (z - 2w)^2$, $F_2(z,w) = z^2$ and if $T_1 : \pi^*T_1 = dd^cG_1$ we find that $\operatorname{supp} T_1 = \mathbb{P}^2$ but that the Julia set is not $\mathbb{P}^2$. Hence a hypothesis on the map R is necessary in order to prove Theorem 4.16.*

Remark 4.20. *Let $f(z,w) = (z^2+c+aw, z)$ be the standard Hénon map. One defines $G^+(z,w) = \lim_n \frac{1}{2^{n+1}} \log^+[|f_1^n|^2 + |f_2^n|^2]$. Let $R[z : w : t] = [z^2 + ct^2 + awt : zt : t^2]$ be the corresponding map on $\mathbb{P}^2$ and G the associated Green function on $\mathbf{C}^3$. We have $G^+(z,w) = G(z,w,1)$ and $G(z,w,0) = \log |z|$. If $\mu^+ = dd^cG^+$ considered as a current in $\mathbf{C}^2$, it was observed in [FS2] that μ^+ has a closed positive extension to $\mathbb{P}^2$ and it is easy to check that this extension is just T. Convergence results for the current μ^+ where obtained in [BS1], [BS2], [FS1], more recently the structure of μ^+ has been studied in [BLS].*

5. Connectedness of Julia sets

One of the main developments in the theory of several complex variables is the solution of the Levi problem. Here we show a dynamical consequence of this fundamental result.

Definition 5.1. *We say that a compact subset X of $\mathbf{C}^2$ satisfies the local maximum principle if for every $p \in X$, all small enough $r > 0$ and all complex polynomials h, $|h(p)| \leq \operatorname{Max} |h|_{\partial \mathbb{B}(p,r) \cap X}$.*

Theorem 5.2. *Suppose f is a normal generic meromorphic map on $\mathbb{P}^2$. Then the Fatou components are domains of holomorphy. The Julia set J is connected and satisfies the local maximum principle. If J is a C^1 manifold in a neighborhood of a point on J, then J is laminated on that neighborhood by Riemann surfaces.*

Proof: We first show that $\mathbf{C}^3\backslash(\Pi^{-1}(J) \cup (0))$ is Stein. Observe that by Theorem 2.12 the function G is plurisubharmonic in $\mathbf{C}^3$ (or $\equiv -\infty$) and pluriharmonic precisely on $\mathbf{C}^3\backslash(\Pi^{-1}(J) \cup (0)$.

The following result is due to Cegrell [Ce], for the reader's convenience we give a proof.

Lemma 5.3. *Let M be a complex manifold and u a plurisubharmonic function on M. If Ω is the maximal open set where u is pluriharmonic, then Ω is pseudoconvex.*

Proof of lemma. Fix $0 < r < 1$. It is enough to show that if $H :=\{(z,w), z \in \mathbf{C},\ w = (w_1, \cdots, w_{n-1}) \in \mathbf{C}^{n-1};\ |\ z\ | < 1,\ \|\ w\ \| < r$ or $r < \|\ z\ \| < 1,\ \|\ w\ \| < 1\}$ and if u is p.s.h. in a neighborhood of the closed unit polydisc $\bar{D}^n$ and pluriharmonic on H, then u is pluriharmonic on D^n. Since u is pluriharmonic on H, $u = Re\ h$, where h is holomorphic on each vertical disc, h being unique after normalization. Then h is holomorphic on H. Let $\tilde{h}$ be the holomorphic extension of h to D^n. We clearly have $u \leq Re\ \tilde{h}$ on D^n and $u - Re\ \tilde{h} = 0$ on H. Hence $u = Re\ \tilde{h}$ on D^n. $\square$

Continuation of the proof of the theorem. Since $\mathbf{C}^3\backslash(\Pi^{-1}(J)\cup(0))$ is pseudoconvex it follows that any Fatou component is locally pseudoconvex. Hence by the solution of the Levi Problem in $\mathbb{P}^2$, the Fatou components are domains of holomorphy (or all of $\mathbb{P}^2$, which cannot happen).

That the Julia set is connected follows from the Hartogs extension phenomenon : If $K \subset \Omega$ is a compact subset of a domain of holomorphy, then $\Omega\backslash K$ has exactly one unbounded connected component Ω' and all holomorphic functions on Ω' extend across K, hence $\Omega\backslash K$ cannot be a domain of holomorphy. So if J is not connected, we can write $J = K_1 \cup K_2$ for disjoint nonempty compact sets, and let $\Omega = \mathbb{P}^2\backslash K_1$ and $K = K_2$ to obtain a contradiction to the solvability of the Levi problem.

If J is C^1 near a point, since the complement is pseudoconvex, then J cannot be a real curve. If J is two dimensional, J must be a Riemann surface. If J is three dimensional, then J is laminated by Riemann surfaces, (Scherbina [Sch]), and if J is four dimensional we can also laminate a neighborhood locally by Riemann surfaces. That J satisfies the local maximum principle is a theorem by Wermer [We]. All these result use just the fact that the complement of J is a domain of holomorphy. $\square$

Proposition 5.4. *For a normal generic meromorphic map the Julia set J does not have a Stein neighborhood. Hence J intersects the support of any positive closed $(1,1)$ current, and in particular any compact complex curve.*

Proof: If $G \equiv -\infty$, then $E = \mathbb{P}^2$, so $J = \mathbb{P}^2$ and we are done since $\mathbb{P}^2$ is not Stein. So assume $G \not\equiv -\infty$. In that case T is well defined. Suppose $U \supset J$ is a Stein neighborhood. Let ρ be a strictly p.s.h. function in U and let θ be a test function supported in U, with value 1 in a neighborhood of J. Since T is closed we have

$$< T, dd^c \rho >=< T, dd^c(\theta \rho) >= 0.$$

But $< T, dd^c \rho >$ bounds the mass of T. So U does not even have p.s.h. functions, strictly plurisubharmonic near a point of J. The complement of the support of a nonzero, positive closed $(1,1)$ current is Stein, as can be deduced from Theorem 4. 1 and Lemma 5. 3. Recall that a compact complex hypersurface is the support of a positive, closed $(1,1)$ current [Le]. $\square$

The following result shows that an open set is in the Fatou set if a subsequence of (R^n) is equicontinuous.

Proposition 5.5. *Let R be a generic meromorphic map on $\mathbb{P}^2$. Let N be the open set of nice points. Assume that on an open set $\Omega \subset N$ there is a subsequence R^{n_i} uniformly convergent on compact sets. Then Ω is contained in the Fatou set of R.*

Proof: Let $G = \lim \frac{1}{d^n} \log \| R^n \|$ where $R : \mathbf{C}^3 \to \mathbf{C}^3$ denotes also a lifting of R. We know that locally on Ω there exist λ_{n_i} nonvanishing holomorphic functions such that $\frac{R^{n_i}}{\lambda_{n_i}} \to h$. Write on $\pi^{-1}(\Omega)$,

$$G = \lim \left[\frac{1}{d^{n_i}} \log \| R^{n_i}/\lambda_{n_i} \| + \frac{1}{d^{n_i}} \log | \lambda_{n_i} | \right].$$

Since the first term converges to 0 on $\pi^{-1}(\Omega)$ we have

$$dd^c G = \lim dd^c \left[\frac{1}{d^{n_i}} \log | \lambda_{n_i} | \right] = 0,$$

hence G is pluriharmonic on $\pi^{-1}(\Omega)$. Then Theorem 2.12 shows that Ω is contained in a Fatou component. $\square$

It is natural to introduce the meromorphic Fatou set $F'(R)$.

Definition 5.6. *Let R be a generic meromorphic map on $\mathbb{P}^2$. A point $p \in \mathbb{P}^2$ is in $F'(R)$ if there exists a neighborhood $U(p)$ on which any subsequence of iterates has a convergent subsequence to a meromorphic map. Here we say that a sequence g_k of meromorphic maps converges to a meromorphic map g if there exists an isolated set of points S in U such that for every compact set $K \subset U \backslash S$ there exists a k_0 so that the sequence $\{g_k\}_{k > k_0}$ is equicontinuous on K and converges to g on K.*

Theorem 5.7. *Let R be a generic normal meromorphic map on $\mathbb{P}^2$, with Green function G. The function G is pluriharmonic on $\pi^{-1}(F'(R))$. Hence $F'(R)$ contains no point of indeterminacy and $F'(R) = F(R)$.*

Proof: Fix a convergent subsequence on $U \subset F'(R)$. As in Proposition 5.5, and using the notations of Definition 5.6, the function G is pluriharmonic on $\pi^{-1}(U \backslash S)$. But $\pi^{-1}(S)$ is just a union of complex lines in $\mathbb{C}^3$, hence G is pluriharmonic on $\pi^{-1}(U)$. Consequently $F'(R)$ does not intersect E. If R is normal or if U is contained in the set of nice points of R then $U \subset F(R)$ as follows from Theorem 2.12. $\square$

Let R be a generic map on $\mathbb{P}^2$. If $G \not\equiv -\infty$ and if R is normal, the Julia set $J(R)$, which we also denote J_0, can be described as the support of the current T. From the example $R = [z^2 : w^2 : t^2]$ we see that there is a natural stratification of the Julia set

$$J_0 = \{[z : w : t]; \mid z \mid = \mid w \mid \geq \mid t \mid ; \mid z \mid = \mid t \mid \geq \mid w \mid ; \mid z \mid = \mid t \mid \geq \mid w \mid \}.$$

On the set $\mid z \mid = \mid t \mid = 1$, $\mid w \mid < 1$ the sequence (R^n) is not normal, but this set is foliated by complex discs and on each of them (R^n) is normal. We introduce the following definition.

Definition 5.8. *Let R be a generic meromorphic map on $\mathbb{P}^2$. A point $p \in \mathbb{P}^2$ is in the one dimensional Fatou set F_1 if there is a neighborhood Ω of p and for every point $q \in \Omega$, there exists a (germ of a) complex curve X_q through q such that the family of iterates R^n is equicontinuous on X_q The one dimensional Julia set J_1 is the complement of F_1. The point $p \in \mathbb{P}^2$ is in the one dimensional Fatou set F_1' if there is a neighborhood Ω of p and for every point $q \in \Omega$, a (germ of a) complex curve X_q through q such that the Green function G is harmonic on X_q (after normalization).*

Notice that we don't require that the X_q is the same in the definition of F_1, F_1'. Also note that when restricted to complex curves the finitely many points of indeterminacy of any iterate R^n are always removable.

We denote by F_0 the Fatou set of R and by F_0' the maximal open set in $\mathbb{P}^2$ such that G is pluriharmonic on $\pi^{-1}(F_0')$. In other words F_0' is the complement of the support of T. Recall that N denotes the set of nice points. Using this notation, we can reformulate Theorem 2.12 and the definitions as follows.

Proposition 5.9. *If R is a generic meromorphic map, then $F_0 \subset F_0'$. Moreover $F_0' \cap N \subset F_0$. Equivalently $J_0' \subset J_0$ and $J_0 \cap N \subset J_0'$. Also $F_0 \subset F_1$, $F_0' \subset F_1'$, $J_1 \subset J_0$.*

Proposition 5.10. *If R is a generic meromorphic map on $\mathbb{P}^2$, then $F_1 \cap N = F_1' \cap N$.*

Proof: On compact subsets of N,

$$|1/d^n \log \| F^n \| -G| \leq c/d^n.$$

If G is harmonic on the normalization of a curve X_q, then $G_{|X_q} = \log | h |$ for a holomorphic function $h \neq 0$ there. So $\{F^n/h^{d^n}\}$ is a normal family. Hence $F_1' \cap N \subset F_1 \cap N$. Suppose $\{F^n/h^{d^n}\}$ is a normal family on $X_q \subset N$. Then it follows that G is harmonic on the normalization. So $F_1 \cap N \subset F_1' \cap N$. $\square$

Example. The following map is studied in [FS3]

$$g([z : w : t]) = [(z - 2w)^2 : (z - 2t)^2 : z^2].$$

The point $p = [1 : 1 : 1]$ is fixed and repelling, hence $p \in J_1$. On the other hand it is shown in [FS3] that $\bigcup_{n \geq 0} g^{-n}(p)$ is dense in $\mathbb{P}^2$. Therefore for this map g, we have $J_1 = \mathbb{P}^2$.

6. $\mathbf{T} \wedge \mathbf{T} =: \mu$.

In the theory of dynamical systems, invariant measures are very useful. In this section we will discuss invariant measures for holomorphic maps in $\mathbb{P}^k$, $k = 1, 2$.

Let f be a holomorphic map on $\mathbb{P}^k$ of degree $d \geq 2$. For a continuous function φ on $\mathbb{P}^k$ define

$$f_* \varphi(x) = \sum_{f(y)=x} \varphi(y).$$

If x is a critical value we take into account the multiplicity, this coincides with the direct image of φ considered as a $(0,0)$ current as defined in paragraph 4, see [deRh].

If ν is a measure on $\mathbb{P}^k$, define the measure $f^* \nu$ by the relation

$$< f^* \nu, \varphi > = < \nu, f_* \varphi > .$$

This coincides with the pullback of ν considered as a (k, k) current provided ν has no mass on $f(C)$ where C is the critical set of f. We also define

$$< f_* \nu, \varphi > = < \nu, f^* \varphi > .$$

In $\mathbb{P}^1$ the current T is identified with the probability measure μ such that $\pi^* \mu = dd^c G$. The fact that μ has mass 1 can be checked as follows. Let ω be the standard Kähler form on $\mathbb{P}^1$ such that $\int_{\mathbb{P}^1} \omega = 1$. By the change of variable formula we have

$$\int_{\mathbb{P}^1} d\mu = \lim_n \int \frac{(f^n)^* \omega}{d^n} = 1.$$

Since μ does not give mass to locally polar sets in $\mathbb{P}^1$ (G is continuous and we apply lemma 4.8) $f^* \mu$ defined above coincides with $f^* T$ as defined in paragraph 4.

If ν is a probability measure on $\mathbb{P}^1$ we define in $\mathbb{C}^2$

$$u(z, w) = \int_{\mathbb{P}^1} \log \frac{|z \zeta_2 - w \zeta_1|}{\| \zeta \|} d\nu(\zeta).$$

Clearly $u \in P$ and it is easy to check that $\pi^* \nu = dd^c u$. This makes explicit the correspondance L_2 in Paragraph 4 between probability measures on $\mathbb{P}^1$ and plurisubharmonic functions u in $\mathbb{C}^2$ such that

$$u(z, w) \leq \log^+ |(z, w)| + 0(1)$$

$$u(\lambda(z, w)) = \log |\lambda| + u(z, w).$$

Let f be a holomorphic map of degree d on $\mathbb{P}^1$ and let $F = (P, Q)$ be a lifting of f to $\mathbb{C}^2$. If $a = [a_1, a_2] \in \mathbb{P}^1$ we can assume $\| a \| = (| a_1 |^2 + | a_2 |^2)^{1/2} = 1$. Then the potential u_a associated to the Dirac mass ϵ_a at a

is $u_a(z,w) = \log | za_2 - wa_1 |$. It is easy to check that if a is not a critical value the potential associated to $\frac{f^*(\epsilon_a)}{d}$ is $\frac{u_a \circ F}{d}$, by continuity this holds also if a is a critical value. Similarly we prove that if ν is a probability measure with potential u then the potential associated to $\frac{f^* \nu}{d}$ is $\frac{u \circ F}{d}$.

The results we describe in the following theorem are well known, from a different point of view, see [Ly] [LFM], compare also with [HP]. For background on ergodic theory, see [Wa].

Theorem 6.1. *Let $f \in \mathcal{H}_d$ on $\mathbb{P}^1$. Then*

(i) $f^*\mu = d\mu$ *and hence* $f_*\mu = \mu$, $\operatorname{supp}\mu = J(f)$.
(ii) *If* $u \in P$ *and* $\nu = L_1(u)$, *then* $f^{*n}\nu/d^n \to \mu$ *except if* ν *has positive mass on an exceptional point. Similarly* $\frac{u(F^n)}{d^n} \to G$ *in* L^1_{loc} *with the same exception.*
(iii) μ *is ergodic and of maximal entropy.*
(iv) *Let* $\{z_i\}$ *denote the periodic points of order n (or a factor of n), and let* $\mu_n := \sum 1/d^n \epsilon_{z_i}$. *Then* $\lim \mu_n = \mu$.

Proof: (i) The fact that $f^*\mu = d\mu$ is a special case of Theorem 4.10 and is just a consequence of the functional equation $G(f(z)) = dG(z)$. Since μ does not have mass on points, for every continuous function φ on $\mathbb{P}^1$,

$$d < f_*\mu, \varphi > = d < \mu, \varphi \circ f > = < f^*\mu, \varphi \circ f > =$$

$$< \mu, f_a st(\varphi \circ f) > = < \mu, d\varphi > = d < \mu, \varphi > .$$

Hence $f_a st\mu = \mu$. The fact that the support of μ coincides with the Julia set is clear from the remark after Theorem 2.12.

(ii) Let F be a lifting for f. The sequence of plurisubharmonic functions in $\mathbb{C}^2$, $u_n = \frac{u(F^n)}{d^n}$ is relatively compact in L^1_{loc} :
If u_{n_i} converges to $-\infty$ the sequence v_{n_i} also converges to $-\infty$, where

$$v_n = u(F^n)/d^n - 1/d^n \log \| F^n \| = u(F^n/ \| F^n \|)/d^n .$$

Given $M > 0$ then for $\| z \| = 1$ we have $u(F^{n_i}/ \| F^{n_i} \|) < -M d^{n_i}$ for n_i large enough. This contradicts the surjectivity of $F^n/ \| F^n \|$ from $\| z \| = 1$ to $\| z \| = 1$ (u cannot be arbitrarily small on all of $\| z \| = 1$).

So assume $u(F^{n_i})/d^{n_i} \to \varphi$ in L^1_{loc}. Since $u \leq \log \| z \| + O(1)$ then $\varphi \leq G$. If $\varphi = G$ everywhere, we are done. Otherwise since G is continuous $\varphi < G$ is open. Let $\omega \subset\subset \pi\{\varphi < G\} =: \Omega$. By Hartogs lemma, for i large enough

$$1/d^{n_i} u(F^{n_i}) < G - 2\delta \quad \text{on} \quad \pi^{-1}(\omega)$$

i.e.

$$1/d^{n_i} u(F^{n_i} / \| F^{n_i} \|) < -\delta \quad \text{on} \quad \pi^{-1}(\omega)$$

$$u(F^{n_i} / \| F^{n_i} \|) < -\delta d^{n_i}.$$

This implies that the image under f^{n_i} of ω avoids a fixed set of positive measure in $\mathbb{P}^1$, i. e. f^{n_i} is a normal family and hence by Proposition 5.5, ω is contained in a Fatou component. Hence G is pluriharmonic on $\pi^{-1}(\Omega)$. So $\psi := \varphi - G$ is a strictly negative subharmonic function of Ω and is zero on $\partial\Omega$. Hence, by the maximum principle, Ω is a Fatou component.

Let σ be the measure on $\mathbb{P}^1$ such that $\pi^*\sigma = dd^c\varphi (= dd^c\psi \text{ on } \Omega)$. We have identified $(1,1)$ positive currents on $\mathbb{P}^1$ and positive measures on $\mathbb{P}^1$. We want to show that $\sigma = 0$ on Ω.

Let $\theta \geq 0$ be a smooth function with compact support in (any connected component of) Ω

$$< \sigma, \theta > = \lim_{n_i} 1/d^{n_i} \int \sum_{j; f^{n_i}(z_j) = z} \theta(z_j) d\nu(z) \leq C_\theta \nu(f^n(\Omega)).$$

If $(f^{n_i}(\Omega))$ are pairwise disjoint then $< \sigma, \theta > = 0$. So we can assume Ω is preperiodic. If $f^{-1}(\Omega)$ has other components than Ω we also have $< \sigma, \theta > = 0$ because the number of points in $\Omega \supset \text{supp}\,\theta$ grows at most like $(d-1)^n$. So it follows that $f^{-1}(\Omega) = \Omega$, i.e. f is a map of degree d on Ω. Then $f^{-n}(z_0) \to \partial\Omega$ see [Be] or [Mi], except for a z_0 which is exceptional. Since θ is of compact support in Ω we find $< \sigma, \theta > = 0$ except when ν has mass on an exceptional point, i.e. a point such that $f^{-1}(z_0) = \{z_0\}$. So σ is supported on $\pi(\{\varphi = G\})$. Hence φ is continuous on support of $\Delta\varphi$. Therefore φ is continuous [Ts]. So $\varphi = G$ by the maximum principle.

(iii) Let $E \subset \mathbb{P}^1$, a totally invariant set. Assume $\mu(E) = c > 0$. Define $\nu = \chi_E \mu/c$. Then ν is a probability measure and $\frac{L^*\nu}{d} = \nu$. But since $\frac{(f^n)^*\nu}{d^n} \to \mu$ this implies that $\mu = \nu$, hence $c = 1$, therefore μ is ergodic. It is known [Gr], [Ly] that the entropy of f is $\log d$. Since $f^*\mu = d\mu$ the conditional measures are just Dirac masses of mass $\frac{1}{d}$ in almost every fiber of f. It follows by classical arguments [Ro] that μ is of maximal entropy, see [Ly] or [FLM].

(iv) One has to show that $\mu_n := \frac{1}{d^n} \sum \epsilon_{z_i^n} \to \mu$, where z_i^n are the periodic points of order n. Define

$$v_n(z, w) = \frac{1}{d^n} \log | P^n(z, w)w - Q^n(z, w)z |$$

where $F^n = (P^n, Q^n)$.

We have $\pi^* \mu_n = dd^c v_n$. We want to show that $v_n \to G$. Since

$$| P^n(z, w)w - Q^n(z, w)z | \leq \| (z, w) \| \| F^n \|,$$

any limit φ of any subsequence v_{n_i} satisfies $\varphi \leq G$. Assume $\varphi < G$ somewhere. Since G is continuous, this set is open. Let Ω be a component of $\{\varphi < G\}$. For every compact $K \subset \Omega$, there is, by Hartogs Theorem, a $\delta > 0$ so that $v_{n_i} < G - \delta$ on K for all sufficiently large i. Hence

$$| P^{n_i}w - Q^{n_i}z | \leq e^{-\delta d^{n_i}} (| P^{n_i} | + | Q^{n_i} |),$$

i.e. $f^{n_i} \to Id$ u. c. c. on $\pi(\Omega)$. So, there is at most one periodic point in $\pi(\Omega)$ and consequently $dd^c \varphi = 0$ on Ω. Hence φ and G are both pluriharmonic on Ω and agree on $\partial \Omega$ where G is continuous. It follows that $\varphi = G$ on Ω, a contradiction. Hence $\frac{1}{d^n} \sum \epsilon_{z_i^n} \to \mu$. $\square$

Observe that we already proved an analogue of Theorem 6.1 for $\mathbb{P}^2$ i.e. Theorem 4.16, for the current T instead of μ.

We want now to consider the case of a holomorphic map $f \in \mathcal{H}_d$ on $\mathbb{P}^2$ and construct a measure of maximal entropy. Let ω denote the standard Kähler form on $\mathbb{P}^2$ such that $\int_{\mathbb{P}^2} \omega^2 = 1$.

Proposition 6.2. *Let $f : \mathbb{P}^2 \to \mathbb{P}^2$ be a holomorphic map in $\mathcal{H}_d$, $d \geq 2$. Define μ by the identity $\pi^* \mu = (dd^c G)^2$. Then μ is a probability measure which satisfies the equations $f^* \mu = d^2 \mu$ and $f_* \mu = \mu$. Moreover μ does not charge locally pluripolar sets.*

Proof: Let $\nu = (dd^c G)^2$ in $\mathbb{C}^3 \backslash (0)$, where $(dd^c G)^2 := dd^c (G \, dd^c G)$. The definition makes sense since G is continuous. Moreover, the $(2, 2)$ current ν extends to $\mathbb{C}^3$ as a positive, closed current [HaP]. As for $(1, 1)$ currents, we can define μ by the functional equation $\pi^* \mu = \nu$. Say, in the chart $(t \neq 0)$, $\mu = (dd^c G(z, \omega, 1))^2$. Since G is bounded, it follows that μ has no mass on locally pluripolar sets ([BT1], [CLN] or [Kl]).

We show that $\mu = \lim\limits_{n\to\infty} \frac{(f^n)^*\omega^2}{d^{2n}}$. We have

$$\frac{\pi^*[(f^n)^*\omega^2]}{d^{2n}} = \frac{(F^n)^*\pi^*\omega^2}{d^{2n}}$$

$$= (F^n)^* \frac{(dd^c\log\|z\|)^2}{d^{2n}}$$

$$= \left(\frac{dd^c\log\|F^n\|}{d^n}\right)^2$$

$$= (dd^c G_n)^2.$$

So in the chart $t \neq 0$, $\frac{(f^n)^*\omega^2}{d^{2n}} = (dd^c G_n(z, w, 1))^2$.

By the change of variable formula, $\{(f^n)^*\omega^2/d^{2n}\}$ have uniformly bounded mass. Let ν be any weak limit. Since G_n converge uniformly on compact sets in $\mathbf{C}^3\backslash 0$ to G, by ([CLN]), $(dd^c G_n)^2 \to (dd^c G)^2$ in $\mathbf{C}^3\backslash 0$. Hence $\nu = (dd^c G)^2 = \mu$, for example in the chart $t \neq 0$, so $\nu = \mu$.

It follows from the change of variable formula, since f is a d^2 to 1 map, that μ is a probability measure.

We prove that $f^*\mu = d^2\mu$. On $\mathbb{P}^2\backslash f^{-1}(f(C))$, f is a submersion. We then have

$$\pi^*(f^*\mu) = F^*(\pi^*\mu)$$

$$= F^*(dd^c G)^2$$

$$= (dd^c(G \circ F))^2$$

$$= d^2(dd^c G)^2$$

$$= d^2\pi^*\mu.$$

So $f^*\mu = d^2\mu$ on $\mathbb{P}^2\backslash f^{-1}(f(C))$. Since μ does not charge complex curves, then $f^*\mu = d^2\mu$ everywhere.

To prove that $f_*\mu = \mu$, observe that $f_*(\varphi\circ f)(x) = d^2\varphi(x)$ if $x \notin f(C)$ which is a set of μ measure 0. Hence,

$$< f_*\mu, \varphi > = < \mu, \varphi \circ f >$$

$$= \frac{1}{d^2} < f^*\mu, \varphi \circ f >$$

$$= \frac{1}{d^2} < \mu, f_*(\varphi \circ f) >$$

$$= < \mu, \varphi >,$$

so $f_*\mu = \mu$. $\square$

We recall first some results in pluripotential theory, see [AT], [BT1], [Kl].

Let B be the unit ball in $\mathbf{C}^k$ and K a compact set in $B(0, r)$, $r < 1$. Following [BT1] define

$$C(K, B) = \sup \left\{ \int_B (dd^c u)^k, \quad 0 \le u \le 1, \quad u \text{ p.s.h.} \right\}.$$

It is shown in [BT1] that C extends to a Choquet capacity whose zero sets are the pluripolar sets.

Define also the Siciak function u_K of the compact K as follows

$$u_K(z) = \sup\{v(z), \ v \le 0 \text{ on } K, \ v(z) \le \log |z| + 0(1) \text{ at infinity}\}.$$

The uppersemicontinuous regularization of u_K is a p.s.h. function of logarithmic growth iff K is not pluripolar.

Alexander and Taylor [AT] proved the following estimate. If $K \subset B(0, r)$, $r < 1$ then, there exists a constant $A(r)$ such that

$$(1) \qquad\qquad m_k := \sup_B u_K \le \frac{A(r)}{C(K, B)}.$$

Using maximum principle one shows easily that

$$\log^+ \| z \| \le u_K(z) \le m_K + \log^+ \| z \|.$$

Now we prove the following result.

Theorem 6.3. sl Let $f \in \mathcal{H}_d$ on $\mathbb{P}^2$. The measure μ is mixing and of maximal entropy.

Proof: It is enough, [Wa], to show that given two non negative smooth test functions φ, ψ we have

$$\lim_{n \to \infty} \int \psi(f^n) \varphi \, d\mu = \left(\int \varphi \, d\mu\right)\left(\int \psi \, d\mu\right).$$

Define

$$\lambda_n(a, \varphi) := \frac{(f^n)_* \varphi(a)}{d^{2n}} = \frac{1}{d^{2n}} \sum_i \varphi(f_i^{-1}(a)).$$

We show first a lemma.

Lemma 6.4. There exists a constant M such that

$$\mu(|\lambda_n(a, \varphi) - c| \ge s) \le \frac{M \, |\varphi|_2}{s \, d^n}.$$

Here $c = \int \varphi \, d\mu$ and $|\varphi|_2$ denotes the C^2 norm of φ.

Proof: It is enough to prove the above estimate locally in $\mathbb{P}^2$. So fix coordinates, say $t = 1$ and define

$$K_s = \{a = (z, w) \in B(0, \tfrac{1}{2}), \ \lambda_n(a, \varphi) - c \geq s\}.$$

Let u_s be the Siciak function for K_s in $\mathbb{C}^2$ and define v_s in $\mathbb{C}^3$ by

$$v_s(z, w, t) = u_s\left(\frac{z}{t}, \frac{w}{t}\right) + \log |t| \, .$$

Let S be the closed $(1,1)$ current in $\mathbb{P}^2$ such that $\pi^* S = dd^c v_s$. Recall that $\nu_s := S \wedge S$ is a probability measure, Theorem 4.4 in [FS4]. Since ν_s is supported where $\lambda_n(a, \varphi) - c \geq s$, we have

$$s \leq \int (\lambda_n(a, \varphi) - c) d\nu_s = \int \lambda_n(a, \varphi) d\nu_s - c$$

$$= \int \lambda_n(a, \varphi) d\nu_s - \int \lambda_n(a, \varphi) d\mu.$$

We have used that $\frac{(f^n)^* \mu}{d^{2n}} = \mu$, hence $c = <\mu, \varphi> = <\mu, \frac{f_*^n \varphi}{d^{2n}}>$. So

$$s \leq \int \lambda_n(a, \varphi)[S \wedge S - T \wedge T] = \int \lambda_n(a, \varphi)[S - T] \wedge [S + T]$$

$$= \int \frac{\varphi}{d^n}(f^n)^*[S - T] \wedge \frac{(f^n)^*(S + T)}{d^n}.$$

Now we have

$$\pi^*(f^{n*})(S - T) = (F^n)^* \pi^*(S - T) = dd^c[(v_s - G) \circ F^n].$$

The function $v_s - G$ is well defined in $\mathbb{P}^2$ so

$$(f^n)^*(S - T) = dd^c[(v_s - G) \circ f^n].$$

We then have

$$s \leq \int \frac{dd^c \varphi}{d^n}(v_s - G)(f^n) \wedge \frac{(f^n)^*(S + T)}{d^n}$$

$$\leq \frac{|\varphi|_2}{d^n} \sup |v_s - G| \int \omega \wedge \frac{(f^n)^*(S + T)}{d^n},$$

since the last integral equals 1 we get

(2) $$s \leq \frac{|\varphi|_2}{d^n} \sup |v_s - G|.$$

We now estimate $\sup |v_s - G|$. Let $m(s) = \sup_B u_s$. Since

$$\log^+ \|z, w\| \leq u_s(z, w) \leq m(s) + \log^+ \|z, w\|,$$

we get

$$u_s\left(\frac{z}{t}, \frac{w}{t}\right) - G(\frac{z}{t}, \frac{w}{t}, 1) \leq m(s) - G\left(\frac{z/t, w/t, 1}{\|(z/t, w/t)\| \vee 1}\right)$$
$$\leq m(s) + M$$

where M is a constant independent of s. Similarly

$$G\left(\frac{z}{t}, \frac{w}{t}, 1\right) - u_s\left(\frac{z}{t}, \frac{w}{t}\right) \leq G\left(\frac{z/t, w/t, 1}{\|z/t, w/t\| \vee 1}\right) \leq M.$$

So relation (2) gives

(3) $$s \leq \frac{|\varphi|_2}{d^n} \sup |v_s - G| \leq \frac{|\varphi|_2}{d^n}(m(s) + M).$$

Using the Alexander-Taylor inequality (1) we get

$$m(s) + M \leq \frac{A(1/2)}{C(K_s, B)} + M \leq \frac{A(1/2) + MC(K_s, B)}{C(K_s, B)} \leq \frac{M'}{C(K_s, B)}$$

since $C(K_s, B) \leq C(B, B)$.
  So using (3)

$$C(K_s, B) \leq \frac{1}{s} |\varphi|_2 \frac{M'}{d^n}.$$

In the chart $t \neq 0$, $\mu = (dd^c G(z, w, 1))^2$. The function

$$\lambda = \frac{1}{2}\left(\frac{G(z, w, 1)}{\sup_B |G(z, w, 1)|} + 1\right)$$

is p.s.h. on B and $0 \leq \lambda \leq 1$. So by the very definition of $C(K_s, B)$ we have the existence of a constant α such that

$$C(K_s, B) \geq \int_B (dd^c \lambda)^2 \geq \mu(K_s)\alpha^{-1}.$$

Finally

$$\mu(K_s) \leq \alpha\, C(K_s, B) \leq \frac{\alpha}{s}\frac{M'}{d^n}\mid \varphi\mid_2 \ .$$

A similar computation with the set $H_s = \{a/c - \lambda_n(a, \varphi) \geq s\}$ finishes the proof of the lemma.

End of proof of Theorem 6.3. Define

$$I_n := <\mu, \psi(f^n)\varphi> - <\mu, \psi><\mu, \varphi> \ .$$

Observe that

$$<\mu, \psi(f^n)\varphi> \; =<\frac{(f^n)^*}{d^{2n}}\mu, \psi(f^n)\varphi>$$

$$=<\mu, \frac{(f^n)_*}{d^{2n}}\psi(f^n)\varphi>$$

$$=<\mu, \psi \cdot \frac{f^n_* \varphi}{d^{2n}}> \ .$$

So

$$I_n \; =<\mu, \psi[\lambda_n(a, \varphi) - c]> \ .$$

Let $q > 1$ and p such that $\frac{1}{p} + \frac{1}{q} = 1$. Using Hölder's inequality we have, if $L = 2\sup\mid\varphi\mid$,

$$\mid I_n\mid \leq \left(\int \psi^q d\mu\right)^{1/q}\left(\int \mid\lambda_n(a, \varphi) - c\mid^p d\mu\right)^{1/p}$$

$$\leq\parallel\psi\parallel_q\left(\int_0^L p\, s^{p-1}\mu(\mid\lambda_n(a, \varphi) - c\mid \geq s)ds\right)^{1/p}$$

$$\leq\parallel\psi\parallel_q\left(\int_0^L p\, s^{p-2} M\frac{\mid\varphi\mid_2}{d^n}ds\right)^{1/p}$$

$$\leq\parallel\psi\parallel_q\left(\frac{p}{p-1}\right)^{1/p}(2\parallel\varphi\parallel_\infty)^{\frac{p-1}{p}}\mid\varphi\mid_2^{1/p}d^{-n/p}M^{1/p}$$

$$\leq C_p\parallel\psi\parallel_q\parallel\varphi\parallel_\infty^{(1-1/p)}\mid\varphi\mid_2^{1/p}d^{-n/p}.$$

So $\lim I_n = 0$ and μ is mixing.

Remark 6.5. Observe that we have given in the proof of Theorem 6.3 an estimate of the decay of the coefficient of correlation. Indeed we have shown that if φ is $\mathcal{C}^2$ and ψ is bounded then for $\epsilon > 0$ there exists C_ϵ such that

$$\left| \int \psi(f^n)\varphi \, d\mu - \left(\int \psi \, d\mu\right)\left(\int \varphi \, d\mu\right) \right| \leq C_\epsilon \mid \varphi \mid_2 \mid \psi \mid_\infty \cdot d^{-n(1-\epsilon)}.$$

Given $f \in \mathcal{H}_d$ on $\mathbb{P}^2$ we define J_1' as the support of μ. We don't know whether J_1' is equal to the Julia set of order one as defined in Definition 5.8. It has however some properties showing that it is the right analogue of the Julia set in one variable.

Theorem 6.6. *Let* $f \in \mathcal{H}_d$, $f : \mathbb{P}^2 \to \mathbb{P}^2$. *Let* U *be an open set intersecting* J_1'. *Define* $E := \mathbb{P}^2 \backslash \bigcup_{n=0}^\infty f^n(U)$. *Then* E *is a closed locally pluripolar set in* $\mathbb{P}^2$.

Proof: Let $W := \bigcup_{n=0}^\infty f^n(U)$ and let χ be the characteristic function of W. Since $f(W) \subset W$, we have that $\chi \leq \chi \circ f$. The ergodic theorem [Wa] applied to μ implies that

$$\frac{1}{N} \sum_0^{N-1} \chi \circ f^n(x) \to \int \chi \, d\mu, \qquad \mu \ a. \ e.$$

So μ a.e., $\chi \leq \int \chi \, d\mu$ hence $\int \chi \, d\mu = 1$, so W is an open set of full measure for μ.

Assume there is a small ball B such that the closed set $K := E \cap \bar{B}$ is not pluripolar. We can consider that K is contained in the chart $t \neq 0$, identified with $\mathbf{C}^2$, let

$$U_K(z, w) = (\sup\{v(z, w); v \leq 0 \text{ on } K, \ v \leq \log \| (z, w) \| +0(1) \text{ at } \infty\})^*,$$

where * denotes the uppersemicontinuous regularization. The fact that K is not pluripolar is equivalent to the fact that U_K is not $+\infty$ and hence is a locally bounded p.s.h. function, such that

$$U_K(z, w) \leq \log \| (z, w) \| +O(1)$$

at infinity. Let $\nu := (dd^c U_K)^2$. It is known that ν is supported on K see [Kl]. Define $v(z, w, t) = U_K(\frac{z}{t}, \frac{w}{t}) + \log | t |$. The function v belongs to P.

We have : $\frac{v(F^n)}{d^n} - \frac{1}{d^n}\log \| F^n \| = \frac{1}{d^n} v\left(\frac{F^n}{\|F^n\|}\right) = O\left(\frac{1}{d^n}\right)$. Hence $v_n := \frac{v(F^n)}{d^n}$ converges uniformly on compacts subsets of $\mathbf{C}^3 \backslash 0$ to G. So by [CLN] or [BT1] $(dd^c v_n)^2 \to (dd^c G)^2$ in the sense of currents. Let v_n be the probability measures such that $\pi^* v_n = (dd^c v_n)^2$. We get that $v_n \to \mu$. So $v_n(\chi) \to 1$, hence support of $v_n \subset f^{-n}(K)$ intersects W, contradicting that $K \subset E$. Therefore E is locally pluripolar.

Corollary 6.7. *Let $f \in \mathcal{H}_d$, $f : \mathbb{P}^2 \to \mathbb{P}^2$. If J_1' contains a nonempty open set then $J_1' = \mathbb{P}^2$.*

Proof: If U is an open set in J_1', then $\bigcup_{n=0}^{\infty} f^n(U)$ is dense in $\mathbb{P}^2$ and contained in J_1', so $J_1' = \mathbb{P}^2$.

Remark 6.8. *Let $f : \mathbb{P}^2 \to \mathbb{P}^2$ be a meromorphic, non holomorphic, generic map. When the product $\mu = T \wedge T$ is defined, it turns out that $\mu = 0$. This is clear from the functional equation $f^* \mu = d^2 \mu$ if we apply the change of variable formula since f is generically a d' to 1 map with $d' < d^2$, as shown in Proposition 1.3. We will consider the problem of constructing interesting invariant measures for meromorphic maps in a forthcoming paper.*

As we have said, we don't know whether J_1' is equal to the Julia set of order one. But we have the following result, that we will use when we discuss hyperbolicity.

Proposition 6.9. *Let $f \in \mathcal{H}_d$, $f \, \mathbb{P}^2 \to \mathbb{P}^2$. Then $J_1' \subset J_1$. In particular J_1 is non empty.*

We first prove a lemma.

Lemma 6.10. *Let u be a continuous function on a closed ball $\bar{B}$ in $\mathbf{C}^k$. Assume that u is p.s.h. in B and that through any point p in B, there is a holomorphic disc Δ_p such that $u_{|\Delta_p}$ is harmonic. Then $(dd^c u)^k = 0$ in B.*

Proof: A holomorphic disc is by definition the image of the unit disc under a non constant holomorphic map φ. That $u \,|_{\Delta_p}$ is harmonic means that $u \circ \varphi$ is harmonic on the unit disc. Let v be a continuous function on $\bar{B}$, p.s.h. in B. Assume $v \le u$ on ∂B. We show first that $v \le u$ on B. Suppose not. Let $K = \{z/(v - u)(z) = M\}$ where M is the positive maximum of $v - u$ on $\bar{B}$. Let p be a peak point for a function $h \in \mathcal{C}(K)$ which is a uniform limit on K of holomorphic polynomials, i.e., $h(p) = 1$ and $| \, h \, | < 1$ on $K \backslash \{p\}$. Since on Δ_p, $v \le u + M$ and $v - u$ reaches its maximum at p : we get that $v - u = M$ on Δ_p, hence $\Delta_p \subset K$, contradicting that p is a peak point. It follows that u is the solution of the Bremerman Dirichlet problem with boundary data $u \mid_{\partial B}$. Hence by a result due to Bedford-Taylor [BT2] $(dd^c u)^k = 0$.

Proof of Proposition 6.9. We have to show that μ vanishes on F_1. Let B be a ball $B \subset\subset F_1$. Given any point in B, there is an analytic variety of dimension one through p, X_p, such that $f^n_{|X_p}$ is normal. This means that we can find holomorphic functions λ_{n_i} on $\pi^{-1}(X_p)$ such that $\frac{F^{n_i}}{\lambda_{n_i}}$ is normal on $\pi^{-1}(X_p)$. So if $G_n = \frac{1}{d^n}\log \| F^n \|$ we have

$$G_{n_i} = \frac{1}{d^{n_i}}\log \| \frac{F^{n_i}}{\lambda_{n_i}} \| + \frac{1}{d^{n_i}}\log \mid \lambda_{n_i} \mid .$$

The first term in the sum converges uniformly to zero so $\frac{1}{d^{n_i}}\log \mid \lambda_{n_i} \mid$ converges uniformly to G on $\pi^{-1}(X_p)$, hence G is pluriharmonic on $\pi^{-1}(X_p)$. If σ is a holomorphic section of π and $G_\sigma = G \circ \sigma$ we get that G_σ is harmonic on X_p, consequently by Lemma 6.10 we have $(dd^c G_\sigma)^2 = 0$ on B. So $\mu(B) = 0$.

7. Hyperbolicity.

In this section we consider $f : \mathbb{P}^2 \to \mathbb{P}^2$, $f \in \mathcal{H}_d$.

We define first hyperbolicity (See Ruelle [Ru]). Let $K \subset \mathbb{P}^2$ be a compact set. We assume that K is *surjectively forward invariant*, that is $f(K) = K$. The space $\hat{K} = K^N$ of orbits $\{x_n\}^0_{n=-\infty}$, $f(x_n) = x_{n+1}$, is compact in the product topology. By the tangent bundle T_K of $\hat{K}$ we mean the space of (x, ξ) where $x = \{x_n\} \in \hat{K}$ and $\xi \in T_{\mathbb{P}^2}(x_0)$ is a tangent vector. We give this tangent bundle the obvious topology. Then f lifts to a homeomorphism $\hat{f} : \hat{K} \to \hat{K}$ and f' lifts to a map $\hat{f}'$ on T_K in the obvious way.

Definition 7.1. *Let $K \subset \mathbb{P}^2$ be a compact surjectively forward invariant set. Then f is hyperbolic on K if there exists a continuous splitting $E^u \oplus E^s$ of the tangent bundle of $\hat{K}$ such that $\hat{f}'$ preserves the splitting and for some constants $C, c > 0, \lambda > 1, \mu < 1$ depending on the choice of a Hermitian metric on $\mathbb{P}^2$,*

$$\mid D\hat{f}^n(\xi) \mid \geq c\lambda^n \mid \xi \mid, \quad \xi \in E^u$$

$$\mid D\hat{f}^n(\xi) \mid \leq C\mu^n \mid \xi \mid, \quad \xi \in E^s, \quad n = 1, 2, \cdots$$

Theorem 7.2. *Let $f : \mathbb{P}^2 \to \mathbb{P}^2$ be a holomorphic map of degree $d \geq 2$. Then f cannot be hyperbolic on $\mathbb{P}^2$ nor on J_0.*

Proof: Assume f is hyperbolic on $\mathbb{P}^2$. Since the critical set is nonempty the fibre dimension of E^u is ≤ 1. If $\dim E^s = 2$ then all periodic orbits are attractive. Pick one, p, with immediate basin of attraction Ω. Since f is surjective, $\partial\Omega$ is a non empty, compact, forward invariant subset of $\mathbb{P}^2$. Hyperbolicity implies that orbits of points $q \in \Omega$ close to $\partial\Omega$ converge to $\partial\Omega$ contradicting that they are in the attractive basin of p. Hence $\dim E^s = 1$. Then we have a lamination of $\mathbb{P}^2$ by stable curves, and on each curve the family (f^n) is equicontinuous, so $\mathbb{P}^2 \subset F_1$. We get $J_1 = \emptyset$. This contradicts Proposition 6.9.

Assume J_0 is hyperbolic. Necessarily J_0 intersects C, the critical set (Proposition 5.4). Hence the fibre dimension of $E^u = 1$, so $\dim E^s = 1$. This implies that through every point p in $\mathbb{P}^2$ there exists an analytic disc Δ_p on which $f^n \mid \Delta_p$ is equicontinuous (clearly this is true for points not in J_0, and for points p in J_0, we consider the stable manifold through p). So $F_1 = \mathbb{P}^2$ and this is again impossible since $J_1 \neq \emptyset$. $\square$

Next we consider the question whether there exist maps which are hyperbolic on the nonwandering set.

Definition 7.3. *The nonwandering set of a map $f : \mathbb{P}^2 \to \mathbb{P}^2$ is the set of points p such that for every open neighborhood $U(p)$, there exists a positive integer n such that $f^n(U) \cap U \neq \emptyset$.*

It is clear that the nonwandering set is compact, surjectively forward invariant.

Theorem 7.4. *Let S_l be a compact hyperbolic, surjectively forward invariant set of unstable dimension l. For $l = 0$, $S_0 \subset F_0$ and S_0 is a finite union of attractive periodic orbits. If $l = 1$, $S_1 \subset J_0$ and if $l = 2$, $S_2 \subset J_1$.*

Proof: There is an arbitrarily small finitely connected neighborhood V of S_0 such that $f^n(V) \subset\subset V$ for all n large enough and f^n is strictly distance decreasing on each component of V. Hence $f^n \mid V$ converges to attractive periodic orbits.

Case $\ell = 1$. We need to show that $S_1 \subset J_0$. Let $x \in S_1 \cap F_0$. Then f^n is equicontinuous on some neighborhood of x. Let $\xi \neq 0$ be an unstable tangent vector at x. The iterates $(f^n)'(x)(\xi)$ have to blow up, a contradiction.

Case $\ell = 2$. We need to show that $S_2 \subset J_1$. Suppose $x \in S_2 \cap F_1$. Then there is a complex curve X through x so that $\{f^n \mid X\}$ is equicontinuous. We can assume that X is irreducible at x. If x is a regular point, let ξ

be a nonzero tangent vector to X at x. Then this is an unstable tangent vector, a contradiction. So it remains to consider the case when x is a singular point of X.

We parametrize X, $\varphi : \Delta \to X$, $t \to (t^p, t^q + \cdots) = (z, w)$ in local coordinates, $q > p$. We assume $x = O$. The sequence $\{f^n \circ \varphi\}$ is equicontinuous and $f^n(\varphi(t)) - f^n(\varphi(0)) = O(t^p)$ independently of n. This contradicts that 0 is an unstable point in all directions. $\square$

Assume from now on that the nonwandering set Ω is hyperbolic. We divide
$\Omega = \Omega_0 \cup \Omega_1 \cup \Omega_2$ where Ω_j has unstable dimension j.

Theorem 7.5. *If the nonwandering set Ω of a holomorphic map f of degree at least 2 is hyperbolic, then all Fatou components are preperiodic to finitely many attractive periodic basins.*

Proof: Pick a Fatou component U. Assume at first that $f^n \mid U$ does not converge u. c. c. to the Julia set. So for some compact subset $K \subset U$ and some subsequence f^{n_k}, the iterates converge uniformly to a holomorphic map with values outside some neighborhood of J. These values must then be in a periodic Fatou component, which we may assume is U. Replacing f by an iterate, we may assume that U is fixed. Hence we may assume that for some other subsequence $f^{n_{k+1}-n_k}$ the iterates converge to a holomorphic map with a fixed point in U. This point is then necessarily nonwandering so by the above theorem is an attractive periodic point for f.

On the other hand assume that the iterates $f^n \mid U$ converge u. c. c. to J. Pick $q \in U$. Let p be any cluster point of the iterates $f^n(q)$. Then p must be nonwandering. Any such cluster point belongs to $S_1 \cup S_2$. Note however because of the repelling nature of S_2, it is impossible to only cluster at S_2 without also clustering at other points arbitrarily close to S_2. Since S_1 and S_2 are disjoint compact sets, the cluster set must be contained in S_1. However in a small neighborhood of S_1 we can use the hyperbolicity on sectors in the tangent space to conclude that the iterates of $f \mid U$ must be diverging. Since unstable sectors are mapped to corresponding unstable sectors, the derivatives blow up since we always stay in a neighborhood of S_1, contradicting that we are in a Fatou component. So this case is impossible. We have shown then that all Fatou components are preperiodic to a finite number of attractive basins. $\square$

Question 7.6. *Let S_1 be a compact hyperbolic, surjectively forward invariant set of unstable dimension 1. Is $S_1 \subset F_1 \backslash F_0$?*

Example 7.7. *Consider the map $f = [z^2 : w^2 : t^2]$. Then this map has three superattractive fixed points, $[0 : 0 : 1]$, $[0 : 1 : 0]$, $[1 : 0 : 0]$. The complement of the three basins of attraction is the Julia set. More precisely, $J_0 = \{[z : w : t]$; such that two coordinates have modulus one and the third has modulus at most one\}. In addition to the set of superattractive fixed points S_0 the nonwandering set contains the set S_1 consisting of the three circles where one homogeneous coordinate is 0 and the other two have modulus one and also the set S_2 consisting of the totally real torus where all three coordinates have modulus one. This example is hyperbolic on the nonwandering set.*

Question 7.8. *Assume that X is a closed totally invariant set for f and that X is disjoint from the closure of the forward orbit of the critical set. Is f hyperbolic on X ?*

References

[AT] H. ALEXANDER & B. A. TAYLOR : Comparison of two capacities in $\mathbf{C}^n$. Matematische Zeitschrift 186 (1984), 407-417.

[B] A. F. BEARDON : *Iteration of rational functions.* Springer Verlag (1991).

[BS1] E. BEDFORD & J. SMILLIE : Polynomial diffeomorphisms of $\mathbf{C}^2$. *Inv. Math.* 87 (1990), 69-99.

[BS2] E. BEDFORD & J. SMILLIE : Polynomial diffeomorphisms of $\mathbf{C}^2$. II. *J. Amer. Math. Soc.* 4 (1991), 657-679.

[BLS] E. BEDFORD, M. LYUBICH & J. SMILLIE : Polynomial diffeomorphisms. IV. *Invent. Math.* 112 (1993),77-125.

[BT1] E. BEDFORD & B. A. TAYLOR : A new capacity for p. s. h. functions. *Acta Math.* 149 (1982), 1-39.

[BT2] E. BEDFORD & B. A. TAYLOR : The Dirichlet problem for Monge-Ampere equation. *Invent. Math.* 37 (1976), 1-44.

[Br] H. BROLIN : Invariant sets under iteration of rational functions. *Ark. Mat.* 6 (1965), 103-144.

[Ca] L. CARLESON : Complex dynamics. UCLA course notes (1990).

[Ce] U. CEGRELL : Removable singularities for plurisubharmonic functions and related problems. *Proc. London Math Soc.* 36 (1978), 310-336.

[CLN] S. S. CHERN, H. LEVINE & L. NIRENBERG : Intrinsic norms on a complex manifold, Global analysis. (Papers in honor of K. Kodeira), Univ. Tokyo Press (1969).

[De] J.-P. DEMAILLY : Courants positifs extrêmaux et conjecture de Hodge. *Invent. Math.* 69 (1982), 347-374.

[FLM] A. FREIRE, A. LOPEZ & R. MANE : An invariant measure for rational maps. *Bol. Soc. Bras. Mat.* 6 (1983), 45-62.

[FS1] J.-E. FORNAESS & N. SIBONY : Complex dynamics in higher dimension I. *Astérisque.* 222 (1994), 210-231.

[FS2] J.-E. FORNAESS & N. SIBONY : Complex Hénon mappings in $\mathbb{C}^2$ and Fatou-Bieberbach domains. *Duke Math. J.* 65 (1992), 345-380.

[FS3] J.-E. FORNAESS & N. SIBONY : Critically finite rational maps on $\mathbb{P}^2$, in *Contemporary Mathematics* 137 (1992), 245-260.

[FS4] J.-E. FORNAESS & N. SIBONY : Oka's inequality for currents and applications. To appear.

[Gr] M. GROMOV : Entropy, homology and semi-algebraic Geometry, in Séminaire Bourbaki. *Astérisque* 145-146 (1987), 225-240.

[HaP] R. HARVEY & J. POLKING : Extending analytic objects. *Comm. Pure Appl. Math.* 28 (1975), 701-727.

[HP] J. H. HUBBARD & P. PAPADOPOL : Superattractive fixed points in $\mathbb{C}^n$. *Indiana Univ. Math. J.* 43 (1994), 321-365.

[Kl] M. K. KLIMEK : *Pluripotential theory.* Oxford (1991).

[Le] P. LELONG : *Fonctions plurisousharmoniques et formes différent-ielles positives.* Paris, Londres, New York, Gordon and Breach, Dunod (1968).

[Ly] M. LYUBICH : Entropy properties of rational endomorphisms of the Riemann sphere. *Ergod. Th. & Dynam. Syst.* 3 (1983, 351-385.

[Mi] J. MILNOR : Notes on complex dynamics. Preprint, SUNY Stony Brook.

[Ro1] V. A. ROHLIN : Lectures on entropy theory of measure preserving transformation. *Russian Math. Surveys* 22 (1967).

[deRh] G. de RHAM : *Variétés différentiables.* Paris, Hermann (1955).

[Ru] D. RUELLE : *Elements of differentiable dynamics and bifurcation theory.* Boston, Academic Press (1989).

[Sc] L. SCHWARTZ : *Théorie des distributions.* Paris, Hermann (1966).

[Sch] SCHERBINA : The Levi form for C^1-smooth hypersurfaces and the complex structure on boundary of domains of holomorphy. *Izv. Akad. Nauk SSSR* 45 (1981), 874-895.

[Si] N. SIBONY : Unpublished manuscript. Course at UCLA (1984).

[Sk] H. SKODA : Prolongement des courants positifs, fermés, de masse finie. *Invent. Math.* 66 (1982), 361-376.

[Siu] Y. T. SIU : Analyticity of sets associated to Lelong numbers and extension of closed positive currents. *Invent. Math.* 27 (1974), 53-156.

[T] P. TORTRAT : Aspects potentialistes de l'itération des polynômes. *Sém. Théorie du Potentiel*, Paris, n° 8 (1987). Springer Lecture Notes 1235.

182 John Erik Fornaess and Nessim Sibony

[Ts] TSUJI : *Potential theory in modern function theory,* Mazuren, Tokyo (1959).

[Wa] P. WALTERS : *An introduction to ergodic theory.* Springer Verlag (1981).

[We] J. WERMER : Maximum modulus algebras and singularity sets. *Proc. R. Soc. Edinb.* 86 (1980), 327-331.

John Erik Fornaess
Mathematics Department
The University of Michigan
Ann Arbor, MI 48104, USA

Nessim Sibony
Université Paris-Sud
Mathématiques - Bât. 425
91405 Orsay Cedex, France

SET THEORETICAL
REAL ANALYTIC SPACES

Hans Grauert

Introduction

In the complex case the theory of analytic and meromorphic decompositions works only for normal complex spaces. These are special reduced
complex spaces. They do not have nilpotent elements in their structure
sheaves. The local holomorphic functions are special continuous complex
functions.

The real-analytic spaces were first introduced by H. Cartan and F.
Bruhat (see [CB57]). But their structure sheaf is not a subsheaf of the
sheaf of local complex functions. This makes them useless for (geometric) analytic decompositions. But we need geometric real quotients, for
instance in the case of actions of real Lie groups. Therefore, we shall define here set theoretical real analytic spaces similar to the work of other
authors (see for instance [WB59], [Lo64]).

The theory of complex spaces is always used. Therefore, the reader
should have a look in [GR58], [GR79], [GR84], [GPR94].

1. Real Spaces of Dimension n

First, we shall define local representatives. These are the n-dimensional
real analytic sets. Assume that $G \subset \mathbf{R}^N$ is an open set in the N-
dimensional complex number space and that $f_1, \cdots, f_\ell$ are real analytic
functions in G. We put $F = (f_1, \cdots, f_\ell)$.

Definition 1. F is regular of dimension n in a point $0 \in G$ if

1) $F(0) = 0$;

2) there is an enumeration of the components of F and of the coordinates
$x_1, \cdots, x_N$ of G such that the matrix $(f_{\kappa x_\lambda}(0))_{1 \leq \kappa \leq N-n, 1 \leq \lambda \leq N-n}$ is not
singular;

3) there exists a neighborhood $U(0)$ in G such that $F|\{x \in U : f_1(x) = \cdots = f_{N-n}(x) = 0\}$ is zero.

We put $A = \{x \in G : F(x) = 0\}$. Then, in U the set A is a smooth
n-dimensional real analytic subvariety. We define the set of singularities

$$S = \{x \in A : F \text{ is not regular of dimension } n \text{ in } x\}.$$

The local dimenion of S can be defined very well, since S has a triangulation. We always do assume that the local dimension of S is smaller than n everywhere. Then we define

Definition 2. An n-dimensional analytic set in G is a set obtained as $A^* = $ closure of $(A - S)$. The set $S - A^*$ is called the *stings*.

So we obtain A^* by omitting the lower dimensional stings of A. Nevertheless we have the notion of the local analytic functions on A^* : these are real functions which can locally be continued to a local analytic function in G. So A^* is a topological space with a structure sheaf.

Let us consider the following example:

We take $G = \mathbf{R}^3 = \{(x, y, t)\}$ and for A the set $\{f(x, y, t) = tx^2 + y^2 = 0\}$. Then in generic points of A the function $F = f$ is regular of dimension 2. The singular locus S is $\{(x, y, t) : (x, y) = 0\}$. The set S is a line and we have everywhere dim $S = 1$. So A^* is $A \cap \{(x, y, t) : t \leq 0\}$ and $S - A^* = \{(x, y, t) : (x, y) = 0, t > 0\}$ is the sting. For $t = t_0 < 0$ the set A^* is a pair of lines in the (x, y)-plane which cross in the 0-point.

Now it is immediate to define our basic concept:

Definition 3. Assume that X is a Hausdorff space (always with countable topology) and that X is equipped with a structure sheaf of local real functions. Assume moreover that X is locally isomorphic to an n-dimensional analytic set in an open $G \subset \mathbf{R}^n$. Then X is called an *n-dimensional analytic space*.

It is clear that the set S^* of those points in X where X is not smooth, *i.e.*, X does not have n-dimensional local analytic coordinates, is given locally by analytic equations, is nowhere dense, and has local dimension $< n$. So every analytic space is an n-dimensional analytic manifold outside of S^*. Singular points, *i.e.*, points of S^* can be smooth points of X in the differentiable sense.

2. Complexification

Let us first consider an n-dimensional analytic set A^* in an open set $G \subset \mathbf{R}^N$. Assume that $f_1, \cdots, f_\ell$ are analytic functions in G such that $A^* = \{f_1 = \cdots = f_\ell = 0\}$ - stings (of lower dimension). Then there is an open subset $\tilde{G} \subset \mathbf{C}^N \supset \mathbf{R}^N$ with $\tilde{G} \cap \mathbf{R}^N = G$ and holomorphic functions $\tilde{f}_1, \cdots, \tilde{f}_\ell$ such that $\tilde{f}_\lambda | G = f_\lambda$ for $\lambda = 1, \cdots, \ell$. We put $\tilde{A} = \{\tilde{f}_1 = \cdots =$

$\tilde{f}_\lambda = 0\}$. Without loss of generality we assume that the functions $\tilde{f}_1, \cdots, \tilde{f}_\ell$ generate locally the sheaf of local holomorphic functions which vanish on $\tilde{A}$ and that every irreducible component on $\tilde{A}$ has complex dimension n and is uniquely determined along G. We call $\tilde{A}$ a *complexification* of A^*.

If we map $z \in G$ onto the point with conjugate complex components we get an anti-holomorphic map γ. We may assume that $\tilde{G}$ is mapped onto $\tilde{G}$. Then G is the set of fixed points of γ and the conjugate complex of $\tilde{f}_\lambda(\gamma(z))$ is always $f(z)$. We have $\tilde{A} \cap G = A^* \cup$ stings (of lower dimension). Since $\tilde{A}$ is uniquely determined along G the set of stings is in local neighborhoods of points of A^* the minimal possible.

If X is an n-dimensional analytic space and $x \in X$ is an arbitrary point there are neighborhoods $U = U(x)$ and an isomorphism $U \cong A^*$ of U onto an n-dimensional analytic set $A^* \subset G$. We take the complexification $\tilde{A}$ and put $\tilde{U} = \tilde{A}$. If U_1 is another small neighborhood (of another point of X) we have another complexification $\tilde{U}_1$ and for neighborhoods V of $U \cap U_1$ in $\tilde{U}$ and V_1 of $U \cap U_1$ in $\tilde{U}_1$ an isomorphism $F : V \cong V_1$. We take F for glueing $\tilde{U}$ and $\tilde{U}_1$. We do this for the pairs of elements of a suitable covering on X. By this we obtain an n-dimensional (reduced) complex space $\tilde{X}$ such that X is an analytic subspace of $\tilde{X}$. The germ of $\tilde{X}$ along X is uniquely determined. If $\tilde{X}$ is small enough and suitable we get an anti-holomorphic mapping $\gamma : \tilde{X} \cong \tilde{X}$, which locally coincides with the old $\gamma : \tilde{A} \to \tilde{A}$. The set of fixed points of γ is the union of the analytic space X with a minimal set of stings. If f is a local analytic function on X then f has a unique continuation to a local holomorphic function $\tilde{f}$ on $\tilde{X}$. We have that the conjugate complex of $\tilde{f} \circ \gamma = \tilde{f}$. We call $(\tilde{X}, \gamma)$ a complexification of X. Its germ along X is uniquely determined. It is possible to prove as in the case of real analytic manifolds:

Theorem 1. *There are arbitrary small complexifications $\tilde{X}$ around X which are Stein spaces.*

There are some consequences following directly from this theorem:

1) If X is an n-dimensional analytic space and L a Lie group we have the notion of (principal) fiber bundle V over X with L as the structure group. For simplicity we assume that L has a complex Lie group $\tilde{L}$ for complexification. In this case the Oka principle follows directly: *If V_c is a topological fiber bundle over X with L as the structure group, then V_c is topologically equivalent to an analytic fiber bundle V over X. Any two topologically equivalent analytic fiber bundles over X are also analytically equivalent.*

2) The notion of a *coherent (real) sheaf S* is well defined on an analytic space X. Of course, it always is a sheaf of modules over the (real) structure sheaf of X. It always can be uniquely extended to a complex coherent

sheaf into a small complexification $\tilde{X}$ of X. So we get the *vanishing of the cohomology* $H^\mu(X, S)$ *for dimensions* $\mu = 1, 2, \cdots$.

3) Moreover, we have the *Remmert embedding theorem*. Assume that X is an n-dimensional analytic space, which is compact or, at least, has the property that the local embedding dimension is bounded. Then there is a real number space $\mathbf{R}^N$ such that X can be represented as an analytic set of $\mathbf{R}^N$.

There is an inportant notion, which is obtained by playing back to the complexification.

Definition 2. Assume that $\tilde{X}$ is a complexification of X (such that the antiholomorphic map γ is defined on the whole $\tilde{X}$) and γ has $(X \cup \text{stings})$ for fixed point set. Then $\tilde{X}$ is called a *compactification* of X.

Of course, in this case X itself is compact. The complex space $\tilde{X}$ will not be determined by X, by far. It may be that there are very many completely different compactifications. A complexification of $\mathbf{R}^N$ is $\mathbf{C}^N$ and a compactification of the real projective space $\mathbf{RP}_n$ is $\mathbf{CP}_n$. But compactifications of this kind are somewhat determined by Lie group actions. We need them in this theory.

3. Semianalytic spaces

The notion of proper analytic maps also comes via the complexifications. Assume that X and Y are analytic spaces and that $p : X \to Y$ is an analytic map whose rank is m in a dense subset of X.

Definition 1. We call p a proper analytic map, if there are complexifications $\tilde{X}$ and $\tilde{Y}$ of X and Y and a holomorphic continuation $\tilde{p} : \tilde{X} \to \tilde{Y}$ of p such that $\tilde{p}$ is proper in the ordinary sense.

Of course: if p is proper in our sense, it also is proper in the ordinary sense. But the opposite direction is far from being true.

Unfortunately the image of an analytic space by a proper analytic map will not be an analytic subset, in general. Let us consider the following example: We take the polynomial $\omega(y, x) = y^2 - x$ and denote by X the analytic subspaces $\{(y, x) : \omega(x, y) = 0\} \subset \mathbf{R}^2$ and by p the projection $(y, x) \to x : X \to \mathbf{R}$. Its rank is 1 almost everywhere and it comes from the analogous projection $\mathbf{C}^2 \to \mathbf{C}$. Therefore it is proper. But the image $p(X)$ is $\{x \in \mathbf{R} : x \geq 0\}$, which is not an analytic set.

More general, we may consider a normed analytic polynomial

$$\omega(y; x_1, \cdots, x_n)$$

over a domain $G \subset \mathbf{R}^n$. It also has a complexification $\tilde{\omega}$ over a domain $\tilde{G} \subset \mathbf{C}^n$ such that the projection of $\tilde{X} = \{\tilde{\omega} = 0\} \to \tilde{G}$ is proper. So by definition $p : (y; x_1, \cdots, x_n) \to (x_1, \cdots, x_n)$ of $X = (\{\omega = 0\}$ - stings of dimension smaller than $n) \to G$ is proper. We denote by $\Delta(x_1, \cdots, x_n)$ the discriminant of ω. In general $p(X)$ will be different from G. If $x \in G - p(X)$ is a point we consider the intersection $\{\tilde{\omega} = 0\} \cap \{(y, x) : y \in \mathbf{C}\} \subset \mathbf{C} = \{y\}$. This set consists of pairs of conjugate complex y. If x is in $G - p(X)$ and x approaches the boundary of $p(X)$ at least two conjugate complex y will move together. Then Δ will vanish in the limit point. So the boundary of $p(X)$ is contained in the zero set of Δ. We denote by $Q^{\bullet}$ the union of some connected components of $G - \{\Delta(y; x) = 0\}$. Then $p(X)$ is the closure Q of $Q^{\bullet}$.

We want to conclude that every proper analytic image of an analytic space X is an m-dimensional semi analytic space. These spaces will be more general than the m-dimensional analytic spaces. Our definition will be somewhat different from that of [Lo64]. The local representatives of the semi analytic spaces are defined as follows:

Assume that Y is an n-dimensional analytic space and that $A \subset Y$ is a finite union of analytic subspaces of dimensions $0, \cdots, n - 1$ (even if A is 0-dimensional it may disconnect Y locally). Let us denote by $Y^{\bullet}$ a finite union of some connected components of $Y - A$. Then the closure Y^* of $Y^{\bullet}$ in Y is a *local representative*. Such a Y^* always is a locally compact Hausdorff space and it carries a structure sheaf of locally analytic functions. These are just locally the restrictions of local analytic functions in Y.

Definition 2. An *n-dimensional semi analytic space* X is a Hausdorff space, which has the following properties.

1) X is equipped with a structure sheaf of local real analytic functions,

2) X is locally isomorphic to an n-dimensional representative of a semi analytic space.

We have the *Remmert projection theorem*:

Theorem 3. *Assume that X and Y are analytic spaces and that $p : X \to Y$ is a proper analytic map of rank m. Then $p(X) \subset Y$ is an m-dimensional semi analytic subspace.*

There are some important examples of proper analytic maps.

1) An n-dimensional analytic space X is called *normal* if it has a complexification $\tilde{X}$, which is a normal n-dimensional complex space. If Y, X are n-dimensional normal complex spaces, $p : Y \to X$ is a proper analytic map of rank n and $A \subset X$ is a closed subset of X which locally

is a finite union of analytic subsets of dimensions $0, \cdots, n-1$ such that p maps $Y \to p^{-1}(A)$ bianalytically onto $X - A$, then (Y, p) is called a *proper analytic modification* of X. By a proper modification the set A is replaced by the set $p^{-1}(A)$. The rest is unchanged. If $0 \in X$ is a single point there is the so-called σ-process which always replaces 0 by an $(n-1)$-dimensional analytic set. If X is smooth at 0, this set will be a real $(n-1)$-dimensional projective space.

2) The real linear group $GL(n+1, \mathbf{R})$ of rank $n+1$ operates on the n-dimensional real projective space $X = \mathbf{RP}_n$. We take an algebraic subgroup L. Then L has a unique complexification $\tilde{L}$ which is a complex Lie subgroup of $G(n+1, \mathbf{C}) \supset GL(n+1, \mathbf{R})$. Every connected component of $\tilde{L}$ intersects $GL(n+1, \mathbf{R})$ in a component of L. By orbits to the action of $\tilde{L}$ we get a meromorphic equivalence relation in $\tilde{X} = \mathbf{CP}_n$ (see [Gr83] and [Gr85]) and a complex quotient space $\tilde{Q}$ of $\tilde{X}$. Moreover, we have an anti-holomorphic reflection $\underline{\gamma} : \tilde{Q} \to \tilde{Q}$ which belongs to the reflection $\gamma : \tilde{X} \to \tilde{X}$ lying over it. The set {fixed points - stings} is an analytic space Q. The real space X produces a subset Q^* of Q. Since all employed maps are proper, it is semi analytic space. The analytic functions on Q^* are the invariants of the action of L. So we have the finiteness of these invariants. The theory was started in [Mu65], further important results were obtained by Heinzner, see for instance [He92].

References

[Ca55] H. Cartan: Quotient d'un espace analytique par un groupe d'automorphismes. Algebraic geometry and topology. Princeton University Press.

[CB57] H. Cartan and F. Bruhat: Sur la structure des sous-ensembles analytiques réels. C.R. Acad. Sci. Paris **244**, 988-990.

[Gr83] H. Grauert: Set theoretic complex equivalence relations. Math. Ann. **265**, 137-148.

[Gr85] H. Grauert: On meromorphic equivalence relations. In: Contribution to Several Complex Variables in honor of Wilhelm Stoll. Viehweg, Wiesbaden.

[GR58] H. Grauert and R. Remmert: Komplexe Räume. Math. Ann. **136**, 245-318.

[GR79] H. Grauert and R. Remmert: Theory of Stein spaces. Springer, Heidelberg 1979, p.272.

[GR84] H. Grauert and R. Remmert: Coherent analytic sheaves. Springer, Heidelberg.

[GPR94] H. Grauert, Th. Peternell and R. Remmert: Sheaf theoretic methods. Encycl. Mathem. Sciences, Springer 1994.

[He93] P. Heinzner: Equivariant holomorphic extension of real analytic manifolds, Preprint.

[Mu65] D. Mumford: Geometric invariant theory. Erg. Math. **34**, Springer, Heidelberg 1965.

[Lo64] S. Lojasiewics: Triangulation of semi analytic sets. Ann. Sc. Norm. Sup. Pisa (3), **18**, 449-474.

[WB59] H. Whitney and F. Bruhat: Quelques propriétés fundamentales des ensembles analytiques réels. Comment. Math. Helvet. **33**, 132-160.

Mathematical Institute, University of Göttingen, Göttingen, Germany

AN ISOPERIMETRIC ESTIMATE FOR
THE RICCI FLOW
ON THE TWO-SPHERE

RICHARD S. HAMILTON

1. Consider a metric $G = \{g_{ij}\}$ on the two-sphere S^2 evolving under the Ricci flow, which on a surface takes the simple form

$$\frac{\partial}{\partial t}\, g_{ij} = -R g_{ij}$$

where R is the scalar curvature. Consider any curve Λ of length L on S^2 dividing the total area A into two parts $A_1 + A_2 = A$, and take the curve $\overline{\Lambda}$ of least length $\overline{L}$ on the round sphere of the same total area $\overline{A} = A$ dividing it into two parts $\overline{A}_1 + \overline{A}_2 = \overline{A}$ with $\overline{A}_1 = A_1$ and $\overline{A}_2 = A_2$ the same as before. We form the isoperimetric ratio

$$C_S(\Lambda) = L/\overline{L}$$

and let

$$\overline{C}_S = \inf_{\Lambda} C_S(\Lambda)$$

taking the infimum over all curves Λ on S^2.

1.1 Main Theorem. The isoperimetric ratio $\overline{C}_S$ increases under the Ricci flow on the two-sphere.

The shortest curve $\overline{\Lambda}$ cutting off areas $\overline{A}_1$ and $\overline{A}_2$ on the round sphere is clearly a circle of latitude, and its length $\overline{L}$ is given by

$$\frac{4\pi}{\overline{L}^2} = \frac{1}{\overline{A}_1} + \frac{1}{\overline{A}_2}\,.$$

On the round sphere $\overline{C}_S = 1$ by definition, and in any metric we can come as close to this as we wish by taking a very short curve like a small circle. Hence $\overline{C}_S \leq 1$ in any metric. When $\overline{C}_S < 1$ we will show that the value of $\overline{C}_S$ is attained by $C_S(\overline{\Lambda})$ for some $\overline{\Lambda}$ which is a single simple closed loop of constant curvature. The constant $\overline{C}_S$ is also the Sobolev constant, defined as the best constant in the inequality

$$\left\{\inf_c \int (f - c)^2 da\right\}^{1/2} \leq \overline{C}_S \int |\nabla f| da$$

where the infimum is taken over all constants c. In fact equality is attained precisely when f has a jump discontinuity along the optimal curve $\overline{\Lambda}$ and is constant on either side. (See the nice survey article by Yau [**Y**].)

This estimate gives us a new proof of a theorem by Ben Chow [**C**].

1.2 Corollary. (Chow). Under the Ricci flow on the two-sphere, any metric approaches constant curvature.

PROOF. The injectivity radius can be controlled by the maximum of the curvature and the length of the shortest closed geodesic circle. The isoperimetric ratio controls the length of a short geodesic circle in terms of the maximum curvature. Therefore if we take a sequence of points and times where the curvature is as large as it has ever been and dilate so the curvature there is one, we can extract a convergent subsequence. If the curvature times the area is unbounded, the limit will be complete but not compact, and we can arrange that it is an eternal solution to the Ricci flow. Since the scalar curvature is bounded below, after dilating the limit will have nonnegative curvature. Since it is not flat, the strong maximum principle shows it is strictly positive. It then follows that the limit is the cigar solution

$$ ds^2 = \frac{dx^2 + dy^2}{1 + x^2 + y^2} \quad . $$

But the cigar has isoperimetric ratio zero; hence it cannot occur as a limit of surfaces whose isoperimetric ratio is bounded away from zero.

The other possibility is that the curvature times the area is bounded. In this case the limit of the dilations will be compact. Again it will have strictly positive curvature. Since the normalized entropy

$$ E = \int R \ln(RA) \, da $$

is monotone decreasing for the Ricci flow, it must be constant on the limit flow. But then Ben Chow's proof of the entropy estimate shows that a certain integral of a positive expression is zero, which implies that the limit is a compact homothetically shrinking solution. but we know this can only be the sphere.

2. We now proceed with the proof, which is a straightforward calculation. Start with the optimal curve $\overline{\Lambda}$ at time $\bar{t}$ where $\overline{C}_S = C_S(\overline{\Lambda})$, and construct the one-parameter family of parallel curves Λ_r at distance r from $\overline{\Lambda} = \Lambda_0$ on either side. We take $r > 0$ when the curve moves from A_1 into A_2, and $r < 0$ when it moves the other way. We then regard L,

A_1, A_2 and $C_S = C_S(\Lambda_r)$ as functions of r, and t also by considering the same curves Λ_r at times t near $\bar{t}$.

First we clearly have

$$\frac{dA_1}{dr} = L \quad \text{and} \quad \frac{dA_2}{dr} = -L$$

and by a standard formula

$$\frac{dL}{dr} = \int k\, ds$$

where k is the geodesic curvature of the curve Λ_r. (Of course by a standard variational argument k is constant on $\overline{\Lambda}$, but we don't seem to use this, except to check $\overline{\Lambda}$ is smooth.) Thus we also get

$$\frac{d^2}{dr^2} A_1 = \int k\, ds \quad \text{and} \quad \frac{d^2}{dr^2} A_2 = -\int k\, ds\,.$$

Now we use the Gauss-Bonnet formula

$$\int_1 K\, da + \int k\, ds = 2\pi$$

where the first integral is over the part with area A_1. Differentiating the Gauss-Bonnet formula with respect to r gives

$$\frac{d^2 L}{dr^2} = \frac{d}{dr} \int k\, ds = -\int K\, ds\,.$$

So much for the space derivatives.

Now we compute the time derivatives under the Ricci flow. This gives

$$\frac{dL}{dt} = -\int K\, ds$$

and

$$\frac{dA_1}{dt} = -2 \int_1 K\, da = -4\pi + 2 \int k\, ds$$

$$\frac{dA_2}{dt} = -2 \int_2 K\, da = -4\pi - 2 \int k\, ds$$

and of course $dA/dt = -8\pi$. So much for the time derivatives.

The Sobolev constant of Λ_r is

$$C_S^2 = L^2 \left(\frac{1}{A_1} + \frac{1}{A_2} \right) \Big/ 4\pi$$

and taking logarithms for simplicity

$$\ln C_S^2 = 2\ln L - \ln A_1 - \ln A_2 + \ln A - \ln 4\pi.$$

First we compute

$$\frac{d}{dr}\ln C_S^2 = \frac{2}{L}\frac{dL}{dr} - \frac{1}{A_1}\frac{dA_1}{dr} - \frac{1}{A_2}\frac{dA_2}{dr}$$

which gives

$$\frac{d}{dr}\ln C_S^2 = \frac{2}{L}\int kds - L\left(\frac{1}{A_1} - \frac{1}{A_2}\right).$$

At the infimum $\overline{\Lambda}$ when $t = \overline{t}$ and $r = 0$ we must have the first variation equal to zero. This gives the useful relation

$$\int kds = \frac{L^2}{2}\left(\frac{1}{A_1} - \frac{1}{A_2}\right)$$

which we can substitute in future calculations. The second space derivative is

$$\frac{d^2}{dr^2}\ln C_S^2 = \frac{2}{L}\frac{d^2L}{dr^2} - \frac{2}{L^2}\left(\frac{dL}{dr}\right)^2 - \frac{1}{A_1}\frac{d^2A_1}{dr^2} + \frac{1}{A_1^2}\left(\frac{dA_1}{dr}\right)^2$$

$$-\frac{1}{A_2}\frac{d^2A_2}{dr^2} + \frac{1}{A_2^2}\left(\frac{dA_2}{dr}\right)^2$$

which works out to

$$\frac{d^2}{dr^2}\ln C_S^2 = \frac{2}{L}\frac{d^2L}{dr^2} - L^2\left(\frac{1}{A_1} - \frac{1}{A_2}\right)^2 + L^2\left(\frac{1}{A_1^2} + \frac{1}{A_2^2}\right).$$

The time derivative is

$$\frac{d}{dt}\ln C_S^2 = \frac{2}{L}\frac{dL}{dt} - \frac{1}{A_1}\frac{dA_1}{dt} - \frac{1}{A_2}\frac{dA_2}{dt} + \frac{1}{A}\frac{dA}{dt}$$

which works out to

$$\frac{d}{dt}\ln C_S^2 = \frac{2}{L}\frac{dL}{dt} - L^2\left(\frac{1}{A_1} - \frac{1}{A_2}\right)^2 + 4\pi\frac{A_1^2 + A_2^2}{A_1 A_2(A_1 + A_2)}.$$

Since

$$\frac{dL}{dt} = -\int Kds = \frac{d^2L}{dr^2}$$

(which looks like a "heat" equation!) we get the formula below.

2.1 FORMULA. The Sobolev constant C_S satisfies

$$\frac{d}{dt}\,\ell n\,C_S^2 = \frac{d^2}{dr^2}\,\ell n\,C_S^2 + \frac{4\pi(A_1^2 + A_2^2)}{A_1 A_2 (A_1 + A_2)}\,[1 - C_S^2].$$

2.2 Corollary. If $\overline{C}_S^2 < 1$ it will increase.

PROOF. It suffices to show that $\overline{C}_S^2 + \varepsilon t$ increases for all $\varepsilon > 0$ no matter how small. If not, we can find a time $\bar{t} > 0$ when it is no larger than at previous times, hence at previous times $\overline{C}_S^2 + \varepsilon t$ was no smaller. Hence at previous times $\overline{C}_S^2$ was larger by $\varepsilon(\bar{t} - t)$.

Now pick the optimal $\overline{\Lambda}$ at time $\bar{t}$, and construct the family Λ_r as before. Since $\overline{\Lambda}$ is a minimum over all r at $t = \bar{t}$, we get

$$\frac{d^2}{dr^2}\,\ell n\,C_S^2 \geq 0$$

and hence for $\overline{C}_S^2 < 1$

$$\frac{d}{dt}\,\ell n\,C_S^2 \geq 0$$

which means at times $t < \bar{t}$, C_S^2 was not larger than its value at $\bar{t}$ by more than $\delta(\bar{t} - t)$ where δ is as small as we like for t near $\bar{t}$. When $\delta < \varepsilon$ we get a contradiction, proving the Main Theorem.

3. It remains to discuss the existence and smoothness of the optimal curve $\overline{\Lambda}$.

Theorem. On any surface with Sobolev constant $\overline{C}_S^2 < 1$ (the isoperimetric ratio on the sphere) the minimum is attained on a smooth curve $\overline{\Lambda}$.

PROOF. If the length L is sufficiently short, the curve lies in a single coordinate patch where the metric g_{ij} is bounded above and below as closely as we like by the Euclidean metric δ_{ij}, say for any $\varepsilon > 0$

$$(1 - \varepsilon)\delta_{ij} \leq \delta_{ij} \leq (1 + \varepsilon)\delta_{ij}$$

and hence if A_1 is the enclosed area in the coordinate patch

$$L^2 \geq \frac{1 - \varepsilon}{1 + \varepsilon}\,4\pi A_1$$

and hence

$$C_S^2(\Lambda) = L^2 \left(\frac{1}{A_1} + \frac{1}{A_2}\right) / 4\pi \geq \frac{1 - \varepsilon}{1 + \varepsilon}.$$

Thus for every $\eta > 0$ there is a $\lambda > 0$ such that if $C_S^2(\Lambda) \leq 1 - \eta$ then $L \geq \lambda$. If $\overline{C}_S < 1$, we can approximate it as closely as we wish by $C_S(\Lambda) \leq 1 - \eta$ for some Λ. Then this Λ has length $L \geq \lambda$. Since L is not too small, neither area can be too small either, so we can find an $\alpha > 0$ such that $A_1 \geq \alpha$ and $A_2 \geq \alpha$. Both A_1 and A_2 are no more than the total area A, so they are also bounded above. Then L is also bounded above; in fact

$$L^2 \left(\frac{1}{A_1} + \frac{1}{A_2} \right) \Big/ 4\pi = C_S(\Lambda) < 1$$

and

$$\frac{1}{A_1} + \frac{1}{A_2} \geq \frac{2}{A}$$

so

$$L < \sqrt{2\pi A}\,.$$

This restricts the geometry of any curve Λ with $C_S(\Lambda) < 1$.

The fact that the lengths are bounded does not tell us a whole lot. If we parametrize the curve by arc length s then

$$L = \int \sqrt{g_{ij} \frac{dx^i}{ds} \frac{dx^j}{ds}}\, ds$$

so in a coordinate patch where g_{ij} is comparable to the Euclidean metric, the coordinate functions x^1 and x^2 have one derivative with respect to s in L^1, which makes them continuous, but this doesn't help much. Since the dimension is low, standard techniques will apply. But here we take the occasion to outline a new approach.

The idea is to take our approximating curve and improve it by running the curve shrinking flow on the surface for a short time (see Grayson [**G**]). Here we move each point on the curve in the normal direction with velocity equal to the geodesic curvature k at that point. The enclosed areas A_1 and A_2 evolves at a rate

$$\frac{dA_1}{dt} = - \int k\, ds \qquad \frac{dA_2}{dt} = + \int k\, ds$$

while the length evolves at a rate

$$\frac{dL}{dt} = - \int k^2\, ds\,.$$

By the Gauss-Bonnet theorem

$$\int k\, ds + \int K\, da = 2\pi$$

on A_1 (or A_2) since a circle S^1 on S^2 encloses a disk D^2. Now K and A_1 (or A_2) are bounded, so $\int k ds$ is bounded. The Sobolev constant evolves at a rate

$$\frac{d}{dt} \ln C_S(\Lambda) = \frac{2}{L} \frac{dL}{dt} - \frac{1}{A_1} \frac{dA_1}{dt} - \frac{1}{A_2} \frac{dA_2}{dt} + \frac{1}{A} \frac{dA}{dt}$$

and since L is bounded above and A_1 and A_2 are bounded below, we can find a contant E such that $C_S(\Lambda)$ decreases when Λ moves by the curve shrinking flow unless

$$\int k^2 ds \leq E .$$

Lemma. For every surface and every $\eta > 0$ there exists a constant E such that if $C_S(\Lambda^0) \leq 1 - \eta$ then under the curve shrinking flow Λ_0 evolves into a curve Λ_t with

$$C_S(\Lambda^t) \leq C_S(\Lambda^0)$$

and Λ^t has

$$\int k^2 ds \leq E .$$

PROOF. By the previous argument, as long as $\int k^2 ds \geq E$ the Sobolev contant $C_S(\Lambda_t)$ continues to decrease. By Grayson's theorem, if this continues either the curve shrinks to a point (which it cannot since $L \to 0$ contradicts $L \geq \lambda$) or it converges to a geodesic (which it cannot since then $\int k^2 ds \to 0$). This proves the result.

Now to finish the proof of the theorem, take a sequence of curves Λ_j^0 with

$$C_S(\Lambda_j^0) \to \overline{C}_S$$

as $j \to \infty$. Deform each by the curve shrinking flow to a curve Λ_j with

$$C_S(\Lambda_j) \leq C_S(\Lambda_j^0)$$

and

$$\int k_j^2 ds \leq E$$

where k_j is the curvature of Λ_j. Then we still have

$$C_S(\Lambda_j) \to \overline{C}_S$$

so the Λ_j are a much better approximating sequence.

Since each Λ_j has

$$\int k^2 ds \leq E$$

we see that the curves are reasonably well behaved in local coordinate charts. Take a chart where the metric g_{ij} is comparable to the Euclidean metrics δ_{ij}, say $\delta_{ij} \leq g_{ij} \leq 2\delta_{ij}$, and where the connection forms are bounded, say $|\Gamma_{ij}^k| \leq 2$. This happens for example in geodesic coordinates or in harmonic coordinates. If s is the arc length parameter in the g_{ij} metric and T^i and N^i the unit tangent and normal vectors then

$$\frac{dx^i}{ds} = T^i$$

and

$$\frac{d^2x^i}{ds^2} + \Gamma_{jk}^i \frac{dx^j}{dt}\frac{dx^k}{dt} = kN^i \, .$$

Both T and N have bounded Euclidean length, so in the Euclidean metric

$$\left|\frac{d^2x^i}{ds^2}\right| \leq C(|k|+1) \, .$$

Since the length L is bounded, we get a bound

$$\int \left|\frac{d^2x^i}{ds^2}\right|^2 ds \leq C$$

from the bound on k.

Let ℓ be the Euclidean length parameter. Then

$$\left(\frac{d\ell}{ds}\right)^2 = \delta_{ij} \frac{dx^i}{ds}\frac{dx^j}{ds}$$

and $d\ell/ds$ is bounded above and below since the metrics are comparable. Differentiating again

$$2\frac{d\ell}{ds}\frac{d^2\ell}{ds^2} = 2\delta_{ij}\frac{d^2x^i}{ds^2}\frac{dx^j}{ds}$$

and it follows that $d^2\ell/ds^2$ is bounded. Then by the rule for inverse functions $d^2s/d\ell^2$ is bounded also, and we get

$$\int \left|\frac{d^2x^i}{d\ell^2}\right|^2 d\ell \leq C$$

in our coordinate patch. This is just a bound

$$\int k^2 ds \leq C$$

for the curvature and arc length of the Euclidean metric. Since in the place

$$k = \frac{d\theta}{ds}$$

where θ is the angle of the tangent line, we see that

$$\theta_2 - \theta_1 = \int_1^2 k\,ds \le \left(\int_1^2 k^2\,ds\right)^{1/2}\left(\int_1^2 1\,ds\right)^{1/2} \le C\sqrt{s_2 - s_1}$$

so the angle θ is of class $C^{1/2}$ with respect to arc lengths. Then for a short enough segment we can write the curve as the graph of a function $y = f(x)$ with $|dy/dx| \le 1$. Since

$$k = \frac{d^2y/dx^2}{1 + (dy/dx)^{3/2}}$$

we now have

$$\int \left(\frac{d^2y}{dx^2}\right)^2 dx \le C$$

and we see y is of class L_2^1 and hence $C^{1+\frac{1}{2}}$ with respect to x.

It is now easy to see that since the Λ_j are locally uniformly bounded in L_2^1 and $C^{1+\frac{1}{2}}$ we can extract a subsequence which converges in L_p^1 for $p < 2$ or $C^{1+\alpha}$ for $\alpha < 1/2$. The limit $\overline{\Lambda}$ will be a genuine immersed curve. A limit of embedded curves may not be embedded, but at least it cannot cross itself; at worst it will be self-tangent. The limit will still have bounded norm in L_2^1 or $C^{1+\frac{1}{2}}$.

Moreover the limit $\overline{\Lambda}$ attains the minimal ratio $C_S(\overline{\Lambda}) = \overline{C}_S$. From this it easily follows that it cannot be tangent to itself, by the algebraic Lemma A.1.2 in [**Ha**], for it would be more efficient to take just one part of the curve or the other. Therefore $\overline{\Lambda}$ is embedded. Now the usual first variation argument shows $\overline{\Lambda}$ has constant curvature

$$k = \frac{L}{2}\left(\frac{1}{A_1} - \frac{1}{A_2}\right)$$

and hence $\overline{\Lambda}$ is smooth.

It is interesting to observe that if we work a little harder, we do not even need to use Grayson's theorem. We only need a standard derivative estimate.

Lemma. For the curve shrinking flow on a surface, there exists a constant $C < \infty$ such that if we have a solution for $0 \le t \le 1/M$ with $k^2 \le M$

with any $M \geq 1$ then

$$\left(\frac{\partial k}{\partial s}\right)^2 \leq CM/t\,.$$

PROOF. Apply the maximum principle to

$$t\left(\frac{\partial k}{\partial s}\right)^2 + k^2$$

and the result follows.

We can flow the solution until k becomes unbounded, and hence as long as

$$\sup_p \ tk^2(p,t) \leq 1$$

because all the derivatives also stay bounded by the same sort of argument. Suppose the equality above is reached first at some time $\tau \leq 1$ at a point ξ where

$$\tau k^2(\xi, \tau) = 1\,.$$

[If not, L is bounded and k^2 is bounded and we get $\int k^2 ds \leq E$ as before.] Then by the first derivative estimate we get

$$\tau k^2(p,\tau) \geq 1/2$$

for all p whose distance to ξ is less than $c\sqrt{\tau}$ for some $c > 0$. Then

$$\int k^2 ds \geq c/\sqrt{\tau}$$

for some other $c > 0$. This shows $C_S(\Lambda)$ decreases unless τ is not too small, and then $\int k^2 ds \leq E$ for some E anyway.

References

[Ch] B. Chow, "The Ricci flow on the 2-sphere," *J. Diff. Geom.* **33** (1991), 325-334.

[Ha] R. S. Hamilton, "Isoperimetric estimates for the curve shrinking flow," preprint UCSD.

[Y] S.-T. Yau, Seminar book.

ISOPERIMETRIC ESTIMATES FOR THE CURVE SHRINKING FLOW IN THE PLANE

RICHARD S. HAMILTON

In this paper we present two isoperimetric ratios which improve under the curve shrinking flow in the plane. Recall that under this flow each point P on a smooth simple closed embedded curve $\triangle$ in the plane moves in the normal direction N with a velocity equal to the curvature k of the curve; thus

$$\frac{\partial P}{\partial t}$$

defines the flow. (See Gage and Hamilton [**Ga + Ha**].)

For the first ratio, consider any straight line segment Λ which divides the region D enclosed by $\triangle$ into two pieces D_1 and D_2. Let L be the length of Λ and let A_1 and A_2 be the areas of D_1 and D_2, where $A_1 + A_2 = A$ is the area of D. Define the ratio

$$G(\Lambda) = L^2 \left(\frac{1}{A_1} + \frac{1}{A_2} \right)$$

and let

$$\overline{G} = \inf_{\Lambda} G(\Lambda)$$

be the least possible value of $G(\Lambda)$ for all straight line segments Λ.

Main Theorem A. If $\overline{G} \leq \pi$ then $\overline{G}$ increases under the curve shrinking flow.

Since for the circle $\overline{G} = 16/\pi > \pi$, we see this result is not sharp. To obtain the best result, consider any curve Λ dividing D into two regions D_1 and D_2 with areas A_1 and A_2 where $A_1 + A_2 = A$ is the area of D. Form the circle $\overline{D}$ with the same total area $\overline{A} = A$, and form the shortest curve $\overline{\Gamma}$ dividing the domain $\overline{D}$ enclosed by $\overline{D}$ into the same areas $\overline{A}_1 = A_1$ and $\overline{A}_2 = A_2$ as before. Let $\overline{L}$ be the length of $\overline{\Gamma}$. Form the ratio

$$H(\Lambda) = L/\overline{L}$$

and let

$$\overline{H} = \inf_{\Lambda} H(\Lambda)$$

over all curves Λ. Of course the best Γ and the best $\overline{\Gamma}$ will be arcs of circles.

Main Theorem B. The ratio $\overline{H}$ always increases under the curve shrinking flow.

Theorem A is easier both to state and to prove, and should be easier to generalize to other cases. But Theorem B is more precise. Therefore we give both proofs. Either result suffices to give a new proof of Grayson's theorem **[Gr1]** which we outline in part C.

Corollary C. (Grayson's Theorem). Under the curve shrinking flow, any smooth simple closed embedded curve in the plane shrinks to a point, and its dilations converge to a circle.

We would especially like to express our appreciation to Michael Gage for his help and advice in the proof of these results.

A.1. We begin the proof of Theorem A. Consider the set K of straight line segments Λ which start and end on the curve $\triangle$ and which lie entirely in the closed domain D inside $\triangle$. Clearly K is compact, being a closed subset of $\triangle \times \triangle$. The subset K° of line segments which meet $\triangle$ only at their endpoints and there meet transversally is a relatively open subset of K.

A1.1 Lemma. The infimum of $G(\Lambda)$ is attained by some Λ in K°.

PROOF. If Λ meets $\triangle$ at some point in its interior, then Λ divides into two line segments Λ_1 and Λ_2 of lengths L_1 and L_2 chopping off areas A_1, A_2, and A_3, with L_1 separating A_1 from A_2 and L_2 separating A_2 from A_3. We claim it is not as efficient to use $L_1 + L_2$ to cut off A_2 from $A_1 + A_3$ as it is to use L_1 to cut off A_1 from $A_2 + A_3$ or else to use L_2 to cut off $A_1 + A_2$ from A_3. This is now just algebra.

A.1.2 Lemma. For any positive numbers L_1, L_2, A_1, A_2, A_3 we have

$$(L_1 + L_2)^2 \left(\frac{1}{A_2} + \frac{1}{A_1 + A_3} \right) \geq$$

$$\min \left\{ L_1^2 \left(\frac{1}{A_1} + \frac{1}{A_2 + A_3} \right) , \, L_2^2 \left(\frac{1}{A_1 + A_2} + \frac{1}{A_3} \right) \right\} .$$

PROOF. If not, then rearranging we get

$$\frac{A_1(A_2 + A_3)}{A_2(A_1 + A_3)} \leq \frac{L_1^2}{(L_1 + L_2)^2}$$

and

$$\frac{(A_1 + A_2)A_3}{A_2(A_1 + A_3)} \le \frac{L_2^2}{(L_1 + L_2)^2} \, .$$

Adding these inequalities gives

$$1 + \frac{2A_1 A_3}{A_2(A_1 + A_3)} \le 1 - \frac{2L_1 L_2}{(L_1 + L_2)^2} \, .$$

Since all quantities are weakly positive, we get $A_1 = 0$ or $A_3 = 0$ and $L_1 = 0$ or $L_2 = 0$. Thus one region is not there.

Finally it is not hard to see that the optimal Λ meets Γ transversely at its endpoints. For if Λ is tangent to Γ at one or both ends on either side, it is always possible to find a one parameter variation of Λ where the length L decreases at an infinite rate, while the areas change at only a finite rate. Hence such a Λ is not optimal for $\overline{G}$.

A.2. Now we are ready for the interesting part of the argument, the computation of the rate of change of the Sobolev constant C_S. For any line segment Λ dividing D into D_1 and D_2 at any time t we define

$$G(\Lambda) = L^2 \left(\frac{1}{A_1} + \frac{1}{A_2} \right) \, .$$

Then $\overline{G} = G(\overline{\Lambda})$ for some $\overline{\Lambda}$ where the infimum is attained at time $\overline{t}$. Since everything is invariant under rotation, we assume for simplicity that $\overline{\Lambda}$ is vertical and hence lies on the vertical line over some $\overline{x}$. Now we define a one-parameter family of lines $\Lambda = \Lambda(x, t)$ as being the segment of the vertical line at x at time t that divides D in two and is close to $\overline{\Lambda}$. This is well-defined for x near $\overline{x}$ and t near $\overline{t}$ since $\overline{\Lambda}$ meets Δ transversally. Then we define $G(x, t) = G(\Lambda(x, t))$. The idea of the proof rests on the following result.

A.2.1 Main Lemma. If $G(\overline{\Lambda}) \le \pi$ then

$$\frac{d}{dt} G \le 0$$

at $x = \overline{x}, t = \overline{t}$.

We prove this in the next section. Here let us see how the Main Lemma implies the Main Theorem.

To see that $\overline{G}$ increases when $\overline{G} \le \pi$, it suffices to show that $\overline{G} + \varepsilon t$ increases when $\overline{G} + \varepsilon t \le \pi$ for every $\varepsilon > 0$. Suppose this is not true for some ε. Pick a time $\overline{t}$ when $\overline{G} + \varepsilon t$ is less than it has ever been. Realize

$\overline{G}$ as $G(\overline{\Lambda})$ for $\overline{\Lambda}$ as above. Find the family $\Lambda = \Lambda(x,t)$ with $\overline{\Lambda} = \Lambda(\overline{x},\overline{t})$. Since $\frac{d}{dt}G(\Lambda(x,t)) \leq 0$ at $x = \overline{x}$, $t = \overline{t}$, we get

$$\frac{d}{dt}[G(\Lambda(x,t)) + \varepsilon t] \geq \varepsilon > 0$$

and hence for $x = \overline{x}$ and some $t < \overline{t}$ we have

$$G(\Lambda(\overline{x},t)) + \varepsilon t < G(\Lambda(\overline{x},\overline{t})) + \varepsilon\overline{t}.$$

Now $\overline{G}(t) \leq G(\Lambda(\overline{x},t))$ so

$$\overline{G}(t) + \varepsilon t < \overline{G}(\overline{t}) + \varepsilon\overline{t}$$

so $\overline{G} + \varepsilon t$ was not first that small at $\overline{t}$, contrary to our initial choice. This proves that the Main Lemma implies the Main Theorem.

A.3. Now we prove the Main Lemma. Corresponding to $\Lambda = \Lambda(x,t)$ we have lengths $L = L(x,t)$ and areas $A_1 = A_1(x,t)$ and $A_2 = A_2(x,t)$ with $A_1(x,t) + A_2(x,t) = A(t)$. Then as before $G = G(x,t)$ is given by

$$G = L^2\left(\frac{1}{A_1} + \frac{1}{A_2}\right).$$

The curve $\triangle$ near where it meets the top of Λ is given by the graph $\triangle_+$ of a function

$$y_+ = y_+(x,t)$$

and near where it meets the bottom of Λ is given by the graph $\triangle_-$ of another function

$$y_- = y_-(x,t).$$

The relevant geometrical information is the angles ψ_+ and ψ_- of the tangents to the graphs and the curvatures k_+ and k_- of the graphs at $x = \overline{x}$ and $t = \overline{t}$. We suppose $\triangle_+$ has tangent at an angle θ_+ above the horizontal, while $\triangle_-$ has graph at an angle θ_- below the horizontal. Since in the mean curvature flow the curvature is that of the inward normal, our sign convention has y_+ concave when $k_+ > 0$ and y_- convex when $k_- > 0$.

A.3.1 Lemma. We have $\psi_+ = \psi_-$.

PROOF. To see this, consider the one-parameter family of line segments $\widetilde{\Lambda}(u)$ going from the point over $x = \overline{x} - u$ on the bottom curve $\triangle_-$ to the point $x = \overline{x} + u$ on the top curve $\triangle_+$. Its length is $\widetilde{L}$ where

$$\widetilde{L}^2 = 4u^2 + [y_+(\overline{x} + u) - y_-(\overline{x} - u)]^2$$

so that at $u = 0$ we have

$$\frac{\partial \tilde{L}}{\partial u} = \tan \psi_+ - \tan \psi_- .$$

On the other hand, the area $\tilde{A}_1$ to the left of $\tilde{\Lambda}$ is given by

$$\tilde{A}_1 = \int_{\bar{x}}^{\bar{x}+u} y_+(x)dx + \int_{\bar{x}-u}^{\bar{x}} y_-(x)dx - u[y_+(\bar{x}+u) - y_-(\bar{x}-u)]$$

so that at $u = 0$

$$\frac{\partial \tilde{A}_1}{\partial u} = 0 \quad \text{and likewise} \quad \frac{\partial \tilde{A}_2}{\partial u} = 0$$

if $\tilde{A}_2$ is the area to the right. If

$$\tilde{G} = \tilde{L}^2 \left(\frac{1}{\tilde{A}_1} + \frac{1}{\tilde{A}_2} \right)$$

then $\tilde{G}$ by assumption has a minimum at $u = 0$. Hence at $u = 0$

$$\frac{\partial}{\partial u} \tilde{G} = 0$$

which makes $\partial \tilde{L} / \partial u = 0$ which makes $\psi_+ = \psi_-$. This simplifies our computation.

Returning now to the family Λ of vertical lines, we see that the functions y_+ and y_- have derivatives

$$\frac{\partial y_+}{\partial x} = \tan \psi \quad \text{and} \quad \frac{\partial y_-}{\partial x} = -\tan \psi,$$

$$\frac{\partial^2 y_+}{\partial x^2} = -k_+ \sec^3 \psi \quad \text{and} \quad \frac{\partial^2 y_-}{\partial x^2} = k_- \sec^3 \psi.$$

Moreover both y_+ and y_- satisfy the curve shrinking equation for a graph

$$\frac{\partial y}{\partial t} = \frac{\partial^2 y / \partial x^2}{1 + (\partial y / \partial x)^2}$$

so that

$$\frac{\partial y_+}{\partial t} = -k_+ \sec \psi \qquad \frac{\partial y_-}{\partial t} = k_- \sec \psi .$$

The length of the vertical segment Λ is $L = y_+ - y_-$. If we put $k = k_+ + k_-$ for simplicity, we get

$$\frac{\partial L}{\partial x} = 2 \tan \psi$$

$$\frac{\partial^2 L}{\partial x^2} = -k \sec^3 \psi$$

$$\frac{\partial L}{\partial t} = -k \sec \psi .$$

The area A_1 and A_2 to the left and right of Λ clearly satisfy

$$\frac{\partial A_1}{\partial x} = L \quad \text{and} \quad \frac{\partial A_2}{\partial x} = -L$$

and hence

$$\frac{\partial^2 A_1}{\partial x^2} = \frac{\partial L}{\partial x} = 2 \tan \psi \quad \text{and} \quad \frac{\partial^2 A_2}{\partial x^2} = -\frac{\partial L}{\partial x} = -2 \tan \psi .$$

To see how fast A_1 and A_2 change with t, consider the simpler problem of the area A under the graph of $y = y(x,t)$ between $x = a$ and $x = b$, given by

$$A = \int_a^b y \, dx$$

where y evolves by the curve shrinking equation for a graph, which can also be written

$$\frac{\partial y}{\partial t} = \frac{\partial}{\partial x} \arctan \frac{\partial y}{\partial x} .$$

Then if $\theta = \arctan (\partial y/\partial x)$ is the angle of the tangent line above the horizontal, we get

$$\frac{\partial A}{\partial t} = \int_a^b \frac{\partial y}{\partial t} dx = \theta(b) - \theta(a) .$$

Now consider the easy case where A_1 is enclosed by one graph on top and one on bottom, both coming from a point where the tangent is vertical. The change of angle on the top curve is from $\pi/2$ to ψ, and on the bottom from $-\pi/2$ to $-\psi$. The change from the endpoint moving (where the tangent is vertical) is zero since the height between the two curves there is zero. Thus

$$\frac{\partial A_1}{\partial t} = 2\psi - \pi \quad \text{and} \quad \frac{\partial A_2}{\partial t} = -2\psi - \pi$$

by a symmetrical argument. If the area has a more complicated shape, the formula comes out the same since the result only depends on the angles at the end and the winding number.

The argument now follows from the fact that $G(x,t)$ has a minimum over all x at $x = \bar{x}$ when $t = \bar{t}$, so that at $x = \bar{x}$ and $t = \bar{t}$ we have

$$\frac{\partial}{\partial x} G = 0 \quad \text{and} \quad \frac{\partial^2}{\partial x^2} G \geq 0 .$$

We have now all the formulas we need, and the rest is just a computation. Everything can be expressed in terms k and θ. It is easiest to work with

$$\ell n\, G = 2\ell n\, L - \ell n\, A_1 - \ell A_2 + \ell n\, A$$

where $A = A_1 + A_2$ is independent of x and satisfies $\partial A/\partial t = -2\pi$. Then

$$\frac{\partial}{\partial x}\ell n\, G = \frac{2}{L}\frac{\partial L}{\partial x} - \frac{1}{A_1}\frac{\partial A_1}{\partial x} - \frac{1}{A_2}\frac{\partial A_2}{\partial x}$$

gives

$$\frac{\partial}{\partial x}\ell n\, G = \frac{4}{L}\tan\psi - L\left(\frac{1}{A_1} - \frac{1}{A_2}\right).$$

Since $\partial G/\partial x = 0$ we get

$$\tan\psi = \frac{L^2}{4}\left(\frac{1}{A_1} - \frac{1}{A_2}\right)$$

and this determines the angle ψ in terms of L, A_1 and A_2.

Next we compute

$$\frac{\partial^2}{\partial x^2}\ell n\, G = \frac{2}{L}\frac{\partial^2 L}{\partial x^2} - \frac{2}{L^2}\left(\frac{\partial L}{\partial x}\right)^2$$
$$-\frac{1}{A_1}\frac{\partial^2 A_1}{\partial x^2} + \frac{1}{A_1^2}\left(\frac{\partial A_1}{\partial x}\right)^2 - \frac{1}{A_2}\frac{\partial^2 A_2}{\partial x^2} + \frac{1}{A_2^2}\left(\frac{\partial A_2}{\partial x}\right)^2$$

which makes

$$\frac{\partial^2}{\partial x^2}\ell n\, G = -\frac{2k}{L}\sec^3\psi - \frac{8}{L^2}\tan^2\psi$$
$$-2\left(\frac{1}{A_1} - \frac{1}{A_2}\right)\tan\psi + L^2\left(\frac{1}{A_1^2} + \frac{1}{A_2^2}\right).$$

Using the formula for $\tan\psi$, the middle terms can be simplified and we get

$$\frac{\partial^2}{\partial x^2}\ell n\, G = -\frac{2k}{L}\sec^3\psi + \frac{2L^2}{A_1 A_2}.$$

Finally we compute

$$\frac{\partial}{\partial t}\ell n\, G = \frac{2}{L}\frac{\partial L}{\partial t} - \frac{1}{A_1}\frac{\partial A_1}{\partial t} - \frac{1}{A_2}\frac{\partial A_2}{\partial t} + \frac{1}{A}\frac{\partial A}{\partial t}$$

which gives

$$\frac{\partial}{\partial t}\ell n\, G = -\frac{2k}{L}\sec\psi - 2\psi\left(\frac{1}{A_1} - \frac{1}{A_2}\right)$$

$$+\pi \left[\frac{A_1^2 + A_2^2}{A_1 A_2 (A_1 + A_2)} \right] .$$

We are now tempted to write this as a "heat equation" for G;

$$\frac{\partial}{\partial t} \ell n\, G = \cos^2 \theta \frac{\partial^2}{\partial x^2} \ell n\, G - \frac{2L^2}{A_1 A_2} \cos^2 \psi$$

$$-2\psi \left[\frac{1}{A_1} - \frac{1}{A_2} \right] + \pi \left[\frac{A_1^2 + A_2^2}{A_1 A_2 (A_1 + A_2)} \right] .$$

Now we use $\cos^2 \psi \leq 1$ and for ψ in the range $0 \leq \psi < \pi/2$ we have $\psi \leq \tan \psi$ and

$$\tan \psi = \frac{L^2}{4} \left(\frac{1}{A_1} - \frac{1}{A_2} \right)$$

so $A_1 \leq A_2$. [If $-\pi/2 < \psi \leq 0$ we get the opposite inequalities with the same result, as should be clear by the symmetry of the situation switching A_1 and A_2.] This gives an approximation for the two middle terms on the right hand side, which can then be combined to give

$$\frac{2L^2}{A_1 A_2} \cos^2 \psi + 2\psi \left[\frac{1}{A_1} - \frac{1}{A_2} \right] \leq \frac{L^2}{2} \left(\frac{1}{A_1} + \frac{1}{A_2} \right)^2 .$$

The last term can be approximated using

$$A_1^2 + A_2^2 \geq \frac{1}{2} (A_1 + A_2)^2$$

to give

$$\frac{A_1^2 + A_2^2}{A_1 A_2 (A_1 + A_2)} \geq \frac{1}{2} \left(\frac{1}{A_1} + \frac{1}{A_2} \right) .$$

Then using

$$\overline{G} = L^2 \left(\frac{1}{A_1} + \frac{1}{A_2} \right)$$

we get finally that at $x = \overline{x}$, $t = \overline{t}$

$$\frac{\partial}{\partial t} \ell n\, G \geq \cos^2 \psi \frac{\partial^2}{\partial x^2} \ell n\, G + \frac{1}{2} \left(\frac{1}{A_1} + \frac{1}{A_2} \right) (\pi - \overline{G}) .$$

At $x = \overline{x}$ and $t = \overline{t}$ the function G has

$$\frac{\partial^2}{\partial x^2} \ell n\, G \geq 0$$

and hence

$$\frac{\partial}{\partial t} \ell n\, G \geq 0$$

when $\overline{G} \leq \pi$. Therefore if $\overline{G} \leq \pi$ it increases, and we have proven the Main Theorem.

B.1. We turn to the proof of Theorem B. Since $\overline{H} \leq 1$ always, while $H(\Gamma) \to 1$ if $L(\Gamma) \to 0$, we can easily check that $\overline{H}$ cannot drop suddenly, and it suffices to show H increases when $\overline{H} < 1$. Then we do not need to consider curves Γ which are too short. For any given division of area $A = A_1 + A_2$, there will be a shortest curve (or collection of curves) effecting this division, and the curve (or each component curve) will have constant curvature and be perpendicular to the boundary. (If there are several components, they will all have the same curvature. This prevents any component from being too short, and hence also limits the number of components.) It is among such curves that we see one Γ of minimum length $L(\Gamma)$, and since this set is compact we can surely find one.

Now we claim the best Γ will have only one component. Suppose, for example, that Γ has two components Γ' and Γ'' of lengths L' and L'', dividing A into regions of area $A' + A'' + A''' = A$ as shown.

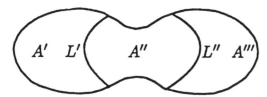

Taking $L = L' + L''$ and $A_1 = A' + A'''$ and $A_2 = A''$ gives a ratio

$$(L' + L'') / \overline{L}(A' + A''', A'')$$

and we claim one of the ratios

$$L' / \overline{L}(A', A'' + A''') \quad \text{or} \quad L'' / \overline{L}(A' + A'', A''')$$

will be smaller, corresponding to using only one part of the curve for the division. Now suppose this is not true; then each is larger. This gives

$$\frac{L'}{L' + L''} \geq \frac{\overline{L}(A', A'' + A''')}{\overline{L}(A'', A' + A''')}$$

and

$$\frac{L''}{L' + L''} \geq \frac{\overline{L}(A''', A' + A'')}{\overline{L}(A'', A' + A''')} \, .$$

Adding this and simplifying gives

$$\overline{L}(A'', A' + A''') \geq \overline{L}(A', A'' + A''') + \overline{L}(A''', A' + A'') \, .$$

This would imply that the shortest curve to cut off an area $A' + A'''$ from an area A'' in a circle of area A is not a single arc, which is absurd. This proves the following result.

B.1.1 Lemma. If $\overline{H} < 1$, the minimum $H(\Gamma)$ is attained by a single smooth curve Γ of constant curvature perpendicular to $\triangle$.

B.2. Next we need to compute the first and second variation of the length $L(\Gamma)$ and the areas $A_1(\Gamma)$ and $A_2(\Gamma)$ for any one-parameter family of curves Γ_μ with parameter μ. We assume Γ_0 is our arc of constant curvature at $\mu = 0$. It is easiest therefore to work in polar coordinates, where Γ_μ is given by the graph of $r = r(\theta, \mu)$ between $\theta = \theta_-(\mu)$ and $\theta = \theta_+(\mu)$. At $\mu = 0$ the function r is constant so $\partial r/\partial \theta = 0$ and $\partial^2 r/\partial \theta^2 = 0$, while the curvature k of Γ is $k = 1/r$. Since Γ is perpendicular to $\triangle$ at $\mu = 0$, $\partial \theta_+/\partial \mu = 0$ and $\partial \theta_-/\partial \mu = 0$ at $\mu = 0$. The boundary $\triangle$ has curvatures K_+ at $\theta = \theta_+$ and K_- at $\theta = \theta_-$ at $\mu = 0$ which can be computed as the curvatures of the graph of

$$\theta = \theta_+(\mu) \quad \text{and} \quad r = r(\theta_+(\mu), \mu)$$

for K_+ and the same for K_-, except with a sign change because we keep the sign convention for K in the mean curvature flow. If a parametrized curve $P = P(\mu)$ has velocity vector $V = dP/d\mu$ and acceleration vector $Z = dV/d\mu$, then its curvature is

$$K = \frac{\det (V\ Z)}{|V|^3}.$$

Applying this and simplifying at $\mu = 0$ gives

$$\left(r \frac{d^2\theta}{d\mu^2} \right)_+ = -K_+ \left(\frac{\partial r}{\partial \mu} \right)_+^2$$

and

$$\left(r \frac{d^2\theta}{d\mu^2} \right)_- = +K_- \left(\frac{\partial r}{\partial \mu} \right)_-^2.$$

where subscripts $+$ or $-$ denote the value of the function at $\theta = \theta_+$ or $\theta = \theta_-$.

The result is most easily expressed in terms of the velocity v and acceleration z of Γ, where

$$v = \frac{\partial r}{\partial \mu} \quad \text{and} \quad z = \frac{\partial^2 r}{\partial \mu^2}$$

at $\mu = 0$. The length is given by

$$L = \int_{\theta_-}^{\theta_1} \sqrt{r^2 + \left(\frac{dr}{d\theta}\right)^2}\, d\theta$$

A straightforward calculation at $\mu = 0$ gives the following result.

B.2.1 Lemma.

$$\frac{dL}{d\mu} = \int_{\theta_-}^{\theta_+} v\, d\theta$$

$$\frac{d^2 L}{d\mu^2} = \int_{\theta_-}^{\theta_+} z\, d\theta + k \int_{\theta_-}^{\theta_+} \left(\frac{dv}{d\theta}\right)^2 d\theta - (K_+ v_+^2 + K_- v_-^2).$$

Similarly we can calculate the rate of change of the areas A_1 and A_2. If $A_1(\mu)$ denotes the area on the origin side of Γ_μ then

$$A_1(U) - A_1(0) = \int_{\mu=0}^{U} \int_{\theta=\theta_-(\mu)}^{\theta_+(\mu)} r \frac{\partial r}{\partial \mu}\, d\theta d\mu$$

since in polar coordinates the area element is

$$da = dx dy = r dr d\theta,$$

and in μ and θ coordinates

$$r = r(\mu, \theta)\quad \theta = \theta$$

the Jacobian determinant is

$$J = \det \begin{pmatrix} \partial r/\partial \mu & \partial r/\partial \theta \\ \partial \theta/\partial \mu & \partial \theta/\partial \theta \end{pmatrix} = \det \begin{pmatrix} \partial r/\partial \mu & \partial r/\partial \theta \\ 0 & 1 \end{pmatrix} = \frac{\partial r}{\partial \mu}$$

so $da = r\frac{\partial r}{\partial \mu} d\mu d\theta$. Then it is straightforward to compute the first and second derivatives of A_1 with respect to μ at $\mu = 0$. Of course $A_1 + A_2 = A$ is independent of μ, so the derivatives of A_2 are the negatives.

B.2.2 Lemma.

$$\frac{dA_1}{d\mu} = \frac{1}{k} \int_{\theta_-}^{\theta_+} v\, d\theta$$

and

$$\frac{d^2 A_1}{d\mu^2} = \frac{1}{k} \int_{\theta_-}^{\theta_+} z\, d\theta + \int_{\theta_-}^{\theta_+} v^2\, d\theta.$$

B.3. Now we can write down the condition that $H(\Gamma)$ attains its minimum at Γ_0. As usual this says that the first variation is zero, and the second variation is positive. Recall that

$$H(\Gamma) = L/\overline{L}$$

where $L = L(\Gamma)$ and $\overline{L} = L(\overline{\Gamma})$ where $\overline{\Gamma}$ is an arc in the round ball having the same areas $A_1 = \overline{A}_1$ and $A_2 = \overline{A}_2$ on each side as Γ does. When $H(\Gamma)$ attains its minimum at Γ_0, then for any one-parameter family of paths Γ_μ with velocity v and acceleration z we can find a one-parameter family of paths $\overline{\Gamma}_\mu$ with velocity $\overline{v}$ and acceleration $\overline{z}$ such that $A_1(\mu) = \overline{A}_1(\mu)$ and

$$L(\mu) \,/\, \overline{L}(\mu) \geq L(0) \,/\, \overline{L}(0)\,.$$

The last is better written in terms of logarithms; thus

$$\log L(\mu) - \log \overline{L}(\mu) \geq \log L(0) - \log \overline{L}(0)\,.$$

Hence at $\mu = 0$ we must have

$$\frac{dA_1}{d\mu} = \frac{d\overline{A}_1}{d\mu} \quad \text{and} \quad \frac{d^2 A_1}{d\mu^2} = \frac{d^2 \overline{A}_1}{d\mu^2}$$

and

$$\frac{1}{L}\frac{dL}{d\mu} = \frac{1}{\overline{L}}\frac{d\overline{L}}{d\mu} \quad \text{and} \quad \frac{1}{L}\frac{d^2 L}{d\mu^2} \geq \frac{1}{\overline{L}}\frac{d^2 \overline{L}}{d\mu^2}\,.$$

Applying the formulas for the first and second variation from the last section gives the following criterion.

B.3.1 Lemma. At the minimal Γ_0, for every v and z we can find $\overline{v}$ and $\overline{z}$ such that

(a)
$$\frac{1}{k}\int_{\theta_-}^{\theta_+} v\,d\theta = \frac{1}{\overline{k}}\int_{\overline{\theta}_-}^{\overline{\theta}_+} \overline{v}\,d\overline{\theta}$$

(b)
$$\frac{1}{k}\int_{\theta_-}^{\theta_+} z\,d\theta + \int_{\theta_-}^{\theta_+} v^2\,d\theta = \frac{1}{\overline{k}}\int_{\overline{\theta}_-}^{\overline{\theta}_+} \overline{z}\,d\overline{\theta} + \int_{\overline{\theta}_-}^{\overline{\theta}_+} \overline{v}^2\,d\overline{\theta}$$

(c)
$$\frac{1}{L}\int_{\theta_-}^{\theta_+} v\,d\theta = \frac{1}{\overline{L}}\int_{\overline{\theta}_-}^{\overline{\theta}_+} \overline{v}\,d\overline{\theta}$$

(d)
$$\frac{1}{L}\int_{\theta_-}^{\theta_+} z\,d\theta + \frac{k}{L}\int_{\theta_-}^{\theta_+}\left(\frac{dv}{d\theta}\right)^2 d\theta - \frac{K_+ v_+^2 + K_- v_-^2}{L}$$
$$\geq \frac{1}{\overline{L}}\int_{\overline{\theta}_-}^{\overline{\theta}_+} \overline{z}\,d\overline{\theta} + \frac{\overline{k}}{\overline{L}}\int_{\overline{\theta}_-}^{\overline{\theta}_+}\left(\frac{d\overline{v}}{d\overline{\theta}}\right)^2 d\overline{\theta} - \frac{\overline{K}_+ \overline{v}_+^2 + \overline{K}_- \overline{v}_-^2}{\overline{L}}\,.$$

B.3.2 Corollary. $k/L = \bar{k}/\bar{L}$.

PROOF. Compare (a) and (c).

B.3.3 Corollary. For every v we can find $\bar{v}$ such that

$$\int_{\theta_-}^{\theta_+} \left[\left(\frac{dv}{d\theta} \right)^2 - v^2 \right] d\theta - \frac{K_+ v_+^2 + K_- v_-^2}{k}$$

$$\geq \int_{\bar{\theta}_-}^{\bar{\theta}_+} \left[\left(\frac{d\bar{v}}{d\bar{\theta}} \right)^2 \bar{v}^2 \right] d\bar{\theta} - \frac{\bar{K}_+ \bar{v}_+^2 + \bar{K}_- \bar{v}_-^2}{\bar{k}} .$$

PROOF. Multiply (d) by $L/k = \bar{L}/\bar{k}$ and substract (b).

Now it is convenient to divide the variation v into odd and even parts, primarily because the quantity we want to estimate is $K_+ + K_-$. By translating θ we can arrange that $\theta_- = -\theta_+$; to simplify the notation we let the common value be denoted by λ. Likewise let

$$K = \frac{K_+ + K_-}{2} \quad \text{and} \quad K_\# = \frac{K_+ - K_-}{2} .$$

Write $v = x + y$ where x is an even function of θ on $-\lambda \leq \theta \leq \lambda$ and y is an odd function; and write

$$X = x_+ = x_- \quad \text{and} \quad Y = y_+ = -y_-$$

for the boundary values of x and y.
Now $kL = \theta_+ - \theta_- = 2\lambda$. Then since $L \leq \bar{L}$ and $k/L = \bar{k}/\bar{L}$ we get this result.

B.3.4 Lemma. We have $\lambda/k^2 = \bar{\lambda}/\bar{k}^2$ and $\lambda \leq \bar{\lambda}$.

We use this to reinterpret (a) in odd and even parts. The odd parts integrate to zero. We are left with the following.

B.3.5 Lemma.

$$\frac{1}{\sqrt{\lambda}} \int_0^\lambda x d\theta = \frac{1}{\sqrt{\bar{\lambda}}} \int_0^{\bar{\lambda}} \bar{x} d\bar{\theta} .$$

Likewise if we reinterpret Corollary 4.3 we get the following.

B.3.6 Corollary.

$$\int_0^\lambda \left[\left(\frac{dx}{d\theta} \right)^2 - x^2 \right] d\theta + \int_0^\lambda \left[\left(\frac{dy}{d\theta} \right)^2 - y^2 \right] d\theta - \frac{K}{k}(X^2 + Y^2) - \frac{K_\#}{k}(2XY)$$

$$\geq \int_0^{\overline{\lambda}} \left[\left(\frac{d\overline{x}}{d\overline{\theta}} \right)^2 - \overline{x}^2 \right] d\overline{\theta} + \int_0^{\overline{\lambda}} \left[\left(\frac{d\overline{y}}{d\overline{\theta}} \right)^2 - \overline{y}^2 \right] d\overline{\theta}$$

$$- \frac{\overline{K}}{\overline{k}} (\overline{X}^2 + \overline{Y}^2) - \frac{\overline{K}_\#}{\overline{k}} (2\overline{XY}).$$

Now the barred half is easier to understand because $\overline{K}_\# = 0$ for the circle. Our quantifiers tell us that for every x and y we can find $\overline{x}$ and $\overline{y}$ such that Lemma 4.5 and Corollary 4.6 hold. Since there is no restriction on $\overline{y}$, we had better have

$$\int_0^{\overline{\lambda}} \left[\left(\frac{d\overline{y}}{d\overline{\theta}} \right)^2 - \overline{y}^2 \right] d\overline{\theta} \geq \frac{\overline{K}}{\overline{k}} \overline{Y}^2$$

for every choice of $\overline{y}$. In fact we get equality. For a fixed value of $\overline{Y}$ the best choice of an odd $\overline{y}$ to minimize the left hand side is to take (up to a constant multiple) $y = \sin \theta$. This gives

$$\int_0^{\overline{\lambda}} (\cos^2 \overline{\theta} - \sin^2 \overline{\theta}) d\overline{\theta} \geq \frac{\overline{K}}{\overline{k}} \sin^2 \overline{\lambda}.$$

Doing the integral gives

$$\frac{\overline{K}}{\overline{k}} \leq \frac{1}{\tan \overline{\lambda}}.$$

But in fact we have equality.

B.3.7 Lemma. For an arc perpendicular to a circle

$$\frac{\overline{K}}{\overline{k}} = \frac{1}{\tan \overline{\lambda}}.$$

PROOF. The circle has radius $1/\overline{K}$, the arc has radius $1/\overline{k}$, the arc goes through an angle $\pm\overline{\lambda}$, and $\tan \overline{\lambda}$ is the ratio of the radii, as we see from the right triangle in the picture below.

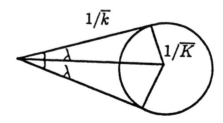

Now choose any x, and choose $y = 0$. Then we can find $\bar{x}$ and $\bar{y}$ to satisfy all our inequalities, and $\bar{y}$ doesn't help. So we may as well take $\bar{y} = 0$ anyway. This gives us the following.

B.3.8 Corollary. For every x we can find $\bar{x}$ with

$$\frac{1}{\sqrt{\lambda}} \int_0^\lambda x \, d\theta = \frac{1}{\sqrt{\bar{\lambda}}} \int_0^{\bar{\lambda}} \bar{x} \, d\bar{\theta}$$

and

$$\int_0^\lambda \left[\left(\frac{dx}{d\theta} \right)^2 - x^2 \right] d\theta - \frac{K}{k} X^2 \geq \int_0^{\bar{\lambda}} \left[\left(\frac{d\bar{x}}{d\bar{\theta}} \right)^2 - \bar{x}^2 \right] d\bar{\theta} - \frac{\overline{K}}{k} \bar{x}^2 .$$

B.4 There is an obvious homogeneity in Corollary 4.8 which allows us to scale x and $\bar{x}$ by a common factor. We can then interpret the first condition as

$$\int_0^\lambda x \, d\theta = \sqrt{\lambda} \quad \text{and} \quad \int_0^{\bar{\lambda}} \bar{x} \, d\bar{\theta} = \sqrt{\bar{\lambda}} .$$

Then subject to this, the best strategy for us is to pick x minimizing

$$\int_0^\lambda \left[\left(\frac{dx}{d\theta} \right)^2 - x^2 \right] d\theta - \frac{K}{k} \lambda^2$$

and then the worst $\bar{x}$ would be the one minimizing

$$\int_0^{\bar{\lambda}} \left[\left(\frac{d\bar{x}}{d\bar{\theta}} \right)^2 - \bar{x}^2 \right] d\bar{\theta} - \frac{\overline{K}}{k} \overline{X}^2 .$$

If we minimize the quantity

$$Z = \int_0^\lambda \left[\left(\frac{dx}{d\theta} \right)^2 - x^2 \right] d\theta - \frac{K}{k} X^2$$

with

$$\int_0^\lambda x \, d\theta = \sqrt{\lambda}$$

the usual calculus of variations argument gives us this result.

B.4.1 Lemma. We have

$$\frac{d^2 x}{d\theta^2} + x = a$$

for some constant a and

$$\left(\frac{dx}{d\theta}\right)_{\theta=\lambda} = \frac{K}{k}X.$$

PROOF. For any variation $\tilde{x}$ in x with

$$\int_0^\lambda \tilde{x}\,d\theta = 0$$

we must have

$$\int_0^\lambda \left[\frac{dx}{d\theta}\frac{d\tilde{x}}{d\theta} - X\tilde{X}\right] d\theta - \frac{K}{k}X\tilde{X} = 0$$

and if we integrate by parts we get

$$\left[\left(\frac{dx}{d\theta}\right)_{\theta=\lambda} - \frac{K}{k}X\right]\tilde{X} - \int_0^\lambda \left(\frac{d^2x}{d\theta^2} + x\right)\tilde{x}\,d\theta = 0$$

for all $\tilde{x}$, which proves the result.

B.4.2 Corollary.

$$\frac{K}{k}X + \sqrt{\lambda} = a\lambda.$$

PROOF. Integrate $d^2x/d\theta^2 + x = a$ over $0 \leq \theta \leq \lambda$.

B.4.3 Corollary. At the minimizing choice of x

$$Z = -a\sqrt{\lambda}.$$

PROOF. Apply the same idea to

$$\int_0^\lambda x\left(\frac{d^2x}{d\theta^2} + x\right) d\theta.$$

The same principles apply to the optimal choice of $\bar{x}$. This gives the following result.

B.4.4 Corollary. We have

$$\frac{K}{k}\frac{X}{\sqrt{\lambda}} \leq \frac{\overline{K}}{\overline{k}}\frac{\overline{X}}{\sqrt{\overline{\lambda}}}.$$

PROOF. We know $Z \geq \overline{Z}$, so $a\sqrt{\lambda} \leq \bar{a}\sqrt{\overline{\lambda}}$. But from the previous Corollary

$$a\sqrt{\lambda} = \frac{K}{k}\cdot\frac{X}{\sqrt{\lambda}} + 1$$

and the result follows.

Now x solves

$$\frac{d^2x}{d\theta^2} + x = a$$

and x also satisfies

$$\int_0^\lambda x\, d\theta = \sqrt{\lambda}$$

and

$$\left(\frac{dx}{d\theta}\right)_{\theta=\lambda} = \frac{K}{k}X\,.$$

The latter conditions determine x uniquely. Since x is even

$$x = a - b\cos\theta$$

for some a and b which satisfy

$$a\lambda - b\sin\lambda = \sqrt{\lambda}$$

and

$$b\sin\lambda = \frac{K}{k}(a - b\cos\lambda)\,.$$

We can solve these equations for a and b. To simplify the answer we write

$$p = K/k\,.$$

By Lemma 3.7, we have

$$\bar{p} = 1/\tan\bar{\lambda}$$

Then

$$a = \frac{\sqrt{\lambda}(\sin\lambda + b\cos\lambda)}{\lambda\sin\lambda - p\sin\lambda + b\lambda\cos\lambda}$$

$$b = \frac{p\sqrt{\lambda}}{\lambda\sin\lambda - p\sin\lambda + p\lambda\cos\lambda}$$

and

$$X = \frac{\sqrt{\lambda}}{X - p + b\lambda\tan\lambda}\,.$$

Then the previous result simplifies to

B.4.5 Corollary.

$$\frac{p}{\lambda - p + p\lambda\tan\lambda} \leq \frac{\bar{p}}{\bar{\lambda} - \bar{p} + \bar{p}\bar{\lambda}\tan\bar{\lambda}}\,.$$

Next we invert and add 1 to get

B.4.6 Corollary.

$$\frac{\lambda}{p} + \frac{\lambda}{\tan \lambda} \le \frac{\overline{\lambda}}{\overline{p}} + \frac{\overline{\lambda}}{\tan \overline{\lambda}}.$$

B.5 Now we digress to prove a useful estimate.

B.5.1 Lemma. If $0 < \lambda \le \overline{\lambda} < \pi/2$ then

$$\frac{\lambda}{\tan \lambda} - \frac{\overline{\lambda}}{\tan \overline{\lambda}} \le (\overline{\lambda} - \lambda) \tan \overline{\lambda}.$$

PROOF. Let

$$y = \frac{x}{\tan x}.$$

Then

$$\frac{dy}{dx} = \frac{1}{\tan x} - \frac{x \sec^2 x}{\tan^2 x}$$

and hence for $0 < x < \pi/2$

$$\frac{dy}{dx} + \tan x = \frac{\sec^2 x}{\tan^2 x}[\tan x - x] \ge 0.$$

By the mean value theorem

$$y(\overline{\lambda}) - y(\lambda) = y'(\xi)(\overline{\lambda} - \lambda)$$

for some ξ in $\lambda \le \xi \le \overline{\lambda}$. Since

$$-y'(\xi) \le \tan \xi \le \tan \overline{\lambda}$$

we get

$$\frac{\lambda}{\tan \lambda} - \frac{\overline{\lambda}}{\tan \overline{\lambda}} \le (\overline{\lambda} - \lambda) \tan \overline{\lambda}$$

as desired.

B.5.2 Lemma. If $0 < \lambda \le \overline{\lambda} < \pi/2$, $p > 0$, $\overline{p} > 0$ and

$$\frac{\lambda}{p} + \frac{\lambda}{\tan \lambda} \ge \frac{\overline{\lambda}}{\overline{p}} + \frac{\overline{\lambda}}{\tan \overline{\lambda}}$$

and $\overline{p} \le 1/\tan \overline{\lambda}$ then $p \le 1/\tan \overline{\lambda}$.

PROOF. The previous result tells us

$$\lambda \left(\frac{1}{p} - \tan \overline{\lambda} \right) \geq \overline{\lambda} \left(\frac{1}{\overline{p}} - \tan \overline{\lambda} \right)$$

so if $1/\overline{p} \geq \tan \overline{\lambda}$ then $1/p \geq \tan \overline{\lambda}$ also.

Applying this to Corollary 4.6 gives

B.5.3 Corollary. We have $p \leq \overline{p}$, so $K/k \leq \overline{K}/\overline{k}$.

B.6 Now at last we examine the curve shrinking flow. We evolve the boundary curve $\triangle$ by

$$\frac{\partial P}{\partial t} = KN$$

where K is the curvature of $\triangle$, P the position and N the unit normal. If we fix the arc Γ, the areas A_1 and A_2 on either side may change in a way we could calculate. But the easy way to compute this is to also evolve Γ by the curve shrinking flow

$$\frac{\partial P}{\partial t} = kN$$

where k is the curvature on Γ. We also impose the boundary condition that Γ remains perpendicular to $\triangle$. This boundary value problem is of Neumann type and is easily seen to have a unique smooth solution; but this is not important, as we only need the first order variation for our argument. The enclosed areas A_1 and A_2 now change at a constant rate

$$\frac{dA_1}{dt} = -\pi \quad \text{and} \quad \frac{dA_2}{dt} = -\pi$$

since this is the integral of the curvature over the boundary. (We lose two angles of $\pi/2$ at the corners.) The same is true for $\overline{A}_1$ and $\overline{A}_2$ if we evolve both $\overline{\triangle}$ and $\overline{\Gamma}$ by the curve shrinking flow in the same way. Then $A_1 = \overline{A}_1$ and $A_2 = \overline{A}_2$ are identities preserved under the flow.

The lengths L of Γ and $\overline{L}$ of $\overline{\Gamma}$ evolve by

$$\frac{dL}{dt} = -(K_+ + K_-) - \int_{\theta_-}^{\theta_+} k^2 ds$$

and

$$\frac{d\overline{L}}{dt} = -(\overline{K}_+ + \overline{K}_-) - \int_{\overline{\theta}_-}^{\overline{\theta}_+} \overline{k}^2 d\overline{s} .$$

Since $H = L/\overline{L}$ we have

$$\frac{d}{dt} \log H = \frac{1}{L}\frac{dL}{dt} - \frac{1}{\overline{L}}\frac{d\overline{L}}{dt}.$$

Hence to show H increases we need to show

$$\frac{1}{L}\frac{dL}{dt} \geq \frac{1}{\overline{L}}\frac{d\overline{L}}{dt}$$

in the same way. Then the identities $A_1 = \overline{A}_1$ and $A_2 = \overline{A}_2$ are preserved by the double curve shrinking flow.

The length L of Γ contracts by $2K = K_+ + K_-$ due to the curve shortening motion of its boundary, and also contracts by

$$\int_{\theta_-}^{\theta_+} k^2 ds = 2k\lambda$$

due to its own curve shortening motion. Thus

$$\frac{dL}{dt} = -2(K + k\lambda)$$

and likewise

$$\frac{d\overline{L}}{dt} = -2(\overline{K} + \overline{k\lambda}).$$

Thus to show H is increasing, we need to verify that at the minimizing Γ

$$\frac{K + k\lambda}{L} \leq \frac{\overline{K} + \overline{k\lambda}}{\overline{L}}.$$

But since $k/L = \overline{k}/\overline{L}$, this is equivalent to

$$\frac{K}{k} + \lambda \leq \frac{\overline{K}}{\overline{k}} + \overline{\lambda}.$$

Now since $\lambda \leq \overline{\lambda}$ and $K/k \leq \overline{K}/\overline{k}$, we are done.

It is easy to justify the argument above rigorously. To show $\log H$ is increasing it suffices to show $\varepsilon t + \log H$ is increasing for every $\varepsilon > 0$. If this fails to be true, take some point t where it first reaches a new low at some minimizing arc Γ. The previous argument allows us to construct a one-parameter family of curves Γ backwards in time where along the path

$$\frac{d}{dt} \log H \geq 0$$

and hence $\varepsilon t + \log H$ was strictly smaller at an earlier time. (Note we do not need to solve a backwards heat equation for Γ, we only need to take any path whose first variation is what it should be.)

C.1. It remains to show how our estimate implies Grayson's theorem. Fortunately all the work has been done elsewhere, and we only have to quote it. By Altschuler's work [**A1**] we can estimate the derivatives of the curvature in terms of the maximum curvature, and obtain a sequence of dilations of the solutions which converges to a limit. This limit is also a solution of the curve shrinking flow, and is complete. It may not be compact, but at least it has finite total curvature

$$\int |k|ds < \infty.$$

Moreover since the total curvature decreases under the flow, in the limit solution it must be constant. However, it is strictly decreasing whenever there is a point where $k = 0$ but $\partial k/\partial s \neq 0$. Then by work of Angenent this cannot happen unless k has only one sign, which we may as well take as $k > 0$.

There are two cases for the limit. In case one we have the solution existing for all $t < T$ with k unbounded as $t \to T$, but $(T - t)|k|$ remains bounded. In Case II it is unbounded. In Case I, Huisken's monotonicity principle [**Hu**] shows that the limit is a homothetically shrinking soliton. These are classified by Abresch and Langer [**Ab+La**], and the only embedded one is the circle, which of course is just what we want. In Case II, the limit solution exists for all time with curvature k satisfying $0 < k \le 1$, and $k = 1$ at the origin at $t = 0$. Then the strong maximum principle applied to the Harnack estimate (see [**Ha1**] and [**Ha2**]) shows the limit is a translating soliton. It is then necessarily the graph $y = f(x, t)$ of a function where

$$\frac{\partial y}{\partial t} = \frac{\partial}{\partial x} \arctan \left(\partial y/\partial x \right) = 1$$

which is easily solved to give the grim reaper

$$y = t + \ell n \sec x.$$

In the grim reaper a horizontal line segment has length $L < \pi$, while if it is high enough it encloses an arbitrarily large area A_1, while there is still an arbitrarily large area A_2 on the other side if we go out enough farther. If the grim reaper is to be the limit, then the original curve comes arbitrarily close to it after translating, rotating, and dilating. All of which doesn't affect the constants $\overline{G}$ (or $\overline{H}$). But then we must have

$\overline{G} \to 0$ (or $\overline{H} \to 0$), which is impossible. This proves Grayson's theorem in the plane. It is pretty clear the argument will also work for a curve on a surface as in [**Gr2**], but we haven't worked out the details.

References

[Ab+La] U. Abresch and J. Langer, "The normalized curve shortening flow and homothetic solutions," *J. Diff. Geom.* **23** (1986).

[Al] S. Altschuler, "Singularities of the curve shrinking flow for space curves," thesis UCSD.

[Ga+Ha] M. Gage and R. S. Hamilton, "The shrinking of convex plane curves by the heat equation," *J. Diff. Geom.* **23** (1986), 69-96.

[Gr1] M. Grayson, "The heat equation shrinks embedded plane curves to points," *J. Diff. Geom.* **26** (1987), 285-314.

[Gr2] M. Grayson, "Shortening embedded curves," *Ann. Math.* **129** (1989), 71-111.

[Ha1] R. S. Hamilton, "The Harnack estimate for the Mean Curvature flow," preprint. [Ha2] R. S. Hamilton, "Eternal Solutions to the Mean Curvature flow," preprint.

[Hu] G. Huisken, "Asymptotic behavior for singularities of the mean curvature flow," *J. Diff. Geom.* **31** (1990), 285-299.

[Y] S.-T. Yau, Seminar book.

THE ABEL-RADON TRANSFORM AND SEVERAL COMPLEX VARIABLES

G. M. Henkin

Contents

Introduction

The connection between abelian integrals and several complex variables is well known. We give here two classical examples. First of all we remind, that the first profound problems of several complex variables, the problems of Weierstrass, of Poincaré, of Cousin, were posed and solved in connection with the problem of inversion of the Jacobi mapping given by abelian integrals. These main problems have brought, as the final result, to the Oka-Cartan theory of coherent analytical sheaves [22].

Further, we note, that the theory of harmonic forms on complex manifolds, including the Hodge theory and the J. Kohn solution of the $\bar{\partial}$-Neumann-Spencer problem, is constructed as a natural development of the classical Riemann results about abelians differentials on the Riemann surfaces [18], [21].

At present the theory of abelian integrals is considered to be not so much a section of analysis but a relatively completed section of algebraic geometry serving, however, the permanent source of new geometric, topological and even physical constructions [21]. However the direct influence of the theory of abelian integrals on the modern functions theory is also not exhausted. In the present paper having partially survey character we present several new results in several complex variables, where again the classical Abel integrals are both the source of ideas and the method of investigations [1], [2], [20], [21].

First of all in this paper, we deal with the following interconnected problems.

1. Let K be a pseudo-convex compact on n-dimensional projective manifold X and let $H(K)$ be a space of holomorphic functions on K. The problem going back to L. Fantappiè [17] is in description (an analytical representation) of linear continuous functionals on $H(K)$. This problem as A. Martineau showed [34], is equivalent to the problem of analytical description of the infinite dimensional space of $\bar{\partial}$-cohomology $H^{n,n-1}(\mathfrak{D})$ of the pseudo-concave manifold $\mathfrak{D} = X \backslash K$.

2. A closed one-dimensional complex analytic subset in $\mathfrak{D}$ and holomorphic 1-form ψ on V define elements of the space $H^{n,n-1}$ generated by analytical currents $\psi \wedge [V]$ i.e. by $\bar{\partial}$-closed currents of the form:

$$(\psi \wedge [V], \chi) = \int_V \psi \wedge \chi,$$

    where χ is an arbitrary $(0,1)$-form with smooth coefficients and compact support in $\mathfrak{D}$. The problem is to find effective conditions for realization of the element $F \in H^{n,n-1}(\mathfrak{D})$ by some analytical current $\psi \wedge [V]$.

3. The investigation of analytical currents in $\mathfrak{D}$ demands solving the following complex Plateau problem. Let γ be a closed real rectifiable curve in $\mathbb{C}P^n$ and ψ be an integrable $(0,1)$-form on γ. It is required to find necessary and sufficient conditions in order that under suitable orientation the curve γ would be a boundary of some open Riemann surface S (possibly with singularities) in $\mathbb{C}P^n$, and the form ϕ would be a boundary value on γ of some holomorphic $(1,0)$-form ψ on S.

As far as the first problem is concerned the basic results were obtained by A. Martineau [35] for the case of concave domains in $\mathbb{C}P^n$. For the third problem the basic results for the case of γ and S in $\mathbb{C}^n$, were obtained by J. Wermer [46]. The leading idea in these and in the next papers [3], [36], [45] was to use Cauchy and (or) the Fantappiè transforms.

Based on the papers [13], [14], [16], [19], [25], [27] we show in §§1-4, that the comparatively complete solution of the three formulated problems may be obtained due to the Abel-Radon transform closely connected with the Cauchy-Fantappiè transform.

For the holomorphic form $\psi = \sum\limits_{0}^{n} \psi_j(z)dz_j$ on the Riemann surface V in the linearly concave domain $\mathfrak{D} \subset \mathbb{C}P^n$ the Abel transform $\mathfrak{A}\psi$ is determined by the following way. Let $\xi \in \mathfrak{D}^* \subset (\mathbb{C}P^n)^*$ i.e. let the hyperplane $\mathbb{C}P_\xi^{n-1} = \{z \in \mathbb{C}P^n : \xi \cdot z = 0\} \subset \mathfrak{D}$. Let, in addition, the intersection $\mathbb{C}P^{n-1} \cap V$ be transverse and consist of the discrete set of the points $\{z^{(1)}(\xi), \ldots, z^{(N)}(\xi)\}$. We define ([1], [20]), $\mathfrak{A}\psi(\xi) = \sum\limits_{\nu=1}^{N}\sum\limits_{j=0}^{n} \psi_j(z^{(\nu)}(\xi))d_\xi z_j^{(\nu)}(\xi)$, i.e. $\mathfrak{A}\psi$ is $(1,0)$-form in $\mathfrak{D}^*$. For $\bar{\partial}$-closed form $F \in C_{n,n-1}(\mathfrak{D})$ the Radon transform RF, defined by the formula [19]

$$RF(\xi) = \sum_{j=0}^{n}(\int_{\mathfrak{D}} \bar{\partial}_z \frac{1}{\xi \cdot z} \wedge z_j F)d\xi_j,$$

is holomorphic $(1,0)$-form in the domain $\mathfrak{D}^*$. The term "Abel-Radon transform" can be explained by the observation that for the form $F \in C_{n,n-1}(\mathfrak{D})$, $\bar{\partial}$-cohomologous to some current $\psi \wedge [V]$ in $\mathfrak{D}$, we have $\mathfrak{A}(\xi) = RF(\xi)$.

Maybe, the most interesting new fact obtained in connection with discussed problems consists of the possibility of the inversion of the statement of the classical Abel theorem about abelian integrals. Namely, the Abel theorem together with its inversion can be formulated in the following form.

The Main Theorem. *Let V be a closed one-dimensional complex analytic submanifold in the linearly concave domain $\mathfrak{D} \subset \mathbb{C}P^n$ with connected dual set $\mathfrak{D}^* = \{\xi \in (\mathbb{C}P^n)^* : \mathbb{C}P_\xi^{n-1} \subset \mathfrak{D}\}$.*

Let ψ be a meromorphic 1-form on V. Then the following statements are equivalents:

a/ $V = \tilde{V} \cap \mathfrak{D}$, where $\tilde{V}$ is an algebraic subset in $\mathbb{C}P^n$ and $\psi = \tilde{\psi}|_V$, where $\tilde{\psi}$ is a rational form,

b/ the Abel transform $\mathfrak{A}\psi$ is a rational form in $\mathfrak{D}^*$.

The property $a/ \Rightarrow b/$ composes the contents of the Abel theorem [1], [2]. It is proved in a series of papers of S. Lie [33], H. Poincaré [39], L. Darboux [11], W. Blaschke [8], W. Wirtinger [47], B. Saint-Donat [41] and Ph. Griffiths [20] that the condition $\mathfrak{A}\psi \equiv 0$ in $\mathfrak{D}^*$ implies a/. In the mentioned form this theorem is obtained in [25]. Note, that the interest of S. Lie and the successive authors in the inversion of the classical Abel theorem to much extent is connected with geometric applications—the theory of surfaces of translation (see [11]).

From the theory [11], [33], [38], [41], [47], it follows in particularly the characterization of Jacobians of Riemann surfaces as those principally polarized abelian varieties whose theta divisor is doubly of translation type (see [11], [33], [34], [38], [41]).

The formulated problems and connected with them results of §§1–4 admit natural generalizations. As far as the first problem is concerned it is natural to consider q-pseudo-concave domains $\mathfrak{D}$ on n-dimensional projective manifold X and to investigate the space $H^{p,q}(\mathfrak{D})$, $n \geq p \geq q$. For the second problem it is necessary to find suitable conditions of realization of elements of $H^{p,q}(\mathfrak{D})$ by analytical currents $\psi \wedge [V]$, where V is a closed $(n-q)$-dimensional complex subset in $\mathfrak{D}$, and ψ is a holomorphic $(p-q,0)$-form on V. At last, for the third problem it is required to formulate natural conditions for a closed $(2k-1)$-dimensional manifold M in $\mathbb{C}P^n$, necessary and sufficient, in order that M would be a boundary of some complex analytical set T in $\mathbb{C}P^n$.

According to the third problem there are profound results of R. Harvey and of B. Lawson [24] for the case when Γ and T are contained in $\mathbb{C}P^n \backslash \mathbb{C}P^{n-k}$. However, in $\mathbb{C}P^n$, this problem has not been solved completely. According to the first problem there are also suitable results for the case of q-concave domains in $\mathbb{C}P^n$ (see [5], [19], [27]).

It is natural to call the Abel-Radon transform in the general case as the Abel-Radon-Penrose transform. In fact, for the case of 1-concave domains $\mathfrak{D}$ in $\mathbb{C}P^3$, the dual domain $\mathfrak{D}^*$ (the set of projective lines contained in $\mathfrak{D}$) may be considered as a domain in the complex Minkowski space $\mathbb{C}M$. In this case the transform of the considered type was firstly introduced by R. Penrose with the purpose of complex-geometric interpretations of conformally invariant equations of Mathematical physics (see [5], [27]).

The mentioned generalized problems in the present paper are not examined since their investigation has not been finished. However, the generalized analog of the main theorem remains valid if $\mathfrak{D}$ is q-concave domain in $\mathbb{C}P^n$, V is a closed $(n-q)$-dimensional complex subset in $\mathfrak{D}$, ψ is a holomorphic $(n-q,0)$-form on V and $\mathfrak{D}^* = \{\xi \in G_{q+1,q+1} : \mathbb{C}P_\xi^q \in \mathfrak{D}\}$. In this case the theorem is strengthening of the main Ph. Griffiths result of [20]. It was proved in [20], in the general case, that b/ follows from a/ and that a/ follows from $\mathfrak{A}\psi \equiv 0$ in $\mathfrak{D}^*$.

1. The Radon transform for $\bar{\partial}$-cohomology in (pseudo) concave domain

Let $\mathbb{C}P^n$ be a projective space with homogeneous coordinates $Z = (Z_0, Z_1, \ldots, Z_n)$ and $(\mathbb{C}P^n)^*$ be a dual projective space with homogeneous coordinates $\xi = (\xi_0, \xi_1, \ldots, \xi_n)$. To the point $\xi \in (\mathbb{C}P^n)^*$ corresponds the projective hyperplane

$$\mathbb{C}P_\xi^{n-1} = \{Z \in \mathbb{C}P^n : \xi \cdot Z = 0\}, \quad \text{where} \quad \xi \cdot Z = \xi_0 \cdot Z_0 + \cdots + \xi_n Z_n.$$

We denote by $\mathbb{O}(l)$ the line bundle over $\mathbb{C}P^n$, sections of which are homogeneous of degree l functions of the homogeneous coordinates. Following W. Rothstein [40] and A. Martineau [35] we call the domain $\mathfrak{D}$ in $\mathbb{C}P^n$ linearly concave if for any point $Z \in \mathfrak{D}$ there exists a projective hyperplane $\mathbb{C}P_\xi^{n-1}(Z)$ passing through the point Z and contained in $\mathfrak{D}$.

The compact K in $\mathbb{C}P^n$ is called linearly convex if the domain $\mathfrak{D} = \mathbb{C}P^n \setminus K$ is linearly concave.

The set of projective hyperplanes contained in the linearly concave domain $\mathfrak{D}$ forms in the dual space the open set $\mathfrak{D}^*$ called dual to $\mathfrak{D}$.

Let M be analytical or, that is the same due to the Chow theorem [21], algebraic subset of $\mathbb{C}P^n$ of the dimension k. We shall call the domain D_M in the manifold M pseudoconcave if it can be represented in the form $D_M = M \cap \mathfrak{D}$, where $\mathcal{D}$ is a linearly concave domain in $\mathbb{C}P^n$. The corresponding compact $K_M = M \setminus \mathfrak{D}_M$ we shall call pseudoconvex.

Let $C_{p,q}^{(s)}(X)$ denote the space of (p, q)-forms on the complex manifold X with coefficients in the space $C^{(s)}(X)$. Let $H^q(X, \Omega^p)$ denote the space of q-dimensional cohomologies X with coefficients in holomorphic p-forms. The Dolbeault theorem (see [21], [22]) shows that for any $s \in [-\infty, +\infty]$ there is the canonical isomorphism

$$H^q(X, \Omega^p) \simeq H^{p,q}(X)$$

$$= \{\phi \in C_{p,q}^{(s)}(X) : \bar{\partial}\phi = 0\} \Big/ \{\phi \in C_{p,q}^{(s)}(X) : \phi = \bar{\partial}F \in C_{p,q-1}^{(s)}(X)\}$$

The theorems of Andreotti-Grauert, Andreotti-Norguet, Andreotti-Vesentini (see [18], [26]) show that for any non-compact pseudo-concave domain $\mathfrak{D}_M$ the spaces $H^{p,q}(\mathfrak{D}_M)$ are separable and

$$\dim H^{p,q}(\mathfrak{D}_M) \quad < \infty, \text{ if } q \neq k - 1, \text{ and}$$
$$\dim H^{p,q}(\mathfrak{D}_M) \quad = \infty, \text{ if } q = k - 1.$$

The Kohn results [18], [30] allow for any p and q to realize elements of the space $H^{p,q}(\mathfrak{D}_M)$ for strictly pseudoconcave domains by harmonic (in the Fubini-Studi metric) forms on $\mathfrak{D}_M$, satisfying the $\bar{\partial}$-Neumann boundary condition on the boundary $b\mathfrak{D}_M$ of the domain $\mathfrak{D}_M$.

Let $H^*(K_M)$ denote the space of analytical functionals on the space $H(K_M)$ of holomorphic in the neighbourhood K_M functions, where K_M is a pseudo-convex compact on the algebraic manifold M. For the functional $\mu \in H^*(K_M)$ we define the Fantappiè indicatrix (compare with [17], [36]) as the following holomorphic form

$$\mathfrak{F}\mu(\xi) = f(\xi) = \sum_{j=0}^{n} f_j(\xi)d\xi_j, \quad \text{where} \quad f_j(\xi) = \left(\mu, \frac{Z_j}{\xi \cdot Z}\right), \quad \xi \in \mathfrak{D}^*.$$

(1.1)

The coefficients $\{f_j(\xi)\}$ of the form $f(\xi)$ are holomorphic in $\mathfrak{D}^*$ and have the homogeneity of the degree (-1) with respect to ξ. The form $f(\xi)$ itself gives the holomorphic 1-form in $\mathfrak{D}^* \subset (\mathbb{C}P^n)^*$ only under additional condition $\sum_{j=0}^{n} \xi_j f_j(\xi) = 0$, i.e. when $\mu(1) = 0$. For any $\bar{\partial}$-closed form $F \in C_{k,k-1}^{(s)}(\mathfrak{D}_M)$, $s \geq 0$ we define the Radon transform RF by the following formula (compare with [19])

$$RF(\xi) = f(\xi) = \sum_{j=0}^{n} f_j(\xi)d\xi_j, \quad \xi \in \mathfrak{D}^* \tag{1.2}$$

$$\text{where} \quad f_j(\xi) = \int_{\mathbb{C}P_\xi^{n-1} \cap M} \langle \xi \cdot dZ \rangle \rfloor Z_j \cdot F = \int_M \left(\bar{\partial}_z \frac{1}{\xi \cdot Z}\right) \wedge Z_j F;$$

$\langle \xi \cdot dZ \rangle \rfloor Z_j \cdot F$ denotes the residue form of H. Poincaré, J. Leray, Herrera-Lieberman [12], [28], [32] for the form $\frac{Z_j \cdot F}{(\xi \cdot Z)}$ on the manifold $\mathbb{C}P_\xi^{n-1} \cap M$. For $\bar{\partial}$-closed differential form $\Phi \in C_{k-1,k-1}^{(s)}(\mathfrak{D}_M)$, $s \geq 0$ we denote the Radon transform by the following formula (see Andreotti-Norguet [4])

$$\rho_0 \Phi = \phi_0(\xi) = \int_{\mathbb{C}P_\xi^{n-1} \cap M} \Phi. \tag{1.3}$$

Let F be a smooth $\bar{\partial}$-closed $(k, k-1)$-form on $\mathfrak{D}_M$. We denote by μ_F the functional on $H(K_M)$ defined by the equality

$$(\mu_F, h) = \int_{b(K_M)_\epsilon} F \cdot h, \quad \text{where} \quad h \in H(K_M), \tag{1.4}$$

$(K_M)_\epsilon$ is sufficiently small neighbourhood of the compact K_M in M. From the Stokes formula we have: $\mu_F(1) = 0$.

Proposition 1.1 [19] *For any form $F \in C_{k,k-1}^{(s)}(\mathfrak{D}_M)$, $s \geq 0$ and any $\xi \in \mathfrak{D}^*$ the Radon transform $RF(\xi)$ and the Fantappiè indicatrix $\mathfrak{F}\mu_F(\xi)$ are connected by the relation*

$$RF(\xi) = \frac{1}{2\pi i}\mathfrak{F}\mu_F(\xi), \quad \xi \in \mathfrak{D}^*.$$

With the help of this proposition it is easy to prove the following proposition.

Proposition 1.2 [19], [37] *The transforms $F \to RF$ and $\Phi \to \rho_0 \Phi$ define linear continuous mappings of spaces of cohomologies $H^{k,k-1}(\mathfrak{D}_M)$ and $H^{k-1,k-1}(\mathfrak{D}_M)$ in pseudoconcave domain $\mathfrak{D}_M$ respectively to spaces of holomorphic 1-forms $H^{1,0}(\mathfrak{D}^*)$ and holomorphic functions $H^{0,0}(\mathfrak{D}^*)$ in the open set $\mathfrak{D}^*$. If, in addition, $F \in C^{(s)}_{k,k-1}(\mathfrak{D}_M)$, $\Phi \in C^{(s)}_{k-1,k-1}(\mathfrak{D}_M)$ and $F = d\Phi$, then $RF = d\rho_0\Phi$.*

Now we shall formulate the main result of this paragraph.

Theorem 1.1 (The generalized Martineau theorem) *Let $\mathfrak{D}$ be a linear concave domain in $\mathbb{C}P^n$, and M be a k-dimensional algebraic manifold of the form $M = \{Z \in \mathbb{C}P^n : P_1(Z) = \ldots = P_r(Z) = 0\}$, where homogeneous polynomials $P_1, \ldots, P_r$ are such that everywhere in M rank $[\operatorname{grad} P_1, \ldots, \operatorname{grad} P_r] = n - k$.*

Let $\mathfrak{D}_M = M \cap \mathfrak{D}$ be a pseudoconcave domain on M. Then:

(i) *If the open set $\mathfrak{D}^*$ is connected, then the transform $F \to RF$ gives the continuous transform $R : H^{k,k-1}(\mathfrak{D}_M) \to H^{1,0}(\mathfrak{D}^*)$ with finite-dimensional kernel equal to the image of the space $H^{k,k-1}(M)$ in $H^{k,k-1}(\mathfrak{D}_M)$ under restriction of cohomologies from M to $\mathfrak{D}_M$. This kernel is equal to zero, i.e. the transform R is injective when M is a complete intersection in $\mathbb{C}P^n$ and $\dim M > 1$.*

(ii) *If for any $Z \in \mathfrak{D}$ the section of $\mathfrak{D}^*$ by the hyperplane $\{\xi \in \mathfrak{D}^* : \xi \cdot Z = 0\}$ is contractible, then*

$$RH^{k,k-1}(\mathfrak{D}_M) = \left\{ f \in H^{1,0}(\mathfrak{D}^*) : f = d\phi, \ \phi \in H^{0,0}(\mathfrak{D}^*) \quad \text{and} \right.$$

$$\left. P_j\left(\frac{d}{d\zeta}\right)\phi = 0, \ j = 1, 2, \ldots r \right\},$$

where P_j are the polynomials defining the manifold M.

Corollary 1.1 (i). *If under conditions of Theorem 1.1.(i) M is a complete intersection in $\mathbb{C}P^n$ then the transform $\rho_0 : H^{k-1,k-1}(\mathfrak{D}_M) \to H^{0,0}(\mathfrak{D}^*)$ has as a kernel the subspace in $H^{k-1,k-1}(\mathfrak{D}_M)$ of the form*

$$\{\Phi \in H^{k-1,k-1}(\mathfrak{D}_M) : d\Phi = 0 \text{ in } H^{k,k-1}(\mathfrak{D}_M)$$

$$\text{and} \quad \Phi = 0 \text{ in } H^{2k-2}(\mathfrak{D}_M)\}.$$

(ii) Under conditions of Theorem 1.1 (ii) the transform ρ_0 has as the image the space

$$\left\{ \phi \in H^{0,0}(\mathfrak{D}^*) : P_j \left(\frac{d}{d\xi} \right) \phi = 0, \ j = 1, 2, \ldots r \right\}.$$

Remark: In papers of A. Martineau [35], [36] and in refining it paper [19] the theorem, which is equivalent to Theorem 1.1, is proved under assumption that $M = \mathbb{C}P^n$.

The result of Theorem 1.1 (i) for the case when the manifold M is complete intersection was obtained in paper [27]. The result of Theorem 1.1 (ii) and of Corollary 1.1 (ii) is a variant of the Ehrenpreis fundamental principle for the transforms of Fantappiè and Radon instead of the classical Fourier-Laplace transform. These results develop also the construction of the non-linear Fourier transform introduced by Ehrenpreis-Malliavin [16]. In particular, it follows from Theorem 1.1 (ii) and from Proposition 1.1 that under conditions of Theorem 1.1 (ii) any holomorphic of the homogeneity (-1) solution f_0 of the system of equations $P_j \left(\frac{d}{d\xi} \right) f_0(\xi) = 0$ in $\mathfrak{D}^*$ may be represented in the form of suitable sum of elementary solutions $\frac{Z_0}{\xi \cdot Z}$, where $Z \in K_M = M \backslash \mathfrak{D}_M$. More exactly, any holomorphic solution f_0 in $\mathfrak{D}^*$ of this system (and only such function) may be represented in the form of $f_0(\xi) = \left(\mu, \frac{Z_0}{\xi \cdot Z} \right)$, where μ is an analytical functional from $H^*(K_M)$. In §5 we give explicit version for this representation in the spirit of explicit version of the "fundamental principle" in Berndtsson-Passare paper [6].

As Corollary 1.1 (i) shows the transform ρ_0 has infinite dimensional kernel in the space $H^{n-1,n-1}(\mathfrak{D})$. So, for the complete description of the space $H^{n-1,n-1}(\mathfrak{D})$ it is necessary to introduce more general transform of the Radon type for forms from $C_{n-1,n-1}(\mathbb{C}P^n)$.

Let $C_{0,n-1}^{(s)}(\mathfrak{D}, \mathbb{O}(l))$ denote the space of $(0, n-1)$-forms whose coefficients are sections of the class $C^{(s)}$ of the bundle $\mathbb{O}(l)$. In homogeneous coordinates in $\mathbb{C}P^n$ any form $\Phi \in C_{n-1,n-1}^{(s)}(\mathfrak{D})$ is represented by the following way

$$\Phi = \sum_{j_1,\ldots,j_{n-1}=0}^{n} \Phi_{j_1,\ldots,j_{n-1}}(Z) dZ_{j_1} \wedge \ldots \wedge dZ_{j_{n-1}}, \qquad (1.5)$$

where the forms $\Phi_{j_1,\ldots,j_{n-1}} \in C_{0,n-1}^{(s)}(\mathfrak{D}, \mathbb{O}(-n+1))$ are skew-symmetric with respect to indexes $j_1, \ldots, j_{n-1}$ and for any $j_2, \ldots, j_{n-1}$ we have

$$\sum_{j_1=0}^{n} Z_{j_1} \Phi_{j_1,j_2,\ldots,j_{n-1}}(Z) \equiv 0. \qquad (1.6)$$

Following papers [19], [27] we introduce more general, than ρ_0, transforms of the Radon type on the space $H^{n-1,n-1}(\mathfrak{D})$. For the form $\Phi \in C_{n-1,n-1}^{(s)}(\mathbb{C}P^n)$, $s \geq 1$, we suppose that

$$R_{j_1,\ldots,j_{n-1}}\Phi = \phi_{j_1,\ldots,j_{n-1}}(\xi) =$$

$$\frac{1}{\xi_n^2} \int_{\mathbb{C}P_\xi^{n-1}} \frac{\partial \Phi_{j_1,\ldots,j_{n-1}}}{\partial Z_n} Z_0^n \, d\frac{Z_1}{Z_0} \wedge \ldots \wedge d\frac{Z_{n-1}}{Z_0} \qquad (1.7)$$

Let $\wedge^{n-1} H(\mathfrak{D}^*, \mathbb{O}(-2))$ denote the space of holomorphic sections

$$\phi_{j_1,\ldots,j_{n-1}}$$

over $\mathfrak{D}^*$ of the bundle $\mathbb{O}(-2)$, skew-symmetric with respect to indexes $j_1, j_2, \ldots, j_{n-1}$.

Theorem 1.2 *Let $\mathfrak{D}$ be a linear concave domain in $\mathbb{C}P^n$ and for any $Z \in \mathfrak{D}$ the section $\{\xi \in \mathfrak{D}^* : \xi \cdot Z = 0\}$ is contractible. Then the transforms $\Phi \to R\Phi = \{R_{j_1,\ldots,j_{n-1}}\Phi\}$ of the type (1.7) define a linear continuous mapping of the space $H^{n-1,n-1}(\mathfrak{D})$ to the space of all those elements $\phi_{j_1,\ldots,j_{n-1}} \in \wedge^{n-1} H(\mathfrak{D}^*, \mathbb{O}(-2))$ which satisfy the relations*

$$\sum_{j_1=0} \frac{\partial \phi_{j_1,j_2,\ldots,j_{n-1}}(\xi)}{\partial \xi_{j_1}} = 0 \quad \text{for any} \quad j_2,\ldots,j_{n-1}. \qquad (1.8)$$

In addition, the kernel of the transform

$$R : H^{n-1,n-1}(\mathfrak{D}) \to \wedge^{n-1} H(\mathfrak{D}^*, \mathbb{O}(-2))$$

is one-dimensional and consists of classes of cohomologies of forms of the type (const) $(\partial \bar{\partial} ln|Z|)^{n-1}$.

The similar to Theorem 1.2 results have been recently obtained in the theory of the Radon-Penrose transform [5], [27]. Theorem 1.2 is used further in the different form taking into account together with the transform R the transform ρ_0. For the form $\Phi \in C_{n-1,n-1}^{(s)}(\mathbb{C}P^n)$, $s \geq 1$ we suppose that

$$\rho_k \Phi(\xi) = \phi_k(\xi) = \sum_{j=1}^n (-1)^{n-j} \xi_j \frac{\partial}{\partial \xi_k} R_{1,\ldots,\hat{j},\ldots,n} \Phi, \qquad (1.9)$$

where $\xi \in (\mathbb{C}P^n)^*$, $R_{1,\ldots,\hat{j},\ldots,n}\Phi$ is the transform of the type (1.7), $(1,\ldots,\hat{j},\ldots,n)$
$$=$$
$(1,\ldots,j-1,j+1,\ldots,n)$, $k = 1,2,\ldots,n-1$. The transforms $\Phi \to$

$\rho\Phi = \{\rho_k\Phi\}$ of the type (1.3) and (1.9) and $\Phi \to R\Phi = \{R_{j_1,\ldots,j_{n-1}}\Phi\}$ of the type (1.7) are connected by the relations

$$-\frac{1}{(n-1)!}\frac{\partial\rho_0\Phi}{\partial\xi_0} = \sum_{j=1}^{n}(-1)^{n-j}\xi_j R_{1,\ldots,\hat{j},\ldots,n}\Phi$$

$$-\frac{1}{(n-1)!}\frac{\partial^2\rho_0\Phi}{\partial\xi_0\partial\xi_k} = \rho_k\Phi + (-1)^{n-1}R_{1,\ldots,\hat{k},\ldots,n}\Phi \qquad (1.10)$$

for $k = 1, 2, \ldots, n-1$. These relations immediately follow from definitions and detailed writing of the formula for $\rho_0\Phi$:

$$\frac{1}{(n-1)!}\rho_0\Phi = \sum_{j=1}^{n}\frac{\xi_j}{\xi_n}(-1)^{n-j}\int\limits_{\{Z:\zeta\cdot Z=0\}}\Phi_{1,\ldots,\hat{j},\ldots,n}Z_0^{n-1}d\frac{Z_1}{Z_0}\wedge\ldots\wedge d\frac{Z_{n-1}}{Z_0}$$

$$(1.11)$$

where Φ is the form of the type (1.5).

Corollary 1.2 *Under the conditions of Theorem 1.2 the transform $\Phi \to \rho\Phi$ of the form (1.3), (1.9) maps the space $H^{n-1,n-1}(\mathfrak{D})$ isomorphically to the direct sum of spaces*

$$H(\mathfrak{D}) \oplus H(\mathfrak{D}, \mathbb{O}(-2)) \oplus \cdots \oplus H(\mathfrak{D}, \mathbb{O}(-2))$$

Now we give the proofs of the newest formulated statements.

The proof of Theorem 1.1 (ii). Let $F \in H^{k,k-1}(\mathfrak{D}_M)$. We consider the analytical function $\mu_F \in H^*(K_M)$ of the form (1.4). The Fantappiè indicatrix of the functional μ_F has the following form due to (1.1), (1.4)

$$\mathfrak{F}_{\mu_F}(\xi) = \sum_{j=o}^{n}\left(\int\limits_{Z\in b(K_M)_\epsilon}\frac{Z_jF}{\xi\cdot Z}\right)d\xi_j, \quad \xi \in \mathfrak{D}^*.$$

Due to condition (ii) for any $Z \in K_M$ there exists a single-valued branch of logarithm $\ln(\xi\cdot z)$ in the domain $\mathfrak{D}^*$. Therefore

$$\mathfrak{F}_{\mu_F}(\xi) = d\tilde{\phi}(\xi), \quad \text{where} \quad \tilde{\phi}(\xi) = \int\limits_{Z\in b(K_M)_\epsilon}F(Z)\ln(\xi\cdot Z).$$

From $M = \{Z \in \mathbb{C}P^n : P_j(Z) = 0; \ j = 1, \ldots, r\}$ it follows that

$$P_j\left(\frac{d}{d\xi}\right)\tilde{\phi} = \text{const}\int\limits_{Z\in b(K_M)_\epsilon}F(Z)P_j(Z)(\xi\cdot Z)^{-\deg P_j} = 0.$$

From here and from Propositions 1.1, 1.2 it follows that $RF = f = d\phi$, where $\phi = -\frac{1}{2\pi i}\tilde{\phi} = H^{0,0}(\mathfrak{D}^*)$ and $P_j\left(\frac{d}{d\xi}\right)\phi = 0$, $j = 1, 2, \ldots, r$.

Inversely, let $\phi \in H^{0,0}(\mathfrak{D}^*)$ and $P_j\left(\frac{d}{d\xi}\right)\phi = 0$, $j = 1, 2, \ldots, r$. Let $K = \mathbb{C}P^n \backslash \mathfrak{D}$. Consider the functional $\mu \in H^*(K)$ according to the formula

$$(\mu, h) = \int_{Z=bK_\epsilon} hF, \tag{1.12}$$

where

$$F = \frac{(-1)^n}{(2\pi i)^{n-1}} \frac{\partial^n \tilde{\phi}(\eta(Z))}{\partial \xi_0^n} \wedge$$

$$\left(\sum_{k=1}^n (-1)^k \eta_k(Z) d\eta_1(Z) \wedge \ldots \wedge d\hat{\eta}_k(Z) \wedge \ldots \wedge d\eta_n(Z)\right) \bigwedge_{k=1}^n d\frac{Z_k}{Z_0},$$

$$h \in H(\bar{K}_\epsilon), \ Z \mapsto \eta(Z)$$

is a smooth mapping of bK_ϵ to $\mathfrak{D}^*$ such that $\eta(Z) \cdot Z = 0$. The existence of the last mapping is provided by condition of Theorem 1.1 (ii). We have $(\mu, 1) = 0$. Besides, in view of the inversion Martineau formula [36], refined in [19], we have

$$\mathfrak{F}_\mu(\xi) = d\tilde{\phi}(\xi). \tag{1.13}$$

In particular,

$$\tilde{\phi}_0(\xi) = \frac{\partial\tilde{\phi}}{\partial\xi_0} = \int_{Z\in bK_\epsilon} \frac{Z_0}{\xi \cdot Z} F.$$

Since $P_j\left(\frac{d}{d\xi}\right)\tilde{\phi} = 0$, $j = 1, 2, \ldots, r$, then for any $\xi \in \mathfrak{D}^*$ from the previous equality we obtain

$$0 = P_j\left(\frac{d}{d\xi}\right)\tilde{\phi}_0(\xi) = \text{const}\int_{Z\in bK_\epsilon} (\xi \cdot Z)^{-1-\deg P_j} \cdot Z_0 P_j(Z)F(Z)$$

for $j = 1, 2, \ldots, r$. In view of the general Cauchy-Fantappiè formula (5.5), §5 the functions of the form $\left(\frac{1}{\xi \cdot Z}\right)^{1+\deg P_j}$, $\xi \in \mathfrak{D}^*$ are dense in the space of holomorphic functions $H(K)$. Therefore from the last equality we have

$$\int_{Z\in bK_\epsilon} P_j(Z)F(Z) \cdot h(Z) = (\mu, P_j \cdot h) = 0, \quad j = 1, 2, \ldots, r \tag{1.14}$$

for any function $h \in H(K)$.

Due to Oka-Cartan theorem (see [22], [26]) for any sufficiently small $\epsilon > 0$ there exist such linear continuous operators $E : H^\infty(K_\epsilon \cap M) \to$

$H(K)$ and $T_j : H^\infty(K_\epsilon) \to H(K)$, $j = 1, 2, \ldots, r$, that for any function $h \in H^\infty(K_\epsilon)$ the following representation takes place

$$h(Z) = Eh\Big|_{K_\epsilon \cap M}(Z) + \sum_{j=1}^{r} P_j(Z)(T_j h)(Z), \quad Z \in K.$$

Hence and from the equalities $(\mu, P_j h) = 0$, $j = 1, 2, \ldots, r$, $h \in H(K)$ it follows that for any function $h \in H(K)$ and any sufficiently small $\epsilon > 0$ we have

$$(\mu, h) = \left(\mu, Eh\Big|_{K_\epsilon \cap M}\right) \tag{1.15}$$

Therefore, there exists a functional $\mu_M \in H^*(K \cap M)$ such that $\mathfrak{F}_{\mu_M}(\xi) = \mathfrak{F}_\mu(\xi) = d\tilde{\phi}(\xi)$, $\xi \in \mathfrak{D}^*$.

As $(\mu_M, 1) = 0$ then considering μ_M as (k, k)-current $[\mu_M]$ on M with support in K we obtain that $[\mu_M] = \bar{\partial} F$, where $F \in C^{(-\infty)}_{K, K-1}(M)$.

In addition, $(\mu_M, h) = \int_M \bar{\partial} F \wedge h = \int_{b(K_\epsilon \cap M)} F \cdot h$.

Since $\bar{\partial} F = 0$ on $\mathfrak{D}_M = \mathfrak{D} \cap M$ the form F defines an element of the space $H^{k,k-1}(\mathfrak{D}_M)$. Due to Proposition 1.1 we have, at last, $RF = d\phi$. Theorem 1.1 (ii) is proved.

The proof of Theorem 1.2 for $n = 2$: We shall start with the description of the kernel of the transform $\Phi \to R\Phi$. Let for the form $\Phi = \Phi_0 dZ_0 + \Phi_1 dZ_1 + \Phi_2 dZ_2$, $\Phi \in C^{(s)}_{1,1}(\mathfrak{D})$, $s \geq 2$ we have $R\Phi = \{R_j, \Phi\} = 0$. It follows from (1.6), (1.11) that

$$\rho_0 \Phi = \int_{\{Z : \xi \cdot Z = 0\}} Z_0 \left(\Phi_1 - \frac{\xi_1}{\xi_2} \Phi_2\right) d\frac{Z_1}{Z_2}$$

$$= -\int_{Z : \xi \cdot Z = 0} \frac{Z_0^2}{Z_1} \left(\Phi_0 - \frac{\xi_0}{\xi_2} \Phi_2\right) d\frac{Z_1}{Z_0}.$$

Hence and from (1.7) it follows that

$$\frac{\partial \rho_0 \Phi}{\partial \xi_0} = \xi_1 R_2 \Phi - \xi_2 R_1 \Phi \quad \text{and} \quad \frac{\partial \rho_0 \Phi}{\partial \xi_1} = \xi_2 R_0 \Phi - \xi_0 R_2 \Phi.$$

Thus, from condition $R\Phi = 0$ it follows that

$$\frac{\partial \rho_0 \Phi}{\partial \xi_0} = \frac{\partial \rho_0 \Phi}{\partial \xi_1} = 0 \quad \text{i.e.} \quad \rho_0 \Phi \equiv \text{const}.$$

Note, that for the closed $(1, 1)$-form $\omega = \partial \bar{\partial} \ln |Z|$ we also have, firstly, $\rho_0 \omega = \text{const} \neq 0$ and, secondly, $R\omega = 0$. So, there exists such a constant

C that for the form $\tilde{\Phi} = \Phi + C\partial\bar{\partial}\ln|Z|$ we have two equalities simultaneously

$$R\tilde{\Phi} = 0 \qquad \text{and} \qquad \rho_0\tilde{\Phi} = 0.$$

Due to result [19], [27] it follows from the condition $R\tilde{\Phi} = 0$ that there exist such functions $\tilde{\mathfrak{U}}_j \in C^{(s)}(\mathfrak{D}, \mathbb{O}(-1))$, $j = 0, 1, 2$, that

$$\tilde{\Phi}_j = \bar{\partial}_j\tilde{\mathfrak{U}}_j, \quad j = 0, 1, 2, \qquad \text{where} \qquad \tilde{\Phi} = \tilde{\Phi}_0 dZ_0 + \tilde{\Phi}_1 dZ_1 + \tilde{\Phi}_2 dZ_2$$

Hence, and from the relation (1.6) it follows that

$$\bar{\partial}\left(Z_0\tilde{\mathfrak{U}}_0 + Z_1\tilde{\mathfrak{U}}_1 + Z_2\tilde{\mathfrak{U}}_2\right) = 0$$

Due to the Liouville theorem it implies that

$$Z_0\tilde{\mathfrak{U}}_0 + Z_1\tilde{\mathfrak{U}}_1 + Z_2\tilde{\mathfrak{U}}_2 = \text{ const in } \mathfrak{D}$$

We shall show that this constant is equal to zero.

Let, for definiteness, the point $(0, 0, 1) \in \mathfrak{D}^*$. Since $\rho_0\tilde{\Phi} = 0$, we have

$$0 = \int\limits_{Z_2=0} \tilde{\Phi} = \int\limits_{Z_2=0} \tilde{\Phi}_1\left(\frac{Z_1}{Z_0}\right) d\frac{Z_1}{Z_0}$$

$$= -\int\limits_{Z_2=0} \frac{\tilde{\Phi}_1\left(\frac{Z_0}{Z_1}\right) d\frac{Z_0}{Z_1}}{\frac{Z_0}{Z_1}} = \pm 2\pi i\tilde{\mathfrak{U}}_1(0, 1, 0).$$

Hence, it follows that

$$Z_0\tilde{\mathfrak{U}}_0(Z) + Z_1\tilde{\mathfrak{U}}_1(Z) + Z_2\tilde{\mathfrak{U}}_2(Z)$$

$$\equiv 0 \cdot \tilde{\mathfrak{U}}_0(0, 1, 0) + 1 \cdot \tilde{\mathfrak{U}}_1(0, 1, 0) + 0 \cdot \tilde{\mathfrak{U}}_2(0, 1, 0) = 0$$

Thus, the form $\tilde{\Phi}$ is $\bar{\partial}$-exact in $\mathfrak{D} \subset \mathbb{C}P^2$, i.e. $\tilde{\Phi} = \bar{\partial} \sum\limits_{j=0}^{2} \tilde{\mathfrak{U}}_j dZ_j$.

Thereby, it is proved that the kernel of the transform $\Phi \to R\Phi$ consists of one-dimensional space of forms (const. $\partial\bar{\partial}\ln|Z|$).

We describe now the image of the transform $\Phi \to R\Phi$. Let

$$\phi_j = R_j\Phi = \frac{1}{\xi_2^2} \int\limits_{\{Z\,:\,\xi\cdot Z=0\}} \frac{\partial\Phi_j}{\partial Z_2} Z_0^2 d\frac{Z_1}{Z_0},$$

where $\Phi \in C_{1,1}^{(s)}(\mathfrak{D})$ and $\bar{\partial}\Phi = 0$. We have

$$\frac{\partial\phi_j}{\partial\xi_j} = -\frac{1}{\xi_2} \sum \int\limits_{\{Z\,:\,\xi\cdot Z=0\}} Z_j \frac{\partial^2\Phi_j}{\partial(\xi_2 Z_2)^2} Z_0^2 d\frac{Z_1}{Z_0}, \quad j = 0, 1, 2$$

Hence and from (1.6) it follows that

$$\frac{\partial \phi_0}{\partial \xi_0} + \frac{\partial \phi_1}{\partial \xi_1} + \frac{\partial \phi_2}{\partial \xi_2} = 0.$$

Inversely, let $\phi_j(\xi) \in H(\mathfrak{D}^*, \mathbb{O}(-2))$, $j = 0, 1, 2$ and let the previous equality be valid. In view of the result [19], [27] one can find such smooth $\bar{\partial}$-closed forms $\Phi_j \in C_{0,1}(\mathfrak{D}, \mathbb{O}(-1))$ that

$$\phi_j(\xi) = \frac{1}{\xi_2^2} \int\limits_{\{Z : \xi \cdot Z = 0\}} \frac{\partial \Phi_j(Z)}{\partial Z_2} Z_0^2 d\frac{Z_1}{Z_0}, \quad j = 0, 1, 2.$$

Suppose, that $Z_0\Phi_0 + Z_1\Phi_1 + Z_2\Phi_2 = G \in C_{0,1}(\mathfrak{D}, \mathbb{O})$, where $\bar{\partial}G = 0$. We shall prove that the form G is $\bar{\partial}$-exact. For this, due to [19], [27], it is sufficient to verify that for any $\xi \in \mathfrak{D}^*$ the integral

$$\frac{1}{\xi_2^2} \int\limits_{\{Z : \xi \cdot Z = 0\}} \frac{\partial^2}{\partial Z_2^2} G(Z) Z_0^2 d\frac{Z_1}{Z_0}$$

is equal to zero. Substituting into this integral the expression $G = \sum_0^2 Z_j\Phi_j$ we shall have

$$\frac{1}{\xi_2^2} \int\limits_{\{Z : \xi \cdot Z = 0\}} \frac{\partial^2 (\sum Z_j\Phi_j)}{\partial Z_2^2} Z_0^2 d\frac{Z_1}{Z_0}$$

$$= \int\limits_{\{Z : \xi \cdot Z = 0\}} Z_0 \frac{\partial^2 \Phi_0}{\partial(\xi_2 Z_2)^2} Z_0^2 d\frac{Z_1}{Z_0} + \int\limits_{\{Z : \xi \cdot Z = 0\}} Z_1 \frac{\partial^2 \Phi_1}{\partial(\xi_2 Z_2)^2} Z_0^2 d\frac{Z_1}{Z_0}$$

$$+ \int\limits_{\{Z : \xi \cdot Z = 0\}} Z_2 \frac{\partial^2 \Phi_2}{\partial(\xi_2 Z_2)^2} Z_0^2 d\frac{Z_1}{Z_0} + 2\frac{1}{\xi_2} \int\limits_{\{Z : \xi \cdot Z = 0\}} \frac{\partial \Phi_2}{\partial(\xi_2 Z_2)} Z_0^2 d\frac{Z_1}{Z_0}$$

$$= \int\limits_{\{Z : \xi \cdot Z = 0\}} Z_0 \frac{\partial^2 \Phi_0}{\partial(\xi_2 Z_2)^2} Z_0^2 d\frac{Z_1}{Z_0} - \frac{\xi_0}{\xi_2} \int\limits_{\{Z : \xi \cdot Z = 0\}} Z_0 \frac{\partial^2 \Phi_2}{\partial(\xi_2 Z_2)^2} Z_0^2 d\frac{Z_1}{Z_0}$$

$$+ 2\frac{1}{\xi_2} \int\limits_{\{Z : \xi \cdot Z = 0\}} \frac{\partial \Phi_2}{\partial(\xi_2 Z_2)} Z_0^2 d\frac{Z_1}{Z_0}$$

$$= -\xi_2 \frac{\partial}{\partial \xi_0} R_0\Phi - \xi_2 \frac{\partial}{\partial \xi_1} R_1\Phi + \xi_0 \frac{\partial}{\partial \xi_0} R_2\Phi + \xi_1 \frac{\partial}{\partial \xi_1} R_2\Phi + 2R_2\Phi$$

$$= -\xi_2 \left(\sum_{j=0}^{2} \frac{\partial}{\partial \xi_j} R_j\Phi \right)$$

$$= -\xi_2 \left(\sum_{j=0}^{2} \frac{\partial \phi_j}{\partial \xi_j} \right)$$

$$= 0.$$

Thus the form G is $\bar{\partial}$-exact in $\mathfrak{D}$, i.e. there exists $\psi \in C_{0,0}^{(s)}(\mathfrak{D})$ such that $G = \bar{\partial}\psi$, in addition, one can suppose that $\psi(0) = 0$. The function ψ may be represented in the form

$$\psi = \sum_{j=0}^{2} Z_j \psi_j, \qquad \text{where} \qquad \psi_j \in C_{0,0}^{(s)}(\mathfrak{D}, \mathbb{O}(-1)).$$

Suppose, that $\tilde{\Phi} = \tilde{\Phi}_0 dZ_0 + \tilde{\Phi}_1 dZ_1 + \tilde{\Phi}_2 dZ_2$, where $\tilde{\Phi}_j = \Phi_j - \bar{\partial}\psi_j$, $j = 0, 1, 2$. By construction we have $\sum Z_j \tilde{\Phi}_j = 0$, $\tilde{\Phi}_j \in C_{0,1}^{(s)}(\mathfrak{D}, \mathbb{O}(-1))$ and $\bar{\partial}\tilde{\Phi} = 0$.

Thus $(1, 1)$-form $\tilde{\Phi}$ represents some element of the space $H^{1,1}(\mathfrak{D})$. In addition,

$$R_j \tilde{\Phi} = \frac{1}{\xi_2^2} \int\limits_{\{Z : \xi \cdot Z = 0\}} \frac{\partial \tilde{\Phi}_j}{\partial Z_2} Z_0^2 d\frac{Z_1}{Z_0} = \frac{1}{\xi_2^2} \int\limits_{\{Z : \xi \cdot Z = 0\}} \frac{\partial \Phi_j}{\partial Z_2} Z_0^2 d\frac{Z_1}{Z_0} = \phi_j$$

for $j = 0, 1, 2$. Theorem 1.2 (for $n = 2$) is completely proved.

The Abel transform as the Radon transform

Let V be a closed one-dimensional submanifold (an open Riemann surface with singularities) in the domain $\mathfrak{D} \subset \mathbb{C}P^n$.

Let Reg V and Sing V denote respectively the sets of regular and singular points on V. The set Sing V is discrete, locally finite subset in V. Following [20] 1-form ψ on V is called meromorphic, if there exists a discrete locally finite subset $S \subset V$, containing Sing V, such that the form ψ is a usual holomorphic 1-form on $V \backslash S$, satisfying condition of the finite growth near S i.e. for any point $W \in S$ there exist a neighbourhood $\mathfrak{U}_W$ of W and natural N such that

$$\left| \int\limits_{\{Z \in V \cap U_W : |Z - W| \geq r\}} \psi \wedge \bar{\psi} \right| = 0(r^{-N}) \qquad (2.1)$$

The set of meromorphic 1-forms on V we denote through $\mathfrak{M}^{1,0}(V)$.

In view of the residue theory any form $\psi \in \mathfrak{M}^{1,0}(V)$ being holomorphic on $V \backslash S$ defines some current $\psi \wedge [V] \in C_{n,n-1}^{(-\infty)}(\mathfrak{D})$ by the formula

$$(\psi \wedge [V], \chi) = \lim_{\delta \to 0} \int\limits_{\{Z \in V : \text{dist}(Z, \text{sing} V) \geq \delta\}} \psi \wedge \chi, \qquad (2.2)$$

where χ is any form from $C_{0,1}^{(\infty)}(\mathfrak{D})$ with compact support in $\mathfrak{D}$.

Ph. Griffiths [20] calls $(1,0)$-form ψ holomorphic on V if V is a holomorphic 1-form on Reg V and for any compact subset K in $\mathfrak{D}$,

$$\left| \int\limits_{\text{Reg } V \cap K} \psi \wedge \bar{\psi} \right| < \infty$$

However, for the Abel transform theory, it turned out to be more natural another definition: the meromorphic 1-form ψ on V we shall call holomorphic (or otherwise meromorphic form on V of the first kind) if the current $\psi \wedge [V]$ is $\bar{\partial}$-closed i.e. if ψ is holomorphic on Reg V and for any function $\phi \in C^{(\infty)}(\mathfrak{D})$ with compact support in $\mathfrak{D}$ we have

$$\lim_{\delta \to 0} \int\limits_{\{Z \in \text{Reg } V : \text{dist}(Z, \text{sing } V) \geq \delta\}} \psi \wedge \bar{\partial}\phi = 0. \qquad (2.3)$$

The set of holomorphic 1-forms on V we denote through $H^{1,0}(V)$.

Further, an important role plays the following characterization of holomorphic (respectively, meromorphic) 1-forms on complex curves in $\mathbb{C}P^2$ in terms of Poincaré-Leray residues.

Proposition 2.3 [25] *Let V be a closed one-dimensional submanifold in the domain $\mathfrak{D} \subset \mathbb{C}P^2$. 1-form ψ on V is holomorphic on V (respectively, meromorphic on V with singularities on the discrete set $S \subset V$) iff for any point $Z \in V$ there exists such a neighbourhood $\mathfrak{U} \subset \mathfrak{D}$ that*

$$V \cap \mathfrak{U} = \{Z \in \mathfrak{U} : g(Z) = 0\}, \quad \text{where} \quad g \in H(\mathfrak{U}) \quad \text{and} \quad dg \not\equiv 0 \quad \text{on} \quad \mathfrak{U} \cap V \tag{2.4}$$

$$\psi\Big|_{(V \backslash S) \cap \mathfrak{U}} = \text{Res}_{(V \backslash S) \cap U} \frac{f}{g} = dg \rfloor f\Big|_{(V \backslash S) \cap \mathfrak{U}}, \tag{2.5}$$

where $f \in H^{2,0}(U)$ (respectively, $f \in \mathfrak{M}^{2,0}(U)$) with such a polar set V_f, that $V_f \cap V \subset S$).

In addition, it is determined the current $\left[\frac{f}{g}\right] \subset C_{2,0}^{(-\infty)}(\mathfrak{U})$ of the form

$$\left(\left[\frac{f}{g}\right], \chi\right) = \lim_{\epsilon \to 0} \int\limits_{\{Z \in \mathfrak{U} : |g| \geq \epsilon\}} \frac{f}{g} \wedge \chi, \tag{2.6}$$

where χ is any form from $C_{0,2}^{(\infty)}(\mathfrak{U})$ with compact support in $\mathfrak{U}$.
At last, in the domain $\mathfrak{U}\backslash V_f$ the following equality is valid.

$$\bar{\partial}\left[\frac{f}{g}\right]_{\mathfrak{U}\backslash V_f} = 2\pi i \psi \wedge [V \cap U\backslash V_f]. \tag{2.7}$$

Proof. Let $\mathfrak{U}$ be a complex neighbourhood of the point $Z \in \mathfrak{D}$ containing in $\mathfrak{D}$. From the solvability of the Weierstrass problem in $\mathfrak{U}$ (see [22], [26]) it follows that there exists such a function $g \in H(\mathfrak{U})$ that (2.4) is valid. Let form ψ be meromorphic on V with singularities on the discrete set S. Due to (2.3) the current $\psi \wedge [V\backslash S]$ is $\bar{\partial}$-closed in the domain $\mathfrak{D}\backslash S$. Let W be an analytical curve in $\mathfrak{U}$ with the property $W \cap V \cap \mathfrak{U} = S$. In view of Oka theorem (see [18], [22], [26]) the following $\bar{\partial}$-equation:

$$\psi \wedge [(V\backslash S) \cap \mathfrak{U}] = \bar{\partial}X, \qquad \text{where} \qquad X \in C_{2,0}^{(-\infty)}(\mathfrak{U}\backslash W). \tag{2.8}$$

is solvable in the pseudoconvex domain $\mathfrak{U}\backslash W$.

Since $\mathrm{supp}\,\psi \wedge [(V\backslash S) \cap \mathfrak{U}] \subset (V\backslash S) \cap \mathfrak{U}$, then X defines the holomorphic 2-form $\tilde{X}$ in the domain $\mathfrak{U}\backslash(V \cup W)$. From the estimate (2.1) of the meromorphic form ψ on V and from the weighted integral representations for solutions of the $\bar{\partial}$-equation (2.8) in the domain $\mathfrak{U}\backslash W$ (see a survey of such formulas in [6], [26]) it follows that form $\tilde{X}$ is meromorphic in $\mathfrak{U}$ in the sense of (2.1) and, consequently (see [20]), in the usual sense. Therefore, there takes place the irreducible Poincaré representation for $\tilde{X}$ in $\mathfrak{U}\backslash W$ of the form

$$\tilde{X} = \frac{f}{g^k}, \qquad \text{where} \quad f \in \mathfrak{M}^{2,0}(\mathfrak{U}), \quad \text{in addition,} \quad f \in \mathfrak{U}^{2,0}(\mathfrak{U}\backslash W). \tag{2.9}$$

For the neighbourhood $\mathfrak{U} \subset \mathfrak{D}\backslash\mathrm{Sing}\,V$ the equalities (2.8), (2.9) lead to the following: firstly, $k = 1$, and, secondly, there takes place the equality

$$\mathrm{Res}\,\tilde{X}_{(V\backslash S) \cap \mathfrak{U}} = dg \rfloor f = \psi_{(V\backslash S) \cap \mathfrak{U}}, \qquad \text{where} \qquad V \cap \mathfrak{U} \subset \mathrm{Reg}\,V \tag{2.10}$$

Hence, and from irreductibility of the representation (2.9) it follows that $k = 1$ for any convex neighbourhood $\mathfrak{U} \subset \mathfrak{D}$. Thus ,the property (2.5) follows from holomorphy of the form ψ on $V\backslash S$.

Consider now the current $[\tilde{X}] = \left[\frac{f}{g}\right] \in C_{2,0}^{(-\infty)}(\mathfrak{U})$ of the form (2.6). The existence of the limit in the right-hand-side of (2.6) follows from the Herrera-Lieberman result [28], [12].

We shall prove, that (2.7) follows from (2.4), (2.5), (2.6). For any form $\psi \in C_{0,1}^{(\infty)}(\mathfrak{U}\backslash W)$ with compact support in $\mathfrak{U}\backslash W$ we have the equalities

$$\left(\bar{\partial}\left[\frac{f}{g}\right], \psi\right) = \lim_{\epsilon \to 0} \int_{\{Z \in \mathfrak{U}\backslash W : |g| \geq \epsilon\}} \frac{f}{g} \wedge \bar{\partial}\psi = \lim_{\epsilon \to 0} \int_{\{Z \in \mathfrak{U}\backslash W : |g| = \epsilon\}} \frac{f}{g} \wedge \phi.$$

From (2.10) and from the Coleff-Herrera results [10] it follows (see in [10], sections 1.8, 4.2, 4.3) that

$$
\begin{aligned}
\lim_{\epsilon \to 0} \int_{\{Z \in \mathfrak{U}\backslash W : |g| = \epsilon\}} \frac{f}{g} \wedge \phi &= \lim_{\delta \to 0} 2\pi i \int_{\{Z \in (V\backslash S) \cap \mathfrak{U} : |dg| \geq \delta\}} \operatorname{Res} \frac{f}{g} \wedge \phi \\
&= \lim_{\delta \to 0} 2\pi i \int_{\{Z \in (V\backslash S) \cap \mathfrak{U} : |dg| \geq \delta\}} \psi \wedge \phi \\
&= 2\pi i \left(\psi \wedge [(V\backslash S) \cap \mathfrak{U}], \phi\right).
\end{aligned}
$$

The equality (2.7) is proved. Note, that if ψ is any 1-form on V in $\mathfrak{D}$, satisfying (2.4) and (2.5), then, firstly, ψ is meromorphic on V in the sense of (2.1) and, secondly, in the domain $(U\backslash W) \subset \mathfrak{D}$ there takes place the equality (2.7). Therefore, the current $\psi \wedge [(V\backslash S) \cap \mathfrak{U}]$ is a $\bar{\partial}$-closed, i.e. the form ψ is holomorphic on $V\backslash S$. Proposition 2.1 is proved.

We shall define now the Abel transform of meromorphic forms on $V \subset \mathfrak{D}$, where $\mathfrak{D}$ is a linearly concave domain in $\mathbb{C}P^n$.

In $\mathfrak{D}^*$ there exists an open and dense subset $\mathfrak{D}_0^* \subset \mathfrak{D}^*$ such that for every point $\xi \in \mathfrak{D}_0^*$ the intersection of the hyperplane $\mathbb{C}P_\xi^{n-1}$ with V is transverse and it forms the finite set

$$\{Z^{(1)}(\xi), Z^{(2)}(\xi), \ldots, Z^{(N)}(\xi)\} \subset V\backslash S.$$

The number N is constant while ξ belongs to some one component of $\mathfrak{D}_0^*$.

For $Z \in \mathfrak{D} : Z_0 \neq 0$ we can use affine coordinates $Z = (1, Z_1, \ldots, Z_n)$ and represent form ψ on the manifold $\{Z \in V\backslash S : Z_0 = 1\}$, for example, in the form of $\tilde{\psi}(Z_1, \ldots, Z_n)dZ_n$

For each $\xi \in \mathfrak{D}_0^*$, we define

$$\mathfrak{A}\psi(\xi) = \sum_{\nu=1}^{N} \tilde{\psi}(Z^{(\nu)}(\xi))d_\xi Z_n^{(\nu)}(\xi). \tag{2.11}$$

This transform, going back to N. Abel [1], [2], is described in the Ph. Griffiths paper [20] in invariant terms and is called there "Trace mapping" with the designation Trace ψ.

Before than to formulate results relating to the transform (2.11) we introduce following Abel and Griffiths more general transform.

The domain Ω in $\mathbb{C}P^n$ we shall call m-polynomially concave if for any point $Z \in \Omega$ there exists algebraic hypersurface of the degree m passing through Z and being contained in Ω.

Let us denote by Ω^{m^*} the manifold of all hypersurfaces of the degree m contained in Ω.

To the point $\xi \in \Omega^{m^*}$ with the coordinates $\xi = \{\xi_{j_0,...,j_n}; (j_0,...,j_n) \in Z_+^{n+1} : j_0 + j_1 + \cdots + j_n = m\}$ corresponds the hypersurface of the degree m in Ω of the form

$$\Theta_\xi^m = \{Z \in \Omega : \sum_{j_0 + \cdots + j_n = m} \xi_{j_0,...,j_n} Z_0^{j_0} \cdots Z_n^{j_n} = 0\}.$$

Let V be a closed one dimensional submanifold in the pseudoconcave domain $\Omega \subset \mathbb{C}P^n$ and $\psi = \tilde{\psi}(Z_1,...,Z_n)dZ_n$ be a meromorphic 1-form with a discrete set of singularities $S \subset V$.

Let $\Omega_0^{m^*}$ denote the subset in Ω^{m^*} such that for every point $\xi \in \Omega_0^{m^*}$ the intersection of the surface Θ_ξ^m with manifold V is transverse and it forms the finite set $\{Z^{(1)}(\xi),...Z^{(k)}(\xi)\} \subset V \backslash S$.

For $\xi \in \Omega^{m^*}$ let us define

$$\mathfrak{A}(\xi) = \sum_{\nu=1}^k \tilde{\psi}(Z^{(\nu)})(\xi))d_\xi Z_n^{(\nu)}(\xi). \qquad (2.11')$$

Proposition 2.4 [1], [2], [20] *For any meromorphic form ψ on the Riemann surface V in the linearly concave domain $\mathfrak{D}$ (or in m-polynomially concave domain Ω) the transform $\mathfrak{A}\psi$ (2.11) (or (2.11')) defines meromorphic 1-form in the domain $\mathfrak{D}^*$ (or Ω^{m^*}), which is holomorphic in the domain $\mathfrak{D}_0^*$ (or $\Omega_0^{m^*}$).*

Note, that the statement of Proposition 2.2 relating to the general transform (2.11') implies from the statement relating to the transform (2.11). In fact, consider the injective Veronese imbedding of the space $\mathbb{C}P^n$ with homogeneous coordinates $Z = (Z_0,...,Z_n)$ into the projective space of larger dimension $\mathbb{C}P^{N(n,m)}$, $1 + N(n,m) = \binom{m+n}{m}$, with homogeneous coordinates

$$W = \{W_{j_0},...,j_n, j_0 + \cdots + j_n = m\},$$
$$W_{j_0,...,j_m} = g_{j_0,...,j_n}(Z) = Z_0^{j_0} \cdot Z_1^{j_1} \cdot \ldots \cdot Z_n^{j_n},$$
$$j_0 + j_1 + \cdots + j_n = m_0.$$

The mapping $W = g(Z)$ transfers a holomorphic form $\psi(Z)$ on the submanifold V in m-polynomially concave domain $\Omega \subset \mathbb{C}P^n$ to the holomorphic form $\psi(g^{-1}(W))$ on the submanifold $\tilde{V} = g(V)$ in some linearly concave domain $\mathfrak{D} \subset \mathbb{C}P^{N(n,m)}$. In addition, the transform (2.11') for $\psi(Z)$ on V turns into transform (2.11) for $\psi(g^{-1}(W))$ on $\tilde{V}$.

We will show now that the Abel transform (2.11) is a special case of the Radon transform of the form (1.2).

In order to obtain connection between the transform $\psi \to \mathfrak{A}\psi$ of the form (2.11) and $\Phi \to R\Phi$ of the form (1.2) we shall use the fact that the 1-form ψ holomorphic on $V \backslash S$ defines an element of the space $H^{n,n-1}(\mathfrak{D}\backslash S)$ which is represented by $\bar{\partial}$-closed current $\psi \wedge [V \backslash S] \in C_{n,n-1}^{(-\infty)}(\mathfrak{D}\backslash S)$ of the kind (2.2), (2.3). The current of the kind (2.2), (2.3) we shall call analytical in $\mathfrak{D}\backslash S$.

Proposition 2.5 [25] *Let $\mathfrak{D}$ be a linearly concave domain in $\mathbb{C}P^n$, V a closed one-dimensional submanifold in $\mathfrak{D}$, S - a locally finite subset in V, ψ - a holomorphic 1-form on $V \backslash S$. Let F be a $\bar{\partial}$-closed $(n, n-1)$-form of the class $C_{n,n-1}(\mathfrak{D}\backslash S)$, $\bar{\partial}$-cohomologous on $\mathfrak{D}\backslash S$ to the analytical current $\psi \wedge [V \backslash S]$. Then there takes place the equality*

$$RF(\xi) = \mathfrak{A}\psi(\xi) \quad \text{for any} \quad \xi \in (\mathfrak{D}\backslash S)_0^*.$$

Proof. The Radon transform of the kind (1.2) of the form

$$F \in C_{n,n-1}(\mathfrak{D}\backslash S)$$

may be written in the form

$$RF = \sum_{l=0}^{n} f_l(\xi)d\xi_l, \quad \text{where} \quad f_l(\xi) = \int\limits_{\mathbb{C}P^n} [\mathbb{C}P_\xi^{n-1}] \wedge (\xi dZ) \rfloor Z_l F, \quad (2.12)$$

$[\mathbb{C}P_\xi^{n-1}] - (1,1)$-current of integration along hyperplane

$$\mathbb{C}P_\xi^{n-1} = \{Z \in \mathbb{C}P^n : \xi \cdot Z = 0\}, \xi \in (\mathfrak{D}\backslash S)^* \quad \text{(Compare (2.16))}.$$

Let the form F be a $\bar{\partial}$-cohomologous to the current $\psi \wedge [V \backslash S]$, where $[V \backslash S] - (n-1, n-1)$- current of integration of the form (2.16) along manifold $V \backslash S$, and $\psi = \tilde{\psi}(Z)dZ_n$-holomorphic 1-form on $V \backslash S$.

Hence and from (2.12) it follows that the coefficient $f_l(\xi)$ of the form RF may be represented in the form

$$f_l(\xi) = \int\limits_{\mathbb{C}P^n} [\mathbb{C}P_\xi^{n-1}] \wedge [V \backslash S] Z_l(\xi dZ \rfloor dZ_n)\tilde{\psi}(Z) \quad (2.13)$$

$$= \int\limits_{\mathbb{C}P_\xi^{n-1} \cap (V \backslash S)} Z_l(\xi dZ \rfloor dZ_n)\tilde{\psi}(Z),$$

$$\text{where} \quad \mathbb{C}P_\xi^{n-1} \cap (V \backslash S) = \{Z^{(1)}(\xi), \ldots, Z^{(N)}(\xi)\}.$$

We will show now that in the point of intersection

$$Z^{(j)}(\xi) = \{1, Z_1^{(j)}(\xi), \dots, Z_n^{(j)}(\xi)\}$$

of the hyperplane $\mathbb{C}P_\xi^{n-1}$ with manifold $V \backslash S$ there takes place the equality

$$Z_l(\xi dZ \rfloor dZ_n) = \frac{\partial Z_n^{(i)}(\xi)}{\partial \xi_l} \qquad (2.14)$$

In fact, let in the neighbourhood $\mathfrak{U}_j$ of the point $Z^{(j)}(\xi)$ the manifold $V \backslash S$ has the form $(V \backslash S) \cap \mathfrak{U}_j = \{Z \in \mathfrak{U}_j : g_1(Z) = g_2(Z) = \cdots = g_{n-1}(Z) = 0\}$, where $dg_1 \wedge dg_2 \wedge \dots \wedge dg_{n-1} \neq 0$ on $(V \backslash S) \cap \mathfrak{U}_j$, $g_k \in H(\mathfrak{U}_j)$, $k = 1, 2, \dots, n-1$. Then on $(V \backslash S) \cap \mathfrak{U}_j$ we have the relation

$$(\xi dZ) \rfloor dZ_n = \frac{dZ_n}{\xi_1 dZ_1 + \cdots + \xi_n dZ_n}$$

$$= -\left(\det \left| \frac{\partial g_k}{\partial Z_\nu} \right|_{k,\nu=1}^{n-1} \right) \det^{-1} \begin{vmatrix} \frac{\partial g_1}{\partial Z_1} & \cdots & \frac{\partial g_1}{\partial Z_n} \\ \vdots & \ddots & \vdots \\ \frac{\partial g_{n-1}}{\partial Z_1} & \cdots & \frac{\partial g_{n-1}}{\partial Z_n} \\ \xi_1 & \cdots & \xi_n \end{vmatrix}^{-1}$$

On the other hand the functions $Z_l^{(j)}(\xi)$, $l = 1, 2, \dots, n$ satisfy the equalities

$$\frac{\partial g_k}{\partial Z_1} \frac{\partial Z_1^{(j)}(\xi)}{\partial \xi_0} + \cdots + \frac{\partial g_k}{\partial Z_n} \frac{\partial Z_n^{(j)}(\xi)}{\partial \xi_0} = 0, \qquad k = 1, 2, \dots, n-1$$

$$1 + \xi_1 \frac{\partial Z_1^{(j)}(\xi)}{\partial \xi_0} + \cdots + \xi_n \frac{\partial Z_n^{(j)}(\xi)}{\partial \xi_0} = 0.$$

From these equalities it follows that in the point $Z^{(j)}(\xi)$ we have

$$(\xi dZ) \rfloor dZ_n = \frac{\partial Z_n^{(j)}(\xi)}{\partial \xi_0}.$$

Hence, and from Proposition 3.1, §3 it follows that there takes place the equality (2.14) in the point $Z^{(j)}(\xi)$.

From (2.12), (2.13), (2.14) it follows that

$$RF(\xi) = \sum_{j=1}^N \sum_{l=0}^n \tilde{\psi}(Z^{(j)}(\xi)) \frac{\partial Z_n^{(j)}(\xi)}{\partial \xi_l} d\xi_l$$

$$= \sum_{j=1}^{N} \tilde{\psi}(Z^{(j)}(\xi)) d_\xi Z_n^{(j)}(\xi) = \mathfrak{A}\psi(\xi).$$

The proposition is proved.

From Proposition 1.2 and 2.3 follows the proposition.

Proposition 2.6 [25] *For any holomorphic 1-form ψ on the closed one-dimensional submanifold V in the linearly concave domain $\mathfrak{D} \subset \mathbb{C}P^n$ the form $\mathfrak{A}\psi(\xi)$ is a holomorphic 1-form in the dual domain $\mathfrak{D}^*$.*

In fact, formulas (2.2), (2.3) mean that the current $\psi \wedge [V]$ is $\bar{\partial}$-closed in $\mathfrak{D}$. From Proposition 2.3 it follows that $\mathfrak{A}\psi = R(\psi \wedge [V])$ and from Proposition 1.2 it implies that the form $f = R(\psi \wedge [V]$ is holomorphic in $\mathfrak{D}^*$. The following classical result implies from Proposition 2.2 and 2.4.

Theorem 2.3 (i) (N. Abel, 1826 [1], [2]). *Let V be the closed Riemann surface (possibly, with singularities) in $\mathbb{C}P^n$ and ψ-meromorphic form on V. Then the Abel transform (2.11) (or (2.11')) is a rational 1-form on $\mathbb{C}P^n)^*$ (or $(\mathbb{C}P^n)^{m*}$).*

(ii) (K. Jacobi, 1834 [29]). If, in addition, the form ψ is holomorphic on V, then $\mathfrak{A}\psi(\xi) \equiv 0$ on $(\mathbb{C}P^n)^*$ (or $(\mathbb{C}P^n)^{m*}$).

Remark: i) In the original N. Abel papers it was supposed, of course, that V - an algebraic curve in $\mathbb{C}P^2$ and ψ - a rational 1-form. However, due to results of W. Chow, J. -P. Serre and Ph. Griffiths (see [20]) any closed submanifold in $\mathbb{C}P^n$ is algebraic and any meromorphic form on the algebraic manifold is rational. Therefore it is natural to formulate the Abel theorem in the mentioned above form.

ii) The relation $\mathfrak{A}(\xi) = 0$ was obtained by K. Jacobi [29] for any affine algebraic curve $V = \{Z \in \mathbb{C}^2 : q(Z_1, Z_2) = 0\}$ of the degree deg q and rational form $\psi = \frac{p(Z_1,Z_2)dZ_1}{\frac{\partial q}{\partial Z_2}(Z_1,Z_2)}$ under condition deg $p \le \deg q - 3$. However, Proposition 2.1 and computations of Ph. Griffiths [20] show that this Jacobi relation is equivalent to the statement ii) of Theorem 2.1.

Now we give an important refinement of the Abel theorem for the special case when the meromorphic form ψ on $V \subset \mathbb{C}P^n$ is the following one

$$\psi = d\frac{Z_k}{Z_0}, \quad k = 1, 2, \dots, n-1.$$

Corollary 2.3 *If under the conditions of Theorem 1.1 the meromorphic form ψ on the Riemann surface V is equal to $d\frac{Z_k}{Z_0}$, $1 \le k \le n-1$, where $(Z_0, \dots, Z_n)$-homogeneous coordinates of the point $Z \in \mathbb{C}P^n$, then the Abel transform of such a form is the following rational 1-form*

$$\mathfrak{A}\left(d\frac{Z_k}{Z_0}\right)(\xi_0, \xi_1, \dots, \xi_n) = d\phi_k(\xi_0, \xi_1, \dots, \xi_n),$$

where the function $\phi_k(\xi_0, \xi_1, \ldots, \xi_n)$, $1 \leq k \leq n-1$ is the following rational one

$$\sum_{\nu=1}^{N} \frac{b_k^{(\nu)} - \xi_0 \cdot a_k^{(\nu)} + \left(\sum_{j=1}^{n-1} \xi_j b_j^{(\nu)}\right) \cdot a_k^{(\nu)} - \left(\sum_{j=1}^{n-1} \xi_j a_j^{(\nu)}\right) b_k^{(\nu)}}{1 + \sum_{j=1}^{n-1} \xi_j a_j^{(\nu)}}$$

Here $b_j^{(\nu)}$ and $a_j^{(\nu)}$, $j = 1, 2, \ldots, n-1$; $\nu = 1, 2, \ldots, N$ are some fixed complex numbers.

In particular, the function ϕ is linear with respect to ξ_0 and rational with respect to $\xi_1, \ldots, \xi_{n-1}$.

Fixing on irreducible components $\{V_\nu\}$ of the closed one dimensional submanifold V in the concave domain $\mathfrak{D} \subset \mathbb{C}P^n$ the meromorphic forms $\pm d\frac{Z_k}{Z_0}$ and considering their Abel transforms we come, naturally, to the following definition of the Abel transform of holomorphic 1-chain.

Let V be an arbitrary holomorphic 1-chain in the concave domain $\mathfrak{D} \subset \mathbb{C}P^n$, i.e. $V = \sum \epsilon_\nu V_i$, where $\epsilon_\nu = \pm 1$ and irreducible curves $\{V_\nu\}$ are not necessarily all different.

The hyperplane $\mathbb{C}P_\xi^{n-1} = \{Z \in \mathbb{C}P^n : \xi \cdot Z = 0\}$, where $\xi \in \mathfrak{D}_0^*$, intersects V along 0-chain of the form $\mathbb{C}P_\xi^{n-1} \cap V = \sum_{j=1}^{N} \sigma_j \cdot Z^{(j)}(\xi)$, where $\sigma_j = \pm 1$, $Z^{(j)}(\xi)$ is the point with coordinates $\{Z_0^{(j)}(\xi), \ldots, Z_n^{(j)}(\xi)\}$.

We shall call by the Abel transform of the chain V the union of functions on $\mathfrak{D}^*$ of the form

$$\mathfrak{A}_V^{(0)}(\xi) = \sum_{j=1}^{N} b_j; \qquad \mathfrak{A}_V^{(k)}(\xi) = \sum_{j=1}^{N} \sigma_j \frac{Z_k^{(j)}}{Z_0^{(j)}}(\xi), \quad k = 1, 2, \ldots, n-1$$

$$(2.15)$$

Note, that for the case $n = 2$, $k = 1$, $b_j = 1$, $j = 1, 2, \ldots, N$ the relation

$$\frac{\partial^2}{\partial \xi_0^2} \mathfrak{A}_V^{(1)}(\xi) = 0$$

is equivalent to the well-known in algebraic geometry Reiss (see [21]) or Bäcklund (see [9]) relation. This relation for fixed ξ gives necessary and sufficient condition on fixed points $\{Z^{(j)}(\xi)\}$ of the line $\mathbb{C}P_\xi^1$ and on values of the first and of the second derivatives in these points in order that through these points one can draw the algebraic curve V of the degree N with the prescribed behaviour of the second order in these points.

We will show that the Abel transform (2.15) is a particular case of the Radon transform of cohomologies $H^{n-1,n-1}(\mathfrak{D})$ of the form (1.3), (1.9).

We shall remind that due to Lelong's theorem [21], [31] the 1-chain V defines an element of the space $H^{n-1,n-1}(\mathfrak{D})$ which may be represented by $\bar{\partial}$-closed current $[V]$ of the form

$$([V], \chi) = \int_{\text{Reg } V} \chi = \sum \sigma_j \int_{\text{Reg } V_j} \chi, \qquad (2.16)$$

where the form $\chi \in C_{1,1}^{(\infty)}(\mathfrak{D})$ has compact support in $\mathfrak{D}$.

Proposition 2.7 *Let Φ be $\bar{\partial}$-closed form of the class $C_{n-1,n-1}^{(s)}(\mathfrak{D})$ $\bar{\partial}$-cohomologous to the current $[V]$. Then the Radon transform (1.3), (1.9) and the Abel transform (2.15) are connected by the equalities*

$$\rho_k \Phi(\xi) = \frac{\partial^2}{\partial \xi_0^2} \mathfrak{A}_V^{(k)}(\xi), \quad \xi \in \mathfrak{D}_0^*, \quad k = 0, 1, \ldots, n-1.$$

The following proposition implies from Proposition 2.5 and from Corollary 2.1 of Theorem 1.2.

Proposition 2.8 *For any closed in the concave domain $\mathfrak{D} \subset \mathbb{C}P^n$ holomorphic 1-chain V the functions $\frac{\partial^2}{\partial \xi_0^2} \mathfrak{A}_V^{(k)}(\xi)$, $k = 0, 1, \ldots, n-1$, are holomorphic in $\mathfrak{D}^*$, in addition, the function $\mathfrak{A}_V^{(0)}(\xi)$ is constant in every component of $\mathfrak{D}^*$.*

The following theorem implies from Proposition 2.5 and from Corollary 2.1 of the Abel-Jacobi Theorem 2.1.

Theorem 2.4 (M. Reiss, 1837) *Let V be a holomorphic 1-chain in $\mathbb{C}P^n$. Then for the Abel-Radon transform of the current $[V]$ there take place the relations: $\rho_0([V]) = \text{const} \in Z$ and $\rho_k([V]) = 0$, $k = 1, 2, \ldots, n-1$.*

As an immediate result of interpretation of the Abel transform as the Radon transform we give here two results partially inverting Theorems 2.1 and 2.2 and going back to the papers of S. Lie [33], H. Poincaré [38], L .Darboux [21].

Theorem 2.5 [8], [20], [41], [47], [48] *Let V be a holomorphic 1-chain in the line concave domain $\mathfrak{D} \subset \mathbb{C}P^n$ with connected dual set $\mathfrak{D}^*$. If the Abel-Radon transform $\rho_k([V])$ of the form (1.9) and (2.15) of the current $[V]$ is equal to zero for $k = 1, 2, \ldots, n-1$ then 1-chain V is a restriction on $\mathfrak{D}$ of some holomorphic 1-chain in $\mathbb{C}P^n$ i.e. it is a restriction on $\mathfrak{D}$ of algebraic 1-chain.*

Proof. The case of arbitrary n can be easily reduced to the case $n = 2$ by a projection from $\mathbb{C}P^n \backslash \mathbb{C}P^{n-3}$ on $\mathbb{C}P^2$ (see [20]). We give further the proof for the case n=2. For the transform $[V] \to \rho_0([V])$ in view of Proposition 2.5 we have $\rho_0([V]) = l \in Z$. If $l \neq 0$ then instead of 1-chain V we consider 1-chain $\tilde{V} = V - l \cdot \mathbb{C}P_\eta^1$, where $\mathbb{C}P_\eta^1$ is some projective line contained in $\mathfrak{D}$. For 1-chain $\tilde{V}$ we have

$$R([\tilde{V}]) = \{\rho_0([\tilde{V}]), \ \rho_1([\tilde{V}])\} = 0 \text{ in } \mathfrak{D}^*.$$

Based on the consequence of Theorem 1.2 we conclude that for any $\xi \in \mathfrak{D}^*$ the current $[\tilde{V}]$ is $\bar{\partial}$-exact in some linearly concave domain $\mathfrak{D}_\xi$ containing the line $\mathbb{C}P_\xi^1$ i.e.

$$[\tilde{V}]\Big|_{\mathfrak{D}_\xi} = \bar{\partial}X, \quad \text{where} \quad X \in C_{1,0}^{(-\infty)}(\mathfrak{D}_\xi).$$

In view of the Oka-Lelong result (see [23], [26]) $X = \frac{1}{i}\frac{\partial F}{F}$, where F is some meromorphic function in $\mathfrak{D}_\xi$ with divisor of zeros and of poles $[\tilde{V}]_{\mathfrak{D}_\xi}$.

Due to the Hartogs-Levi theorem (see [22], [23]) meromorphic function F in the linearly concave domain $\mathfrak{D}_\xi$ extends up to the meromorphic function on all $\mathbb{C}P^2$. Therefore, the divisor $[\tilde{V}]_{\mathfrak{D}_\xi}$ is a restriction of some holomorphic 1-chain on all $\mathbb{C}P^2$. In view of arbitrariness of $\xi \in \mathfrak{D}^*$ and of connectedness of $\mathfrak{D}^*$ the same is valid and for the divisor $[\tilde{V}]$ in $\mathfrak{D}$ and, consequently, for the divisor $[V] = [\tilde{V}] + l \cdot [\mathbb{C}P_\eta^1]$.

Theorem 2.6 [8], [20], [41], [47] *Let V be a closed one-dimensional submanifold in the linearly concave domain $\mathfrak{D} \subset \mathbb{C}P^n$ with the connected dual set $\mathfrak{D}^*$. Let ψ be a holomorphic 1-form on V, $\psi \neq 0$ on each component of V. If the Abel-Radon transform (2.11) $\mathfrak{A}\psi$ is equal to zero, then, firstly, V is a restriction to $\mathfrak{D}$ of some algebraic curve $\tilde{V}$ in $\mathbb{C}P^n$ and, secondly, the form ψ is a restriction on V of some rational 1-form $\tilde{\psi}$ in $\mathbb{C}P^n$ holomorphic on $\tilde{V}$.*

Proof. The case of arbitrary n can be easily reduced to the case $n = 2$ by a projection from $\mathbb{C}P^n \backslash \mathbb{C}P^{n-3}$ on $\mathbb{C}P^2$ (see [20]). We shall give the proof for the case $n = 2$. We consider $\bar{\partial}$-closed current $\psi \wedge [V]$ in $C_{2,1}^{(-\infty)}(\mathfrak{D})$. For the Radon transform $R(\psi \wedge [V])$ of the current $\psi \wedge [V]$ in view of Proposition 2.3 we have $R(\psi \wedge [V]) = 0$. Based on Theorem 1.1 we conclude that the current $\psi \wedge [V]$ is $\bar{\partial}$-exact in $\mathfrak{D}$, i.e. $\psi \wedge [V] = \bar{\partial}X$, where $X \in C_{2,0}^{(-\infty)}(\mathfrak{D})$. Since $\text{supp}\psi \wedge [V] \subset V$, then X defines holomorphic 2-form $\tilde{X}$ outside of V. From integral representations for solutions of $\bar{\partial}$-equation $\bar{\partial}X = \psi \wedge [V]$ (see [26]) it follows that the form $\tilde{X}$ is meromorphic in $\mathfrak{D}$. Due to the Hartogs-Levi theorem the form $\tilde{X}$

extends to the meromorphic form on all $\mathbb{C}P^2$. Consequently, the form $\tilde{X}$ is rational. Let $\tilde{X} = \frac{P}{Q}$ be an irreducible representation, where P is 2-form on $\mathbb{C}P^2$ with polynomial coefficients of positive homogeneity and Q is a homogeneous polynomial in homogeneous coordinates (Z_0, Z_1, Z_2). It follows from Proposition 2.1 that $dQ \not\equiv 0$ on the surface $\tilde{V} = \{Z \in \mathbb{C}P^2 : Q = 0\}$. We shall consider the current $\left[\frac{P}{Q}\right]$ of the form (2.6) in $\mathbb{C}P^2$. From Proposition 2.1 it follows that in the domain $\mathfrak{D}$ there takes place the equality

$$\bar{\partial}\left[\frac{P}{Q}\right]\Big|_{\mathfrak{D}} = 2\pi i\psi \wedge [V]\|_{\mathfrak{D}}.$$

Let $\psi = \frac{1}{2\pi i}\operatorname{Res}\frac{P}{Q}|_{\tilde{V}}$. Due to Proposition 2.1 we have also $\bar{\partial}[\frac{P}{Q}] = 2\pi i\tilde{\psi} \wedge [\tilde{V}]$ in $\mathbb{C}P^2$. From here it follows that $\tilde{\psi}|_V = \psi|_V$, $V = \tilde{V} \cap \mathfrak{D}$. Besides, due to $\bar{\partial}$- closeness of the current $\tilde{\psi} \wedge [\tilde{V}]$ we conclude that a rational form $\tilde{\psi}$ is holomorphic on $\tilde{V}$. Theorem 2.4 is proved.

Theorem 2.4 may be transformed (see [9]) to the following classical result about surfaces of translation proved at first in a complicated way and for the particular case by S. Lie [33] ($g = 3$), and later, more simply by G. Darboux [11] and in the complete generality ($g \geq 3$) by W. Wirtinger [47].

Theorem 2.7 S. Lie [33], W. Wirtinger [47] *Let S be a germ of the analytical hypersurface in $\mathbb{C}^g$, passing through origin and admitting the double parametric representation of the form*

$$W_\alpha = \sum_{\lambda=1}^{g-1} f_{\alpha\lambda}(Z_\lambda) = \sum_{\lambda=g}^{2g-2} f_{\alpha\lambda}(Z_\lambda).$$

Suppose, that vectors $(f'_{1\lambda}(0), \ldots, f'_{g\lambda}(0))$ for $\lambda = 1, \ldots, 2g-2$ are mutually linearly independent and that for any λ the vector $f''_{1\lambda}(0), \ldots, f''_{g\lambda}(0)$ is not tangent to S. Then $(2g - 2)$ analytical curves in $\mathbb{C}P^{g-1}$: $Z \mapsto (f'_{1\lambda}(Z), \ldots, f'_{g\lambda}(Z))$ are branches of one algebraic curve V in $\mathbb{C}P^{g-1}$ of the degree $2g - 2$.

Theorem 2.5 allows to give the following local characterization of divisors of θ-functions of the Riemann surfaces.

Corollary 2.4 (S. Lie [33], H. Poincaré [38], B. Saint-Donat [41], J. Little [34]). *Under conditions of Theorem 2.5 we suppose that a curve V in $\mathbb{C}P^{g-1}$ is irreducible and without singularities. Then genus of V is equal to g and S coincides with the germ of the divisor of θ-functions of the Riemann surface V and the imbedding $V \subset \mathbb{C}P^{g-1}$ coincides with the canonical imbedding.*

The Corollary 2.2 brings, in turn, to the classical Torelli theorem: a compact Riemann surface is determined up to isomorphism by its polarized Jacobian variety (see [21], [41]).

3. The representation of $\bar\partial$-cohomology by analytical currents and the inverse Abel theorems

Let again $\mathfrak{D}$ be a linearly concave domain in $\mathbb{C}P^n$. The aim of this paragraph is to obtain in terms of the Radon transform the conditions of the representability of elements $H^{n,n-1}(\mathfrak{D})$ and $H^{n-1,n-1}(\mathfrak{D})$ by analytical currents of the form (2.2), (2.3), (2.13) in the domain $\mathfrak{D}$ and to prove in connection with it the more general inverse Abel theorems. Let $\mathfrak{D}^*$ be a dual to $\mathfrak{D}$ open set in $(\mathbb{C}P^n)^*$.

Theorem 3.8 (G. Henkin [25]) *Let for the linearly concave domain* $\mathfrak{D}$ *in* $\mathbb{C}P^n$ *any hyperplane section of the dual set* $\mathfrak{D}^*$ *be contractible. Then the element F from $H^{n,n-1}(\mathfrak{D})$ has a representation in the form of the analytical current $\psi \wedge [V]$ (2.2), (2.3) where V is a closed one-dimensional submanifold in $\mathfrak{D}$ and ψ is a holomorphic 1-form on V iff the Radon transform $f = RF$ (1.2) is represented in the neighbourhood $\mathfrak{D}_\eta^*$ of the fixed point $\eta \in \mathfrak{D}^*$ by the finite sum of the closed holomorphic 1-form in $\mathfrak{D}_\eta^*$*

$$f = \sum_{j=1}^{N} f^{(j)} = \sum_{j=1}^{N}\sum_{k=0}^{n} f_k^{(j)}(\xi)d\xi_k \tag{3.1}$$

such that the functions

$$g_k^{(j)} = f_k^{(j)}(\xi)[f_0^{(j)}(\xi)]^{-1}, \quad k = 1, \ldots, n; \; j = 1, \ldots, N$$

satisfy the equations of the "shock waves"

$$g_k^{(j)} \frac{\partial g_l^{(j)}}{\partial \xi_0} = \frac{\partial g_l^{(j)}}{\partial \xi_k}, \quad k, l = 1, \ldots, n. \tag{3.2}$$

For the case $n = 2$ this result was published in [25] together with applications to the inverse Abel theorem.

Theorem 3.9 *Under the conditions of Theorem 3.1 the element Φ of $H^{n-1,n-1}(\mathfrak{D})$ has a representation in the form of the current $[V]$ (2.16), where V is a holomorphic 1-chain in $\mathfrak{D}$ iff the Radon transform (1.3),*

(1.9): $\phi = \rho\Phi = (\rho_0\Phi, \rho_1\Phi, \ldots, \rho_{n-1}\Phi)$ can be represented in the neighbourhood $\mathfrak{D}^*_\eta$ of the fixed point $(\eta_0, \eta_1, \ldots, \eta_{n-1}, 1) \in \mathfrak{D}^*$ by the finite sum

$$\phi = \sum_{j=1}^{N_+} \phi^{+(j)} - \sum_{j=1}^{N_-} \phi^{-(j)}$$

of the holomorphic vector functions

$$\phi^{\pm(j)} = (\phi_0^{\pm(j)}, \phi_1^{\pm(j)}, \ldots, \phi_{n-1}^{\pm(j)})$$

$$\in H(\mathfrak{D}^*) \oplus H(\mathfrak{D}^*, \mathbb{O}(-2)) \oplus \cdots \oplus H(\mathfrak{D}^*, \mathbb{O}(-2))$$

such that

$$\phi^{\pm(j)}(\xi) = \pm 1, \quad \phi_k^{\pm(j)}(\xi) = \frac{\partial^2}{\partial \xi_j^2} g_k^{\pm(j)}(\xi), \tag{3.3}$$

$$\frac{\partial g_l^{\pm(j)}(\xi)}{\partial \xi_k} = g_k^{\pm}(j) \frac{\partial g_l^{\pm}(j)}{\partial \xi_k},$$

for $k, l = 1, 2, \ldots, n-1$, $\xi \in \mathfrak{D}^*_\eta$.

The current $\psi \wedge [V]$ in $\mathfrak{D}$ of the form (2.2), (2.3) we call semi-algebraic if V is a restriction to $\mathfrak{D}$ of some algebraic curve in $\mathbb{C}P^n$, and ψ is a holomorphic 1-form on V.

The semi-algebraic current $\psi \wedge [V]$ in $\mathfrak{D}$ of the form (2.2), (2.3) we shall call algebraic if the holomorphic 1-form ψ on V is rational.

For the representability of elements of the space $H^{n,n-1}(\mathfrak{D})$ in the form of the semi-algebraic (or algebraic) current there is a criterion noticeably simpler than the formulated above criterion of the analytic representation.

Theorem 3.10 *Under the conditions of Theorem 3.1 the element F of $H^{n,n-1}(\mathfrak{D})$ has a representation in the form of the semi-algebraic (respectively, algebraic) current $\psi \wedge [V]$, where*

$$V = \{Z \in \mathfrak{D} : P_1(Z) = \cdots = P_r(Z) = 0\}, \tag{3.4}$$

$P_1, \ldots, P_r$-homogeneous polynomials and $\mathrm{rank}[\mathrm{grad}\,P_1, \ldots, \mathrm{grad}\,P_r] = n-1$ almost everywhere on V, iff the Radon transform (1.2) RF of the form F may be represented by the following way $RF = d\phi$, where $\phi \in H(\mathfrak{D}^*)$ (respectively $\phi \in H(\mathfrak{D}^*)$ and $d\phi$ is rational), and

$$P_j(\frac{d}{d\xi})\phi = 0, \quad j = 1, 2, \ldots, r, \quad \xi \in \mathfrak{D}^* \tag{3.5}$$

The proof of the necessity in Theorem 3.1, 3.2 is based on the following elementary but very important assertion going back to L. Darboux [11].

Proposition 3.9 *Let V be a one-dimensional analytical submanifold in the concave domain $\mathfrak{D} \subset \mathbb{C}P^n$. Let the hyperplane $\mathbb{C}P_\xi^{n-1}$ transversally intersects V in the points $Z^{(j)}(\xi) = \{1, Z_1^{(j)}(\xi), \ldots, Z_n^{(j)}(\xi)\}$, $j = 1, 2, \ldots, N$. Then for any j we have*

$$Z_k^{(j)} \cdot \frac{\partial Z_l^{(j)}}{\partial \xi_0} = \frac{\partial Z_l^{(j)}}{\partial \xi_k}, \qquad k, l = 1, 2, \ldots, n. \tag{3.6}$$

The detailed proof see in [14].

The proof of the necessity in Theorem 3.1. Let ψ be a holomorphic 1-form on $V \subset \mathfrak{D}$. Then in the neighbourhood of any point $\xi \in \mathfrak{D}_0^*$ the Abel-Radon transform of the current $[\psi_V]$ may be represented in the following way

$$\mathfrak{A}\psi(\xi) = f(\xi) = \sum_{\nu=1}^{N} \psi(Z^{(\nu)}(\xi)) d_\xi Z_n^{(\nu)}(\xi). \tag{3.7}$$

In addition,

$$\xi_0 + \xi_1 Z_1^{(\nu)}(\xi) + \cdots + \xi_{n-1} Z_{n-1}^{(\nu)}(\xi) + Z_n^{(\nu)}(\xi) = 0, \qquad \nu = 1, 2, \ldots, N.$$

From the last equality, it follows that

$$-\frac{\partial Z_n^{(\nu)}(\xi)}{\partial \xi_0} = 1 + \sum_{k=1}^{n-1} \xi_k \frac{\partial Z_n^{(\nu)}(\xi)}{\partial \xi_0}$$

$$-\frac{\partial Z_n^{(\nu)}(\xi)}{\partial \xi_k} = Z_k^{(\nu)} + \sum_{j=1}^{n-1} \xi_j \frac{\partial Z_j^{(\nu)}(\xi)}{\partial \xi_k}.$$

Hence, and from Proposition 3.1 we obtain the equalities

$$-\frac{\partial Z_n^{(\nu)}(\xi)}{\partial \xi_k} = Z_k^{(\nu)} \left(1 + \sum_{j=1}^{n-1} \xi_j \frac{\partial Z_j^{(\nu)}(\xi)}{\partial \xi_0}\right) = -Z_k^{(\nu)} \frac{\partial Z_n^{(\nu)}}{\partial \xi_0},$$

for $\nu = 1, 2, \ldots, N$.

Substituting these equalities in (3.7) we obtain the proof of the necessity in Theorem 3.1.

The proof of the necessity in Theorem 3.2 is yet simpler since for the Radon transform

$$\rho([V]) = \{\rho_0([V]), \rho_1([V]), \ldots, \rho_{n-1}([V])\}$$

we have

$$\rho_k([V]) = \frac{\partial^2}{\partial \xi_0^2} \sum \epsilon_\nu Z_k^{(\nu)}(\xi), \qquad \epsilon_\nu = \pm 1, k \geq 1$$

and the assertion immediately follows from Proposition 3.1.

For the proof of the sufficiency in Theorem 3.1 (for the case $n = 2$) the following two lemmas are very important.

Lemma 3.1 *Let $\mathfrak{D}$ be a linearly concave domain in $\mathbb{C}P^2$, $\mathfrak{D}^*$-a dual domain. If the closed holomorphic 1-form $f(\xi) = f_0 d\xi_0 + f_1 d\xi_1$ satisfies in the neighbourhood $\mathfrak{D}_\eta^*$ of the point $\eta = (\eta_0, \eta_1, 1) \in \mathfrak{D}^*$ the conditions (3.1), (3.2) then in the section $\mathfrak{D}_\eta^{1*} = \{\xi \in \mathfrak{D}_\eta^* : \xi_1 = \eta_1\}$ there take place the equalities*

$$\frac{\partial}{\partial \xi_1} \sum_{j=1}^{N} \left(g_1^{(j)}(\xi_0, \eta_1) \right)^k f_0^{(j)}(\xi_0, \eta_1) = \frac{\partial}{\partial \xi_0} \sum_{j=1}^{N} \left(g_1^{(j)}(\xi_0, \eta_1) \right)^{k+1} f_0^{(j)}(\xi_0, \eta_1)$$

$$(3.8)$$

$$\frac{\partial^k}{\partial \xi_1^k} f_1(\xi_0, \eta_1) = \frac{\partial^k}{\partial \xi_0^k} \left(\sum_{j=1}^{N} \left(g_1^{(j)}(\xi_0, \eta_1) \right)^{k+1} \cdot f_0^{(j)}(\xi_0, \eta_1) \right), \quad k = 0, 1, 2, \ldots$$

$$(3.9)$$

Proof. The equalities (3.9) are consequences of (3.8). For $k = 0$ the equality (3.8) immediately implies from (3.1). Let (3.8) be valid for $k = l$. Then

$$\frac{\partial}{\partial \xi_1} \left(\sum_{j=1}^{N} (g_1^{(j)})^{l+1} f_0^{(j)} \right) = \sum_{j=1}^{N} \left[(l+1)(g_1^{(j)})^l \frac{\partial g_1^{(j)}}{\partial \xi_1} f_0^{(j)} + (g_1^{(j)})^{l+1} \frac{\partial f_0^{(j)}}{\partial \xi_1} \right]$$

(due to (3.2) and to closeness of the form $f^{(j)} = f_0^{(j)} d\xi_0 + f_1^{(j)} d\xi_1$)

$$= \sum_{j=1}^{N} \left[(l+1)(g_1^{(j)})^{l+1} \frac{\partial g_1^{(j)}}{\partial \xi_0} f_0^{(j)} + (g_1^{(j)})^{l+1} \frac{\partial (g_1 f_0^{(j)})}{\partial \xi_0} \right]$$

$$= \frac{\partial}{\partial \xi_0} \left[\sum_{j=1}^{N} (g_1^{(j)})^{l+2} \cdot f_0^{(j)} \right].$$

Lemma is proved.

Lemma 3.2 *Let f be a closed holomorphic (or meromorphic) 1-form in the domain $\mathfrak{D}^* \subset (\mathbb{C}P^2)^*$ and let f satisfy the conditions (3.1), (3.2) in the neighbourhood*

$$\mathfrak{D}_\eta^* = \{\xi \in \mathfrak{D}^* : |\xi_0 - \eta_0|^2 + |\xi_1 - \eta_1|^2 < \epsilon^2\} \quad \text{of the point} \qquad n = (\eta_0, \eta_1, 1).$$

Let the section $\{\xi \in \mathfrak{D} : \xi_1 = \eta_1\}$ be contractible. Consider in the domain

$$\mathfrak{D}_{\eta_1,\epsilon} = \{Z = (1, Z_1, Z_2) \in \mathfrak{D} : |Z_2 + \eta_1 Z_1 + \eta_0| < \epsilon\}$$

the meromorphic function of the form

$$F_{\eta_1}(Z_1, Z_2) = \sum_{\nu=1}^{N} \frac{-f_0^{(\nu)}(-\eta_1 Z_1 - Z_2, \eta_1)}{Z_1 - g_1^{(\nu)}(-\eta_1 Z_1 - Z_2, \eta_1)}. \tag{3.10}$$

Then *i)* the function F_{η_1} extends to the meromorphic function in the domain

$$\mathfrak{D}_{\eta_1} = \{Z = (1, Z_1, Z_2) \in \mathfrak{D} : \xi = (-\eta_1 Z_1 - Z_2, \eta_1, 1) \in \mathfrak{D}^*\},$$

ii) the poles and the residue of the function F_{η_1} do not depend on η_1 and moreover

$$\frac{dF_{\eta_1}(Z_1, Z_2)}{d\eta_1} = +\frac{\partial f_0}{\partial \xi_0}(-\eta_1 Z_1 - Z_2, \eta_1), \quad Z \in \mathfrak{D}_{\eta_1}. \tag{3.11}$$

Proof. We consider the domain

$$\mathfrak{D}_{\eta_1,\epsilon,\mathfrak{R}} = \{Z \in \mathfrak{D}_{\eta_1,\epsilon} : |Z_1| > \mathfrak{R}$$

$$= \sup_{Z \in \mathfrak{D}_{\eta_1,\epsilon}} \sup_{\nu=1,2,\dots,N} |g_1^{(\nu)}(-\eta_1 Z_1 - Z_2, \eta_1)|\}.$$

In this domain the function $F_{\eta_1}(Z_1, Z_2)$ has the following Laurent expansion

$$F_{\eta_1}(Z_1, Z_2) = \sum_{k=0}^{\infty} Z_1^{-k-1} \sum_{\nu=1}^{N} (g^{(\nu)}(-\eta_1 Z_1 - Z_2, \eta_1))^k f_0^{(\nu)}(-\eta_1 Z_1 - Z_2, \eta_1). \tag{3.12}$$

Taking into account, in this expansion, the equalities (3.9) from Lemma 3.1 we obtain the representation of the function $F_{\eta_1}(Z_1, Z_2)$ of the following form

$$F_{\eta_1}(Z_1, Z_2) = \sum_{k=0}^{\infty} -Z_1^{-k-1} \left(\frac{\partial}{\partial \xi_0}\right)^{-k+1} \left(\frac{\partial}{\partial \xi_1}\right)^{k-1} f_1(-\eta_1 Z_1 - Z_2, \eta_1), \tag{3.13}$$

where $\left(\frac{\partial}{\partial \xi_0}\right)^{-k+1}$ is a suitable inverse operator to $\left(\frac{\partial}{\partial \xi_0}\right)^{k-1}$.

Suppose, for the domain G^* being compactly contained in $\mathfrak{D}^*$, that

$$G_{\eta_1} = \{Z = (1, Z_1, Z_2) \in \mathfrak{D} : \xi = (-\eta_1 Z_1 - Z_2, \eta_1, 1) \in G^*\}.$$

For any domain $G^* \subset\subset \mathfrak{D}^*$ one can find such a constant B such that the (3.13) converges and defines a holomorphic function in the domain

$$G_{\eta_1, B} = \{Z = (1, Z_1, Z_2) \in G_{\eta_1} : |Z_1| > B\}.$$

Using now the classical Hartogs-Levi theorem (see [23]) and connectedness of the domain $\mathfrak{D}_{\eta_1}$ we obtain the extension of the meromorphic function $F_{\eta_1}(Z_1, Z_2)$, defined by the equalities (3.10), (3.13) in the domain $\mathfrak{D}_{\eta_1, \epsilon} \cup G_{\eta_1, B}$, to the domain G_{η_1}. Hence, and from the fact that all possible domains G_{η_1} exhaust the domain $\mathfrak{D}_{\eta_1}$ we obtain the assertion (i).

In order to prove (ii) it is sufficient to prove (3.11). Due to the properties of uniqueness it is sufficient to prove (3.11) only in the small domain $\mathfrak{D}_{\eta_1, \epsilon, \mathfrak{R}}$.

In order to compute $\frac{dF_{\eta_1}(Z_1, Z_2)}{\nu}$ we shall use the series (3.12) and the equality

$$\frac{dg(-\eta_1 Z_1 - Z_2, \eta_1)}{d\eta_1} = -Z_1 \frac{\partial g}{\partial \xi_0} + \frac{\partial g}{\partial \xi_1},$$

where $g = g(\xi_0, \xi_1)$, $\xi_0 = -\eta_1 Z_1 - Z_2$, $\xi_1 = \eta_1$.

We shall obtain

$$\frac{dF_{\eta_1}(Z_1, Z_2)}{d\eta_1} = \sum_{k=0}^{\infty} -Z_1^{-k-1} \sum_{\nu=1}^{N} \left(-Z_1 \frac{\partial}{\partial \xi_0} + \frac{\partial}{\partial \xi_1} \right)$$

$$\times \left[g_1^{(\nu)}(\xi_0, \eta_1) \right]^k f_0^{(\nu)}(\xi_0, \eta_1)$$

$$= -\frac{\partial}{\partial \xi_0} \left(\sum_{\nu=1}^{N} -f_0^{(\nu)}(\xi_0, \eta_1) \right) + \sum_{k=1}^{\infty} \frac{1}{Z_1^k} \left[\frac{\partial}{\partial \xi_1} \left(\sum_{\nu=1}^{N} -f_0^{(\nu)}(g_1^{(\nu)})^{k-1} \right) \right.$$

$$\left. - \frac{\partial}{\partial \xi_0} \left(\sum_{\nu=1}^{N} -f_0^{(\nu)}(g_1^{(\nu)})^k \right) \right]$$

Using now the relations (3.1) and (3.8) we obtain

$$\frac{dF_{\eta_1}(Z_1, Z_2)}{d\eta_1} = \frac{\partial}{\partial \xi_0} f_0(-\eta_1 Z_1 - Z_2, \eta_1),$$

where $\xi_0 = -\eta_1 Z_1 - Z_2$.

Lemma 3.2 is proved.

The proof of the sufficiency in Theorem 3.1 for $n = 2$. The case $n \geq 2$ may be reduced to the case $n = 2$. Let, further, for the definiteness, the line Z_0 be contained in $\mathfrak{D}$, i.e. the point $(1, 0, 0) \in \mathfrak{D}^*$. Let the Radon transform $f = RF$ of the $\bar{\partial}$-closed form $F \in C_{2,1}^{(s)}(\mathfrak{D})$ have the properties (3.1), (3.2) in the neighbourhood

$$\mathfrak{D}_\eta^* = \{\xi = (\xi_0, \xi_1, 1) \in \mathfrak{D}^* : |\xi_0 - \eta|^2 + |\xi_1 - \eta_1|^2 \le \epsilon^2\}$$

of the point $\eta = (\eta_0, \eta_1, 1) \in \mathfrak{D}^*$, For any $\xi_1 : |\xi_1 - \eta_1| < \epsilon$ we can on the basis of Lemma 3.2 construct the meromorphic form $F_{\xi_1}(Z_1, Z_2)dZ_1 \wedge dZ_2$ by the starting formula (3.10) (where instead of η_1 one must use ξ_1) in the domain

$$\mathfrak{D}_{\xi_1} = \{Z = (1, Z_1, Z_2) \in \mathfrak{D} : (-\xi_1 Z_1 - Z_2, \xi_1, 1) \in \mathfrak{D}^*\}.$$

This meromorphic form has some polar set V_{ξ_1} and the holomorphic form residue $\psi_{\xi_1}(Z)dZ_2$ on $V_{\xi_1} \subset \mathfrak{D}_{\xi_1}$ (see Proposition 2.1).

The property (ii) in Lemma 3.2 provides with the equalities

$$V_{\xi_1} \cap \mathfrak{D}_{\eta_1} = V_{\eta_1} \cap \mathfrak{D}_{\xi_1} \quad \text{and} \quad \psi_{\xi_1}|_{V_{\xi_1} \cap \mathfrak{D}_{\eta_1}} = \psi_{\eta_1}|_{V_{\eta_1} \cap \mathfrak{D}_{\xi_1}} \qquad (3.14)$$

for all $\xi_1 : |\xi_1 - \eta_1| < \epsilon$.

Consequently, in the domain $G = \bigcup\limits_{\{\xi_1 : |\xi_1 - \eta_1| < \epsilon\}} \mathfrak{D}_{\xi_1}$ the one dimensional submanifold $V = \bigcup\limits_{\{\xi_1 : |\xi_1 - \eta_1| < \epsilon\}} V_{\xi_1}$ is correctly defined.

Note, that the set $\mathbb{C}P^2 \backslash G$ consists of the line $Z_0 = 0$ and of some compact $\mathfrak{K}$ in $\mathbb{C}^2 = \mathbb{C}P^2 \backslash \{Z_0 = 0\}$.

Based on the Wirtinger inequality (see [23]), the area of the set V in the Fubini-Study metric $\mathbb{C}P^2$ in the neighbourhood of the line $Z_0 = 0$ is bounded by the constant depending on N. Therefore, due to the Bishop-Stoll extension theorem [7], [44] the submanifold V extends to the analytical submanifold $\tilde{V}$ in the domain $\tilde{G} = G \cup \{Z_0 = 0\}$.

Consider the 1-form $\psi = \{\psi_{\xi_1}(Z)dZ_2, Z \in V_{\xi_1}\}$ on $\tilde{V}$. Due to Proposition 2.1 this form is holomorphic on V and, in view of elementary estimates, at least meromorphic on $\tilde{V}$ in the sense of (2.1).

From (3.10) it follows that there takes place the equality on $V \cap \mathfrak{D}_{\eta_1, \epsilon}$

$$\psi(Z)|_{V \cap \mathfrak{D}_{\eta_1,\epsilon}} = \text{Res}(F_{\xi_1}dZ_1 \wedge dZ_2)$$

$$= \left(\sum_{\nu=1}^N \frac{-f_0^{(\nu)}(-\xi_1 Z_1 - Z_2, \xi_1)}{\frac{\partial g_1^{(\nu)}(-\xi_1 Z_1 - Z_2, \xi_1)}{\partial \xi_0}}\right)\Bigg|_{V \cap \mathfrak{D}_{\eta_1,\epsilon}} dZ_2$$

This equality, Proposition 3.1 and the definition of the Abel transform (2.11) show, that for $\xi \in \mathfrak{D}_\eta^*$

$$\mathfrak{A}\psi(\xi) = \sum_{\nu=1}^N \frac{f_0^{(\nu)}(\xi_0, \xi_1)d_\xi(-\xi_0 - \xi_1 Z_1^{(\nu)}(\xi))}{-1 - \xi_1 \frac{\partial g_1^{(\nu)}(\xi_0, \xi_1)}{\partial \xi_0}}$$

$$= \left(\sum_{\nu=1}^{N} f_0^{(\nu)}(\xi_0, \xi_1) \right) d\xi_0 + \left(\sum_{\nu=1}^{N} f_1^{(\nu)}(\xi_0, \xi_1) \right) d\xi_1.$$

Let the current $\psi \wedge [\tilde{V}]$ corresponds to the form ψ on $\tilde{V}$. From Proposition 2.3 and Condition (3.1) it follows that $R\psi \wedge [\tilde{V}] = RF = f$ in the neighbourhood $\mathfrak{D}_\eta^*$ of the point $(\eta_0, \eta_1, 1) \in \mathfrak{D}^*$ and, consequently, in view of uniqueness, everywhere in $\tilde{G}^*$.

From the proved necessity in Theorem 3.1 it follows that the properties (3.1), (3.2) for f are held in the neighbourhood of any point $\eta \in \tilde{G}^* = \tilde{G}^* \cup \{(1, 0, 0)\}$.

Therefore, as was mentioned above, in some neighbourhood $\mathfrak{U}_\eta$ of the line $\mathbb{C}P_\eta^1 \subset G$ we can construct the analytical current $\psi' \wedge [V']$ (V' is a complex submanifold in $\mathfrak{U}_\xi$, ψ' is a holomorphic form on V') with the property

$$R\psi' \wedge [V'] = f, \qquad \xi \in \mathfrak{U}_\eta^*.$$

Therefore, the current $\psi \wedge [\tilde{V}] - \psi' \wedge [V']$ satisfies the equality $R(\psi \wedge [\tilde{V}] - \psi' \wedge [V']) = 0$, $\xi \in \mathfrak{U}_\eta$.

Hence, and from Proposition 2.1 and Theorem 2.4 it follows that the current $\psi \wedge [\tilde{V}] - \psi' \wedge [V']$ is a restriction on $\mathfrak{U}_\eta$ of some algebraic (analytical) current in $\mathbb{C}P^2$. Thus, the current $\psi \wedge [\tilde{V}]$ is $\bar{\partial}$-closed, i.e. analytical everywhere in G. Thus, the form ψ is holomorphic on $\tilde{V}$.

Let $\mathbb{C}^2 = \mathbb{C}P^2 \backslash \{Z_0 = 0\}$. Since $\mathfrak{K} = \mathbb{C}^2 \backslash G$ is a compact, then for any $\tilde{\xi}_1 \in \mathbb{C}$ there exists such $\tilde{\xi}_0$ that the line $\mathbb{C}P_\xi^1 = \{Z : \tilde{\xi}_0 + \tilde{\xi}_1 Z_1 + Z_2 = 0\}$ is contained in G, i.e. $(\tilde{\xi}_0, \tilde{\xi}_1, 1) \in G^*$.

From the necessity of conditions (3.1), (3.2) in Theorem 3.1 we conclude that in the neighbourhood of the point $(\tilde{\xi}_0, \tilde{\xi}_1, 1)$ the conditions (3.1), (3.2) are held for the form $f = RF = R\psi \wedge [\tilde{V}]$.

Now, using mentioned above arguments, we can again, and already for all $\xi_1 \in \mathbb{C}$, construct meromorphic forms $\tilde{F}_{\xi_1}(Z_1, Z_2) dZ_1 \wedge dZ_2$ in the domains $\mathfrak{D}_{\xi_1}$ with polar sets $\tilde{V}_{\xi_1}$ and forms-residues $\tilde{\psi}_{\xi_1}(Z) dZ_2$ on $\tilde{V}_{\xi_1} \subset \mathfrak{D}_{\xi_1}$ with the properties (3.14). Consequently, in all domain $\{Z \in \mathfrak{D} : Z_0 = 1\} = \bigcup_{\xi_1 \in \mathbb{C}} \mathfrak{D}_{\xi_1}$ one can correctly define the one-dimensional submanifold $\tilde{V}$ extendible by Bishop-Stoll to submanifold $\tilde{\tilde{V}}$ in $\mathfrak{D}$. In addition, on $\tilde{\tilde{V}}$ one can again correctly define the holomorphic 1-form $\tilde{\psi} = \{\tilde{\psi}_{\xi_1} dZ_2, Z \in \tilde{V}_{\xi_1}\}$ such that $R\tilde{\psi} \wedge [\tilde{V}] = RF(\xi) = f(\xi)$ for any $\xi \in \mathfrak{D}^*$.

Hence, and from Theorem 1.1 it follows that the analytical current $\tilde{\psi} \wedge [\tilde{\tilde{V}}]$ is $\bar{\partial}$-cohomologous to the original form F.

Theorem is proved.

The proof of the sufficiency of the condition (3.3) in Theorem 3.2 is similar, and even simpler, to the proof of the sufficiency of the conditions

(3.1), (3.2) (in Theorem 3.1).

In addition, instead of Lemma 3.2, the following lemma, the proof of which, actually, is contained in [13], will be important.

Lemma 3.3 *Let $\mathfrak{D}$ be linearly concave domain in $\mathbb{C}P^2$. Let the function ϕ_1 belong to the space $H(\mathfrak{D}^*, \mathbb{O}(-2))$ and in the neighbourhood $\mathfrak{D}^*_\eta$ of the point $(\eta_0, \eta_1, 1) \in \mathfrak{D}^*$ it satisfies the condition (3.3). We consider in the domain $\mathfrak{D}_{\eta_1,\epsilon}$ (see Lemma 3.2) the meromorphic function of the form*

$$\psi_{\eta_1}(Z_1, Z_2) = \frac{\prod_{j=1}^{N_+}\left[Z_1 - g_1^{+(j)}(-\eta_1 Z_1 - Z_2, \eta_1)\right]}{\prod_{j=1}^{N_-}\left[Z_1 - g_1^{-(j)}(-\eta_1 Z_1 - Z_2, \eta_1)\right]}.$$

Then the function Φ_{η_1} extends to the meromorphic function in the domain $\mathfrak{D}_{\eta_1}$ (see Lemma 3.2) and the Poincaré-Lelong current $\partial\bar\partial \log |\Phi_{\eta_1}|$ does not depend on η_1 and, moreover,

$$\frac{\partial \log \Phi_{\eta_1}(Z_1, Z_2)}{\partial \eta_1} = -\frac{\partial\phi_1(-\eta_1 Z_1 - Z_2, \eta_1)}{\partial \xi_0}.$$

At the end of the proof of Theorem 3.2, instead of the reference to Theorem 1.1 (i), it is necessary, of course, to use Corollary 1.2 of Theorem 1.2.

The proof of Theorem 3.3 is based, mainly, on Theorem 1.1 but it also uses Theorem 3.1. In fact, the form $\psi \in H^{1,0}(\mathfrak{D} \cap V)$ generates the current $F = \psi \wedge [V] \in H^{n,n-1}(\mathfrak{D})$ and, in addition, the Abel-Radon transforms of ψ and of the current $\psi \wedge [V]$ coincide by Proposition 2.3.

Hence, and from Theorem 1.1 (ii) and from Proposition 2.2 the necessity of the equalities (3.5) in Theorem 3.3 follows immediately.

In order to prove the sufficiency we choose $\eta \in \mathfrak{D}^*$ such that the intersection $\mathbb{C}P^{n-1}_\eta \cap V$ is transverse. Let G be a sufficiently small neighbourhood of the hyperplane $\mathbb{C}P^{n+1}_\eta$ in $\mathfrak{D}$. Further, due to Theorem 1.1 (ii) there exists the form $\psi \in H^{1,0}(G \cap V)$ such that $R\psi \wedge [G \cap V] = RF = d\phi = f$, where ϕ satisfies (3.5) in the neighbourhood of the point η, and V is an algebraic curve, satisfying (3.4).

Due to Theorem 3.1 (necessity) in the neighbourhood of the point η the conditions (3.1), (3.2) for the form f are held. Therefore, using the sufficiency in Theorem 3.1 we can find the representation of the class $F \in H^{n,n-1}(\mathfrak{D})$ by the analytical current $[\tilde\psi[V \cap \mathfrak{D}]]$. If, in addition, $f = d\phi$ is a rational form then $\tilde\psi$ is also a rational form. This fact is a simple consequence of Theorem 3.4 (of the general inverse Abel theorem).

The proof of the sufficiency of the conditions (3.1), (3.2) in Theorem 3.1 brings simultaneously to the following statements which are of an independent interest.

Proposition 3.10 ([25]). *Let a meromorphic 1-form f in the connected domain $\mathfrak{D}^* \subset (\mathbb{C}P^n)^*$ satisfy in the neighbourhood $\mathfrak{U}_\eta$ of the fixed point $\eta = (\eta_0, \eta_1, 1) \subset \mathfrak{D}^*$ the condition (3.1), (3.2) of the Theorem 3.1. Then in the linearly concave domain $\mathfrak{D} = \bigcup_{\xi \in \mathfrak{D}^*} \mathbb{C}P_\xi^{n-1}$ there exist such a closed one-dimensional analytical submanifold V and a meromorphic form ψ on V that for the current $\psi \wedge [V]$ being $\bar{\partial}$-closed outside the discrete subset $S \subset V \subset \mathfrak{D}$ we have $R\psi \wedge [V](\xi) = f(\xi)$ for $\xi \in (\mathfrak{D} \backslash S)^*$.*

Corollary 3.5 ([25]). *Under conditions of Proposition 3.2 the polar set for the form f is a locally finite union of hyperplanes $\bigcup_{z \in S} \{\xi \in \mathfrak{D}^* : \xi \cdot Z = 0\}$, where S is a discrete subset of singularities of ψ on V.*

Proof (of Proposition 3.2). Let $\mathfrak{P}_f$ be a set of poles in $\mathfrak{D}^*$ of the form $f \in \mathfrak{M}^{1,0}(\mathfrak{D}^*)$. The set $\mathfrak{P}_f$ can be represented in $\mathfrak{D}^*$ as the union of two analytic sets $\mathfrak{P}'_f$ and $\mathfrak{P}''_f$, where $\mathfrak{P}'_f$ is a locally finite union of the pieces of hyperplanes, $\mathfrak{P}''_f$ is an analytic set not containing pieces of complex hyperplanes. So, there exists a locally finite set S in $\mathfrak{D}$ such that

(3.14)
$$\mathfrak{P}'_f = \cup_{z \in S}(\mathfrak{D}^* \cap (\mathbb{C}P^{n-1})^*_z),$$

where

$$(\mathbb{C}P^{n-1})^*_z = \{\xi \in (\mathbb{C}P^n)^* : \xi \cdot z = 0\}.$$

Let

$$\mathfrak{D}_0^* = \mathfrak{D}^* \backslash \mathfrak{P}_f, \quad \tilde{\mathfrak{D}}_0^* = \mathfrak{D}^* \backslash \mathfrak{P}'_f$$

$$\mathfrak{D}_0 = \cup_{\xi \in \mathfrak{D}_0^*} \mathbb{C}P_\xi^{n-1} \quad \text{and} \quad \tilde{\mathfrak{D}}_0 = \cup_{\xi \in \tilde{\mathfrak{D}}_0^*} \mathbb{C}P_\xi^{n-1}.$$

We have $\mathfrak{D}_0 \subset \tilde{\mathfrak{D}}_0$. Let us, firstly, prove that, in fact $\mathfrak{D}_0 = \tilde{\mathfrak{D}}_0$. Indeed, if $z^* \in \tilde{\mathfrak{D}}_0 \backslash \mathfrak{D}_0$, then there exists $\xi^* \in \mathfrak{P}''_f \backslash \mathfrak{P}'_f$ such that $z^* \in \mathbb{C}P_{\xi^*}^{n-1}$.

From (3.14) it follows that $z^* \notin S$. Hence, there exists $\xi \in \mathfrak{D}_0^* \cap (\mathbb{C}P^{n-1})^*_{z^*}$, consequently, $z^* \in \mathfrak{D}_0$.

Let us prove, further, that in $\mathfrak{D}_0$ there exists a closed one-dimensional analytic submanifold V_0 and a holomorphic form ψ_0 on V_0 such that for the analytic current $\psi_0 \wedge [V_0]$ in $\mathfrak{D}_0$ we have

$$R(\psi_0 \wedge [V_0])(\xi) = f(\xi), \quad \xi \in fD_0^*.$$

In order to prove this it is natural to apply the scheme of the proof of theorem 3.1 which is based on lemmas 3.1, 3.2. But this proof (see lemma 3.2) uses the condition of the contractibility of hyperplane sections of $\mathfrak{D}_0^*$, which, generally, is not satisfied. In order to overcome this difficulty let us represent $\mathfrak{D}_0^*$ as a locally finite covering $\mathfrak{D}_0^* = \cup_{j=1}^\infty \bar{\mathfrak{D}}_j^*$ of such closed analytic polyhedres $\bar{\mathfrak{D}}_j^*$ that

i) $\eta \in \bar{\mathfrak{D}}_1^*$,

ii) $\bar{\mathfrak{D}}_j^* \cap \bar{\mathfrak{D}}_k^* = \emptyset$ for any j, k,

iii) $\bar{\mathfrak{D}}_j^* \cap \bar{\mathfrak{D}}_{j+1}^* \neq \emptyset, j = 1, 2, \cdots$

iv) any hyperplane section of any $\bar{\mathfrak{D}}_j^*$ is contractible;

v) any smooth piece of any $b\bar{\mathfrak{D}}_j^*$ is filtered by $(n-1)$-dimensional domains in some hyperplanes of $(\mathbb{C}P^n)^*$.

Using lemmas 3.1, 3.2 as in the proof of theorem 3.1 we can construct in the neighborhood $U(\bar{\mathfrak{D}}_1)$ of $\mathfrak{D}_1 = \cup_{\xi \in \mathfrak{D}_1^*} \mathbb{C}P_\xi^n$ such an analytic current $\psi_1 \wedge [V_1]$, that

$$(3.15) \qquad R(\psi_1 \wedge [V_1]) = (\xi) = f(\xi), \xi \in \bar{\mathfrak{D}}_1^*.$$

Choose $\eta^{(1)} \in \bar{\mathfrak{D}}_1^* \cap \bar{\mathfrak{D}}_2^*$. By theorem 3.1 in the neighborhood of $\eta^{(1)}$ the equations (3.1), (3.2) are valid.

Using again lemmas 3.1, 3.2 we can obtain in the neighborhood $U(\bar{\mathfrak{D}}_2)$ of $\mathfrak{D}_2 = \cup_{\xi \in \mathfrak{D}_2^*} \mathbb{C}P_\xi^{n-1}$ the analytic current $\psi_2 \wedge [V_2]$ such that V_2 is the extension of V_1 from $\mathfrak{D}_1$ in $\mathfrak{D}_2$ and ψ_2 is the extension of ψ_1 from V_1 in V_2.

By induction for any $j = 3, 4, \cdots$ we can obtain in the neighborhood $U(\bar{\mathfrak{D}}_j)$ the analytic current $\psi \wedge [V_j]$ such that V_j is the extension of V_{j-1} and ψ_j is the extension of ψ_{j-1}.

Let $G_0^* = \cup_{j=1}^\infty \mathfrak{D}_j^* \cup (\bar{\mathfrak{D}}_j^* \cap \bar{\mathfrak{D}}_{j+1}^*))$. The intersection of the analytic chain $\cup_{j=1}^\infty V_j$ with the domain $G_0 = \cup_{\xi \in G_0^*} \mathbb{C}P_\xi^{n-1}$ gives a closed analytic set $\tilde{V}_0 \subset G_0$. The forms $\psi_j \in H^{1,0}(V_j)$ define the form $\tilde{\psi}_0 \in H^{1,0}(\tilde{V}_0)$. From (3.15) and from the connectedness of G_0^* we obtain the equality

$$(3.16) \qquad R(\tilde{\psi}_0 \wedge [\tilde{V}_0]) = f(\xi) \quad \forall \xi \in G_0^*$$

By construction of G_0^* we have $\mathfrak{D}_0^* = G_0^* \cup (\cup_{\nu=1}^\infty \Gamma_\nu)$, where $\{\Gamma_\nu\}$ are some components of boundaries of $\{\mathfrak{D}_j\}$. By the property v) of the covering $\{\bar{\mathfrak{D}}_j^*\}$ any real hypersurface Γ_ν has the form

$$\Gamma_\nu = \cup_{z \in \gamma_\nu} (b\mathfrak{D}_{j(\nu)}^*) \cap (\mathbb{C}P^{n-1})_z^*,$$

where γ_ν is some real analytic curve in $\mathfrak{D}_0$. Hence $\mathfrak{D}_0 = G_0 \cup (\cap_{\nu=1}^\infty \gamma_\nu)$.

Under small deformations of the covering $\{\mathfrak{D}_j^*\}$ by the automorphism of $(\mathbb{C}P^n)^*$ the construction of the analytic chain $\cup_{j=1}^\infty V_j$ is not changed but the curves $\{\gamma_\nu\}$ can be made transversal to this chain. So, we can suppose that the set $\mathfrak{A} = (\cup_{j=1}^\infty \gamma_\nu) \cap (\cup_{j=1}^\infty V_j)$ is locally finite. Then the analytic set $\tilde{V}_0$ is closed not only in G_0 but also in $\mathfrak{D}_0 \backslash \mathfrak{A}$.

By the Remmert-Stein theorem the set $\bar{\tilde{V}}_0 = V_0$ gives a closed analytic set in $\mathfrak{D}_0$. From the boundedness of $\tilde{\psi}_0$ on $\tilde{V}_0$ it follows that the form $\tilde{\psi}_0$ can be holomorphically extended on V_0. So, we have analytic current $\psi \wedge [V_0]$ in $\mathfrak{D}_0 = \bar{\mathfrak{D}}_0$. By the Abel theorem $R(\psi_0 \wedge [V_0])(\xi)$ is the holomorphic form in $\bar{\mathfrak{D}}_0$. By the construction for $\xi \in \mathfrak{D}_0^*$ we have

$$R(\psi_0 \wedge [V_0])(\xi) = f(\xi), \quad \xi \in \mathfrak{D}_0^*.$$

Hence, the set $\mathfrak{P}''_f$ is empty. Proposition 3.2 and its consequence 3.1 are proved simultaneously.

Proposition 3.11 ([25]). *Let $\mathfrak{D} \subset \tilde{\mathfrak{D}}$ be linearly concave domains in $\mathbb{C}P^n$ with the connected dual set $\mathfrak{D}^* \subset \tilde{\mathfrak{D}}^*$. Let V be a closed one-dimensional analytical submanifold in $\mathfrak{D}$ and ψ - a holomorphic 1-form on V. Let the Abel transform $f(\xi) = \mathfrak{A}\psi(\xi)$ extend to the holomorphic 1-form $\tilde{f}$ in the domain $\tilde{\mathfrak{D}}^*$. Then in the domain $\tilde{\mathfrak{D}}$ there exist a closed one-dimensional analytical submanifold $\tilde{V}$ and a holomorphic 1-form $\tilde{\psi}$ on $\tilde{V}$ such that $\tilde{V} \cap \mathfrak{D} = V$ and $\tilde{\psi}|_V = \psi$.*

Proof. Due to Proposition 2.3 and to Theorem 3.1 (the necessity) the form $f = \mathfrak{A}\psi(\xi) = R\psi \wedge [V]$ has the properties (3.1), (3.2) in the neighbourhood of the fixed point $\eta = (\eta_0, \eta_1, \ldots, \eta_{n-1}, 1) \in \mathfrak{D}^*$. Due to Proposition 3.2 in the domain $\tilde{\mathfrak{D}}$ there exist a closed one-dimensional complex submanifold $\tilde{V}$ and a holomorphic form $\tilde{\psi}$ on $\tilde{V}$ such that $R\tilde{\psi} \wedge [\tilde{V}](\xi) = R(\psi \wedge [V])(\xi)$ for any $\xi \in \mathfrak{D}^*$.

Due to Theorem 2.4 the current $\tilde{\psi} \wedge [\tilde{V}] - \psi \wedge [V]$ is an algebraic, i.e. it has the form $\psi' \wedge [V']$, where V' - an algebraic curve in $\mathbb{C}P^n$, and ψ' - a rational 1-form holomorphic on V'. Thus, for the current $\psi \wedge [V]$ we have the representation $\psi \wedge [V] = \tilde{\psi} \wedge [\tilde{V}] + \psi' \wedge [V']$, i.e. $V = \tilde{V} \cap \mathfrak{D}$ and $\psi = \tilde{\psi}|_V$, where $\tilde{V} = \tilde{V} + V'$ is an analytical one-dimensional submanifold in $\tilde{\mathfrak{D}}$, and $\tilde{\psi}$ is a holomorphic 1-form on $\tilde{V}$ equal, by definition, to $\tilde{\psi}$ on $\tilde{V}$ and to ψ' on V'. Proposition 3.3 is proved.

Now we shall prove more general, than in §2, the inverse Abel theorems as simple consequences of the results obtained in this paragraph.

Theorem 3.11 ([25]). *Let V be a closed one-dimensional submanifold in the linearly concave domain $\mathfrak{D} \subset \mathbb{C}P^n$ with the connected dual set $\mathfrak{D}^* \subset (\mathbb{C}P^n)^*$.*

Let ψ be a holomorphic 1-form on V, $\psi \not\equiv 0$ on each component of V. Then the following statements are equivalent:

a) *$V = \tilde{V} \cap \mathfrak{D}$, where $\tilde{V}$ is a closed algebraic curve in $\mathbb{C}P^n$ and $\psi = \tilde{\psi}|_V$, where $\tilde{\psi}$ is a rational 1-form on $\mathbb{C}P^n$.*

b) *The Abel transform $\mathfrak{A}(\psi)(\xi)$ is a rational 1-form on $\mathfrak{D}^*$.*

Remark: Theorem 2.1 (Abel) proves that the property a) implies b). Theorem 2.4 (Lie-Darboux-Wirtinger-Saint-Donat-Griffiths) proves a), if the condition $\mathfrak{A}(\xi) \equiv 0$ is held.

Theorem 3.4 in the mentioned above form, strengthening Theorem 2.4, is obtained in [25].

Proof. We shall prove that a) follows from b). Due to Theorem 3.1 (the necessity) and to Proposition 2.3 the rational form $f(\xi) = \mathfrak{A}\psi(\xi)$ has the properties (3.1), (3.2) of Theorem 3.1 in the neighbourhood of any point $\eta \in \mathfrak{D}^*$. In view of proposition 3.2 in $\mathbb{C}P^n$ there exists such a closed one-dimensional analytical submanifold V' and a meromorphic form ψ' on V' that for the current $\psi' \wedge [V']$ we have $R\psi' \wedge [V'](\xi) = f(\xi)$ for any $\xi \in \mathfrak{D}^*$. Due to Theorem 2.4 the current $\psi \wedge [V] - \psi' \wedge [V']$ is an algebraic, i.e. it has the form $\tilde{\psi} \wedge [\tilde{V}]$, where $\tilde{V}$ is an algebraic curve in $\mathbb{C}P^n$, and $\tilde{\psi}$ is a rational 1-form holomorphic on $\tilde{V}$. Let $\tilde{\tilde{V}} = \tilde{V} + V'$ and $\tilde{\psi}$- such a meromorphic form on $\tilde{\tilde{V}}$ that $\tilde{\tilde{\psi}} = \tilde{\psi}$ on $\tilde{V}$ and $\tilde{\tilde{\psi}} = \psi'$ on V'. Then we have $V = \tilde{\tilde{V}} \cap \mathfrak{D}$ and $\psi = \tilde{\tilde{\psi}}|_V$. In addition, $\tilde{\tilde{V}}$ is an algebraic curve in $\mathbb{C}P^n$ due to the Chow theorem and $\tilde{\tilde{\psi}}$ is a rational form due to the principle G.A.G.A. (see [20], [42]).

Theorem 3.3 and 3.4 allow to strengthen not only Theorem 2.4 but and Theorem 2.3.

Theorem 3.12 *Let $V = \sum_{\nu=1} \epsilon_\nu V_\nu$ be a holomorphic 1-chain in the linearly concave domain $\mathfrak{D} \subset \mathbb{C}P^n$ with a connected dual set $\mathfrak{D}^*$. Then the following statements are equivalent.*

a) *The chain V is a restriction to $\mathfrak{D}$ of some algebraic 1-chain in $\mathbb{C}P^n$ with support supp V of the form (3.4).*

b) *All the Abel transform (2.12) $\mathfrak{A}_V^{(k)}(\xi)$ of the chain V are linear with respect to $\xi_0; \xi \in \mathfrak{D}^*$, $k = 1, 2, \ldots, n-1$.*

c) *One of the Abel transforms $\mathfrak{A}_V^{(k)}(\xi)$, $k \in \{1, 2, \ldots, n-1\}$ is rational with respect to $\xi = (\xi_0, \ldots, \xi_n) \in \mathfrak{D}^*$.*

d) *One of the Abel transforms $\mathfrak{A}_V^{(k)}(\xi)$, $k \in \{1, 2, \ldots, n-1\}$ satisfies a system of the linear differential equations of the form (3.5).*

The equivalence of the statments a) and b) is the contents of Theorem 2.3. The statment c) follows from a) due to consequence 2.1 of the Abel theorem 2.1. Theorem 3.4 in application to the form $\psi = \epsilon_\nu d\frac{Z_k}{Z_0}$ on V_ν, $\nu = 1, \ldots, N$ proves that a) follows from c). Theorem 3.3 in application to the same form on V proves the equivalence of a) and d).

4. The complex Plateau problem in $\mathbb{C}P^n$

Let γ be a real closed 1-chain in $\mathbb{C}P^n$, i.e. γ - a linear combination, with integer coefficients, of the closed oriented rectifiable curves γ_j in $\mathbb{C}P^n : \gamma = \sum k_j \gamma_j$.

We suppose, that 1-chain γ satisfies the condition:

suppγ contains such a closed subset τ with Hausdorff's zero measure
that supp$\gamma\backslash\tau$ is a submanifold of the class C^1.   (4.1)

In this paragraph we shall give conditions under which 1-chain γ is a
boundary $\gamma = bV$ of some holomorphic 1-chain $V = \sum l_j V_j$ in $\mathbb{C}P^n$. If
γ is a closed curve, then the examined question is a special case of the
Plateau problem, since an analytical set S with $bS = \gamma$ minimizes locally
the area S (in the Fubini-Study metric of the space $\mathbb{C}P^n$) in the set of
all rectifiable surfaces with the given boundary γ (see [23]).

From results of J. Wermer [46], E. Bishop - G. Stolzenborg [45], H.
Alexander [3] it follows (see [49]) that the rectifiable closed curve γ in
$\mathbb{C}^n = \mathbb{C}P^n\backslash\mathbb{C}P^{n-1}$ is a boundary $\gamma = bS$ (in the sense of currents) of some
Riemann surface (maybe with singularities) compactly contained in $\mathbb{C}^n$ iff
for any $k = 1, 2, \ldots, n$ and any polynomial $P(Z)$, $Z = (Z_1, Z_2, \ldots, Z_n) \in$
$\mathbb{C}^n$ there takes place the equality

$$\int_\gamma P(Z)dZ_k = 0.   \qquad (4.2)$$

The following example of T.C. Dinh, based on an idea of B. Lawson
shows that the complex Plateau problem in $\mathbb{C}P^n$ cannot be, in an obvious
way, reduced to the corresponding problem in $\mathbb{C}^n$ by choosing a suitable
hyperplane in $\mathbb{C}P^n$ as an infinite plane. Let $f : \mathbb{C} \to \mathbb{C}P^2$, $\lambda \to (\lambda, \lambda^3 +$
$1, e^\lambda - 1)$. Then for any sufficiently large $R \gg 0$ any projective line
$\mathbb{C}P^1_\xi = \{Z : \xi_0 Z_0 + \xi_1 Z_1 + \xi_2 Z_2 = 0\}$ intersects the Riemann surface
$S = f(B(0, R))$ which has the curve $\gamma = f(bB(0, R))$ as a boundary.
Here, $B(0, R) = \{\lambda \in \mathbb{C} : |\lambda| \leq R\}$.

We shall give a solution of the formulated problem in $\mathbb{C}P^n$ obtained
in [13], [14].

Let $W_0, W_1, \ldots, W_n$ be such homogeneous coordinates in $\mathbb{C}P^n$ that
the hyperplane $\{W_0 = 0\}$ does not intersect suppγ. We introduce affine
coordinates $Z_k = \frac{W_k}{W_0}$, $k = 1, 2, \ldots, n$ in $\mathbb{C}^n \simeq \mathbb{C}P^n\backslash\{Z_0 = 0\}$. Let
$\mathfrak{D} = \mathbb{C}P^n\backslash$supp$\gamma$. Let $\mathfrak{D}^*$ be a set in $(\mathbb{C}P^n)^*$ dual to $\mathfrak{D}$, i.e. $\mathfrak{D}^* = \{\xi \in$
$(\mathbb{C}P^n)^* : \xi \cdot Z \neq 0, Z \in$ supp$\gamma\}$.

Theorem 4.13 (P. Dolbeault, G. Henkin [13], [14]). *Let γ be a
closed 1-chain in $\mathbb{C}P^n$ with the property (4.1). Then the following conditions are equivalent*

  *(i) the chain γ is a boundary of the holomorphic 1-chain V with a finite
       mass in $\mathbb{C}P^n\backslash$supp$\gamma = \mathfrak{D}$*

(ii)   there exists the point $\eta = (\eta_0, \eta_1, \ldots, \eta_{n-1}, \eta_n) \in \mathfrak{D}^*$ that in the neighbourhood $\mathfrak{U}_\eta$ of which the functions of the form

$$G_k(\xi) = \frac{1}{2\pi i} \int\limits_\gamma Z_k \frac{d_z(\xi \cdot Z)}{\xi \cdot Z}, \qquad k = 1, 2, \ldots, n; \qquad (4.3)$$

$$\xi \cdot Z = \xi_0 + \xi_1 Z_1 + \cdots + \xi_n Z_n$$

satisfy the conditions

$$G_k(\xi) = \sum_{j=1}^{N^+} f_{jk}^+(\xi) - \sum_{j=1}^{N^-} f_{jk}^-(\xi), \qquad k = 1, 2, \ldots, n,$$

where the functions $f_{j,k}^\pm \in H(\mathfrak{U}_\eta)$ and for any j they satisfy the equation

$$f_{jk}^\pm \frac{\partial f_{jl}^\pm}{\partial \xi_0} = \frac{\partial f_{jl}^\pm}{\partial \xi_k}; \qquad k, l = 1, 2, \ldots, n. \qquad (4.4)$$

Corollary 4.6 *If under the condition (ii) of Theorem 4.1 we have $G_k(\xi) = 0$, $k = 1, 2, \ldots, n$ in the neighbourhood of the fixed point $\eta = (1, 0, \ldots, 0)$ i.e. if we have $N^+ = N^- = 0$ in (4.4) then this equality turns into moment condition (4.2) in the space $\mathbb{C}^n \simeq \mathbb{C}P^n|\{W_0 = 0\}$ and the condition (i) is reduced to the Wermer-Alexander statement: there exists an open Riemann surface S in $\mathbb{C}^n$ such that $\gamma = bS$.*

We shall state here the connection, not being contained in the explicit form in [13], [14], of Theorem 4.1 with the Abel-Radon transform theory represented in previous paragraphs.

For the fixed 1-chain $\gamma = \sum k_j \gamma_j$ we consider the current $[\gamma] \in C_{n-1,n}^{(-\infty)}(\mathbb{C}P^n)$ by the formula $([\gamma], \chi) = \int_\gamma \chi = \sum_j k_j \int_{\gamma_j} \chi$, where $\chi \in C_{1,0}^{(\infty)}(\mathbb{C}P^n)$. In view of $H^{n-1,n}(\mathbb{C}P^n) = 0$ (see [21]) we can solve $\bar\partial$-equation of the form

$$\bar\partial\Phi = [\gamma] \qquad \text{where} \qquad \Phi \in C_{n-1,n-1}^{(-\infty)}(\mathbb{C}P^n). \qquad (4.5)$$

The formulas from [27] give the following explicit solution of (4.5):

$$\Phi(Z) =$$

$$\frac{(-1)^n}{2(2\pi i)^{n+1}(n-1)!} \int \frac{\det[\bar\xi, \bar Z, \overbrace{d\bar Z, \ldots, d\bar Z}^{n-1}] \wedge \det[d\xi, d\xi, \overbrace{dZ, \ldots, dZ}^{n-1}]}{(1 - \bar\xi Z)(1 - \xi \bar Z)^n},$$

where $Z \in S^{2n+1}$; $\xi \in P^{-1}(\gamma) \cap S^{2n+1}$; $S^{2n+1} = \{\xi \in \mathbb{C}^{n+1} : |\xi| = 1\}$; $\xi \mapsto P(\xi)$ is the canonical projection from $\mathbb{C}^{n+1}\backslash\{0\}$ to $\mathbb{C}P^n$, the column $d\bar{Z}$ is repeated in $\det[\ldots](n-1)$ times. Hence, $\Phi \in C^{(\infty)}_{n-1,n-1}(\mathfrak{D})$, where $\mathfrak{D} = \mathbb{C}P^n \backslash \gamma$, besides, due to (4.5) $\bar{\partial}\Phi = 0$ in $\mathfrak{D}$.

Since $\mathfrak{D}$ is a linearly concave domain it is natural to consider the Radon transform $\rho\Phi$ (1.3), (1.9) of the form Φ.

Proposition 4.12 *For the solution Φ of the equation (4.5) the values of the transform $\rho\Phi = \{\rho_0\Phi, \rho_1\Phi, \ldots, \rho_{n-1}\Phi\}$ in the points $\xi \in \mathfrak{D}^*$ satisfy the relations*

$$\rho_0\Phi \;=\; G_0(\xi) = \frac{1}{2\pi i}\int\limits_{\gamma}\frac{d_z(\xi \cdot Z)}{\xi \cdot Z},$$

$$\rho_k\Phi \;=\; \frac{\partial^2}{\partial\xi_0^2}G_k(\xi) = k = 1,2,\ldots,n-1.$$

From this proposition it follows that for Φ satisfying (4.5) the values $\rho\Phi$ depend only on γ. Hence, and from Corollary 1.2 we obtain that the class of cohomology Φ in $H^{n-1,n-1}(\mathbb{C}P^n \backslash \gamma)$ is uniquely defined by the chain γ. From Proposition 4.1 and from the necessity in Theorem 3.2 it follows that if $\gamma = bV$, where V is a holomorphic 1-chain in $\mathbb{C}P^n \backslash \gamma$, then the functions $\{G_k(\xi)\}$ of the form (4.3) must satisfy the conditions (4.4) = (3.3). Thus, we obtain the necessity of the condition (ii) for the validity of (i). The sufficiency in Theorem 3.2 guarantees the representation of the class $\Phi \in H^{n-1,n-1}(\mathfrak{D})$ by the analytical current V under realization of the conditions (3.3). Therefore, it is natural to suppose that the condition (ii) is not only necessary but also it is sufficient for the validity of (i). However, we cannot immediately derive this statement from Theorem 3.2 since in conditions of this theorem the connectedness of the dual set $\mathfrak{D}^*$ is essential. But in the conditions of Theorem 4.1 the set dual to the domain $\mathfrak{D} = \mathbb{C}P^n \backslash \gamma$ is not connected. It is possible to overcome this difficulty by transfering the condition (3.3) for $\rho([V])$ from the component to the component of $\mathfrak{D}^*$ with help of analysis of the jumps of the integrals (4.3) (see the proof of theorem 4.1 in [13], [14], where the Harvey-Lawson techniques is essentially used).

The simplest variant of the inverse Abel theorems - Theorem 2.3 allows to obtain also the following statement concerning the uniqueness of the complex Plateau problem in $\mathbb{C}P^n$.

Proposition 4.13 ([14]). *Let γ be a closed 1-chain in $\mathbb{C}P^n$ with the property (4.1) and V and $\tilde{V}$ - two holomorphic 1-chains in $\mathbb{C}P^n \backslash \gamma$ with the property $bV = b\tilde{V} = \gamma$. Then the chain $V - \tilde{V}$ is algebraic.*

In fact, due to Proposition 4.1, the Radon transforms $\rho([V])$ and $\rho([\tilde{V}])$ of the currents $[V]$ and $[\tilde{V}]$ are equal. Therefore, $\rho([V] - [\tilde{V}]) = 0$ in $\mathfrak{D}^*$. Hence, and from Theorem 2.3 it follows that the chain $V - \tilde{V}$ is algebraic. This corollary gives the refinement of the Wermer [46] result, where under condition (4.8) meromorphic forms ψ on S with boundary value ϕ on bS was constructed. Theorem 4.1 admits the following natural refinement.

Theorem 4.14 *Let γ be a closed curve in $\mathbb{C}P^n$ with the property (4.1), and ϕ be an integrable $(1,0)$-form on γ with measure coefficient and with the property*

$$\int_\gamma \phi = 0. \tag{4.6}$$

Then the following conditions are equivalent:

(i) *under suitable orientation γ is a boundary $\gamma = bS$ of the Riemann surface S (possibly with singularities) in $\mathbb{C}P^n \setminus \gamma$ and ϕ is a boundary value on bS of some holomorphic $(1,0)$-form ψ on S.*

(ii) *there exists the point $\eta \in \mathfrak{D}^*$ in the neighbourhood of which the correctly defined in $\mathfrak{D}^*$ 1-form*

$$f(\xi) = \frac{1}{2\pi i} \sum_{j=0}^n \left(\int_{z \in \gamma} \frac{Z_j \cdot \phi}{\xi \cdot Z} \right) d\xi_j \tag{4.7}$$

satisfies the conditions (3.1), (3.2) of Theorem 3.1.

Corollary 4.7 *If under the condition (ii) of Theorem 4.2 we have $f(\xi) \equiv 0$ in the neighbourhood of the point $\eta = (1,0,\ldots,0)$, i.e. if $N = 0$ in (3.1) then this equality turns into the following moment condition in $\mathbb{C}^n = \mathbb{C}P^n \setminus \{Z_0 = 0\}$*

$$\int P(Z)\phi = 0 \tag{4.8}$$

for any polynomial $P(Z)$ of affine coordinates $Z_1, \ldots, Z_n$ and the assertion consists of the following: under the condition (4.8) there exists an open Riemann surface S (possibly with singularities) compactly contained in $\mathbb{C}^n$ and holomorphic forms ψ on S such that $\gamma = bS$ and ϕ is a boundary value on bS of the form ψ.

The proof of Theorem 4.2 is similar to the proof of Theorem 4.1. For the curve γ and for the form ϕ on γ we shall introduce the current $\phi \wedge [\gamma] \in C_{n,n}^{(-\infty)}(\mathbb{C}P^n)$ by the formula

$$(\phi \wedge [\gamma], \chi) = ([\gamma], \phi\chi) = \int_\gamma \phi\chi, \qquad \text{where} \qquad \chi \in C^{(\infty)}(\mathbb{C}P^n).$$

One condition (4.6) is sufficient for the solvability of the $\bar\partial$-equation

$$\bar\partial F = \phi \wedge [\gamma], \qquad \text{where} \qquad F \in C^{(-\infty)}_{n,n-1}(\mathbb{C}P^n), \qquad (4.9)$$

since $H^{n,n}(\mathbb{C}P^n) = \mathbb{C}$ (see [21]).

The formulas from [27] give the following explicit solution F of (4.9):

$$F(Z) =$$

$$\frac{(-1)^n}{(2\pi i)^{n+1} n!} \int_{\xi \in P^{-1}(\gamma) \cap S^{2n+1}} \frac{\det[\bar\xi, \bar Z, \overbrace{d\bar Z, \ldots, d\bar Z}^{n-1}] \wedge \det[d\xi, \overbrace{dZ, \ldots, dZ}^{n}]}{(1 - \bar\xi Z)(1 - \xi \bar Z)^n},$$

where $Z \in S^{2n+1}$.

Hence, $F \in C^{(\infty)}_{n,n-1}(\mathfrak{D})$, where $\mathfrak{D} = \mathbb{C}P^n \backslash \gamma$, and, in addition, $\bar\partial F = 0$ in $\mathfrak{D}$.

We consider the Radon transform (1.2) RF of the form F.

Proposition 4.14 *The value $RF(\xi)$ in the points $\xi \in \mathfrak{D}^*$ satisfy the equality*

$$RF(\xi) = f(\xi), \quad \xi \in \mathfrak{D}^*$$

for the solution F of the equation (4.9).

This proposition is, actually, a limit case of Proposition 1.1.

From Proposition 4.3 and Theorem 1.1 it follows that, firstly, the values RF depend only on the curve γ and on the form ϕ on γ, and secondly, the class of cohomologies of the form F in $H^{n,n-1}(\mathfrak{D})$ is uniquely defined by γ and by ϕ.

From Proposition 4.3 and Theorem 3.1 it follows that if $\gamma = bS$ and $\phi = \psi|_\gamma$, where S is the Riemann surface in $\mathbb{C}P^n \backslash \gamma$ and ψ is a holomorphic $(1,0)$-form on S, then the form $f = R\psi \wedge [S]$ satisfies the conditions (3.1), (3.2) of Theorem 3.1.

From here we obtain the necessity of the condition (ii) for the validity (i) in Theorem 4.2. Theorem 3.1 also states that conditions upon the Radon transform RF (3.1), (3.2) are not only necessary but also sufficient for the analytical representation of the element $F \in H^{n,n-1}(\mathfrak{D})$ by an analytical current.

However, the assumption that the dual set $\mathfrak{D}^*$ is connected, is essential in Theorem 3.1. But the dual set for $\mathfrak{D} = \mathbb{C}P^n \backslash \gamma$ is not connected. This difficulty is overcome by transferring the conditions (3.1), (3.2) from the component to the component of $\mathfrak{D}^*$ with help of analysis of jumps of the integral of the Cauchy-Fantappiè type (4.7) on the boundaries of the components of $\mathfrak{D}^*$ (compare with the proof of Theorem 4.1 in [13], [14]).

5. Inversion formulas
for the Abel-Radon-Fantappiè transforms.

We shall show here that the integral formulas of the Cauchy-Leray type [26], [32], [36] allow to give, under certain conditions, explicit inversion formulas both for the Fantappiè transform and for the Abel-Radon transform.

Let $\mathfrak{D}$ be such a linearly concave domain in $\mathbb{C}P^n$ that $\mathfrak{D} = \cup \mathfrak{D}_\epsilon$, where $\{\mathfrak{D}_\epsilon\}$-linearly concave domains with boundaries of the class C^2 compactly supported in $\mathfrak{D}$, where $\mathfrak{D}_{\epsilon_1} \supset \bar{\mathfrak{D}}_{\epsilon_2}$ if $\epsilon_1 < \epsilon_2$.

Let M be k-dimensional algebraic subset which is a complete intersection of the form

$$M = \{Z \in \mathbb{C}P^n : \tilde{P}_1(Z) = \tilde{P}_2(Z) = \ldots = \tilde{P}_{n-k}(Z) = 0\}, \qquad (5.1)$$

where $\{\tilde{P}_j(Z)\}$-homogeneous polynomials of the homogeneous coordinates $Z = (Z_0, Z_1, \ldots, Z_n)$ such that we have $d\tilde{P}_1 \wedge \ldots \wedge d\tilde{P}_{n-k} \neq 0$ almost everywhere on M.

Suppose, $K = \mathbb{C}P^n | \mathfrak{D}$, $K_\epsilon = \mathbb{C}P^n | \mathfrak{D}_\epsilon$. Suppose, that the hyperplane $\{Z \in \mathbb{C}P^n : Z_0 = 0\}$ is contained in $\mathfrak{D}$. We shall use affine corrdinates in the domain $\mathbb{C}^n = \{Z \in \mathbb{C}P^n : Z_0 \neq 0\}$, i.e. $Z_0 = 1$ and $Z = (Z_1, Z_2, \ldots, Z_n)$.

We introduce the functions $\rho_\epsilon \in C^2(\mathbb{C}^n)$ such that

$$K_\epsilon = \{Z \in \mathbb{C}P^n : \rho_\epsilon(Z) < 0\}, \qquad (5.2)$$

where $d\rho_\epsilon \neq 0$ on bK_ϵ.

Suppose, further, that

$$\eta(Z) = (\eta_0(Z), \eta_1(Z), \ldots, \eta_n(Z)); \quad \eta'(Z) = (\eta_1(Z), \ldots, \eta_n(Z)), \quad (5.3)$$

where $\eta_\nu(Z) = \frac{\partial \rho_\epsilon(Z)}{\partial Z_\nu}$, $\nu = 1, 2, \ldots, n$ and

$$\eta_0(Z) = -Z \cdot \eta'(Z) = \sum_{\nu=1}^{n} Z_\nu \frac{\partial \rho_\epsilon(Z)}{\partial Z_\nu}.$$

We introduce polynomials of the affine coordinates

$$P_j(Z) = \tilde{P}_j(1, Z_1, \ldots, Z_n), /// j = 1, 2, \ldots, n - k$$

and fix the vector-polynomials

$$Q^{(j)}(W, Z) = \{Q_1^{(j)}(W, Z), \ldots, Q_n^{(j)}(W, Z)\}, \qquad W, Z \in \mathbb{C}^n$$

such that

$$P_j(W) - P_j(Z) = (W - Z) \cdot Q^{(j)}(W, Z) = \sum_{\nu=1}^{n} (W_\nu - Z_\nu) \cdot Q_\nu^{(j)}(W, Z). \quad (5.4)$$

Let $\mathfrak{D}^*$ be a domain dual to $\mathfrak{D}$ in the dual projective space $(\mathbb{C}P^n)^*$ with the coordinates $(\xi_0, \xi_1, \ldots, \xi_n)$.

Here we shall have need of two formulas of the Cauchy-Leray type: one—for holomorphic functions in the domain $\mathfrak{D}^*$, another—for holomorphic functions on the manifold $M \cap K_\epsilon$. The first formula, obtained in [36], represents any functions $f_0 \in H(\mathfrak{D}^*, \mathbb{O}(-1))$ in the following form

$$f_0(\xi) = \frac{(-1)^{n-1}}{(2\pi i)^{n-1}(n-1)!} \int\limits_{Z \in b\mathfrak{D}_\epsilon} \frac{\partial^{n-1} f_0(\eta(Z))}{\partial \eta_0^{n-1}} \frac{\det[\eta'(Z), \bar{\partial}\eta'(Z)] \wedge dZ}{\xi_0 + \xi' Z},$$

(5.5)

where $\xi \in (\mathfrak{D}_\epsilon)^*$, and the superdeterminant

$$\det[\eta'(Z), \bar{\partial}\eta'(Z)] \overset{\text{def}}{=} \det[\eta'(Z), \bar{\partial}\eta'(Z), \ldots, \bar{\partial}\eta'(Z)]$$

is determined according to usual rules, except that the components of the column $\bar{\partial}\eta'(Z)$, repeated $(n-1)$ times, are multiplied by the exterior way. The second formula, arising from [10], [26], [43], represents any functions $h \in H(M \cap K_\epsilon)$ in the following form

$$(2\pi i)^k h(Z) = \int\limits_{W \in M \cap bK} h(W) \frac{\det[Q(W,Z), \eta'(W), \bar{\partial}\eta'(W)] \wedge (dP \rfloor dW)}{(\eta_0(W) + \eta'(W) \cdot Z)^k},$$

(5.6)

where $Z \in M \cap K$; the column $\bar{\partial}\eta'(W)$ is repeated $(k-1)$ times in

$$\det[Q(W,Z), \eta'(W), \bar{\partial}\eta'(W)]$$

$$\overset{\text{def}}{=} \det[Q^{(1)}(W,Z), \ldots, Q^{(n-k)}(W,Z), \eta'(W), \bar{\partial}\eta'(W), \ldots, \bar{\partial}\eta'(W)],$$

$$dP \rfloor dW = \det{}^{-1} \left| \frac{\partial P_\alpha}{\partial Z_\beta} \right|_{\alpha,\beta=1}^{n-k} (W) dW_{n-k+1} \wedge \ldots \wedge dW_n$$

$$= \text{Res}(P_1, \ldots, P_{n-k})^{-1} dW_1 \wedge dW_2 \wedge \ldots \wedge dW_n.$$

Yet, we shall need of the well known Ehrenpreis-Malgrange-Martineau result [36] bringing, under made assumptions, to the unique solvability for $l > 0$ of the equation

$$\frac{\partial \psi}{\partial \xi_0} = f, \quad \text{where} \quad f \in H(\mathfrak{D}^*, \mathbb{O}(l-1)), \quad \psi \in H(\mathfrak{D}^*, \mathbb{O}(l)).$$

Therefore in such spaces the operators of the following form are defined

$$D = \left\{ \left(\frac{\partial}{\partial \xi_0}\right)^{-1} \frac{\partial}{\partial \xi_1}, \ldots, \left(\frac{\partial}{\partial \xi_0}\right)^{-1} \frac{\partial}{\partial \xi_n} \right\}.$$

(5.7)

Under made assumptions there takes place the following strengthening of Theorem 1.1.

Theorem 5.15 *Let $f_0 \in H(\mathfrak{D}^*, \mathbb{O}(-1))$ and the function f_0 satisfies the system of differential equations*

$$\tilde{P}_j\left(\frac{d}{d\xi}\right) f_0 = 0, \qquad j = 1, 2, \ldots, n-k, \tag{5.8}$$

where $\{\tilde{P}_j\}$-polynomials of the form (5.1). Then the function f_0 may be represented by the following "sum" of elementary solutions $\frac{1}{\xi_0 + \xi'W}$: $(2\pi i)^k (k-1)! \cdot f_0(\xi)$ is the integral of

$$\frac{det[Q(W, D), \eta'(W), \bar{\partial}\eta'(W)]\frac{\partial^{k-1} f_0(\eta(W))}{\partial \eta_0^{k-1}} \wedge dP \rfloor dW}{(\xi_0 + \xi'W)} \tag{5.9}$$

over $\{W \in bK_\epsilon : P_1(W) = \ldots = P_{n-k}(W) = 0\}$, where D is an operator of the form (5.7), $\eta(W)$ is a vector-function of the form (5.3), $Q(W, Z) = \{Q^{(j)}(W, Z)\}$ is a matrix of the polynomials of the form (5.4).

Remark: In addition to Theorem 1.1, Theorem 5.1 gives the inversion formula for the Fantappiè transform $\mu \to f_0(\xi) = \left(\mu, \frac{1}{\xi_0 + \xi'Z}\right)$ of the analytical functional μ with the support on the intersection of the algebraic manifold M with the linearly convex compact $K \subset \mathbb{C}P^n$.

The result of Theorem 5.1 may be also interpreted as the inversion formula for the "non-linear" Fantappiè transform

$$\mu \to f_0(\xi) = \left(\frac{1}{\sum\limits_{j_0 + \cdots + j_n = n} \xi_{j_0, j_1, \ldots, j_n} Z_1^{j_1} \cdots Z_n^{j_n}}\right)$$

for the analytical functional with the support in the m-polynomial convex compact in $\mathbb{C}P^n$ (compare with the "non-linear" Fourier transform introduced in [16]).

The proof of Theorem 5.1 is based on the appropriate combination of the formulas (5.5), (5.6). In fact, due to formula (5.6) and to the Oka-Cartan theorem [22], [26] for any $\xi \in \mathfrak{D}^*$ the holomorphic function $\frac{1}{\xi_0 + \xi'Z}$ on K is represented in the form

$$\frac{1}{\xi_0 + \xi'Z} + P_1(Z)h_1(Z) + \cdots + P_{n-k}(Z)h_{n-k}(Z)$$
$$= \frac{1}{(2\pi i)^k} \int\limits_{W \in M \cap B_\epsilon} \frac{det[Q(W,Z), \eta'(W), \bar{\partial}\eta'(W)]dP \rfloor dW}{(\xi_0 + \xi'W)(\eta_0(W) + \eta'(W) \cdot Z)^k}, \tag{5.10}$$

where the function $\{h_j(Z)\}$ belong to $H(K)$.

We substitute the formula (5.10) to the formula (5.5) and we use the equality (1.14). Changing the integration order we obtain

$$(k-1)!(2\pi i)^k f_0(\xi) = \frac{(-1)^{n-1}(k-1)!}{(2\pi i)^{n-1}(n-1)!} \int\limits_{W \in bK_\epsilon \cap M} \frac{dP \rfloor dW}{(\xi_0 + \xi'W)} \times$$

$$\int_{Z \in b\mathfrak{D}_\epsilon} \frac{\det[Q(W,Z),\eta'(Z),\bar\partial\eta'(Z)]}{(\eta_0(W)+\eta'(W)\cdot Z)^k} \frac{\partial^{n-1}}{\partial\eta_0^{n-1}} f_0(\eta(Z))\det[\eta'(Z),\bar\partial\eta'(Z)]dZ =$$

(taking into account the equalities $\left(\frac{\partial}{\partial\eta_0}\right)^{-1} \frac{\partial}{\partial\eta_\nu} \frac{1}{\eta_0+\eta'Z} = \frac{Z_j}{\eta_0+\eta'Z}$)

$$= \frac{(-1)^{n-1}}{(n-1)!} \int_{W \in M \cap bK_\epsilon} \frac{dP\rfloor dW}{(\xi_0+\xi'W)}\det[Q(W,D),\eta'(W),\bar\partial\eta'(W)]$$

$$\times \frac{\partial^{k-1}}{\partial\eta_0^{k-1}} \int_{Z \in b\mathfrak{D}_\epsilon} \frac{1}{\eta_0(W)+\eta'(W)Z} \frac{\partial^{n-1}}{\partial\eta_0^{n-1}} f_0(\eta(Z))\det[\eta'(Z),\bar\partial\eta'(Z)]dZ =$$

(using the formula (5.5))

$$= \int_{W \in M \cap bK_\epsilon} \frac{(dP\rfloor dW)\det[Q(W,D),\eta'(W),\bar\partial\eta'(W)]}{(\xi_0+\xi'W)} \frac{\partial^{k-1}}{\partial\eta_0^{k-1}} f_0(\eta(W)).$$

Theorem is proved.

As a consequence of Theorem 5.1 one can also obtain an explicit version of "fundamental principle" of Euler-Ehrenpreis for solutions of the system of the equations of the form

$$P_j\left(\frac{d}{d\xi}\right)\psi(\xi) = 0, \qquad j = 1,2,\ldots,n-k, \tag{5.11}$$

in the class of the entire functions $\psi(\xi)$, $\xi \in \mathbb{C}^n$, of the exponential type.

Let K be a convex compact in $\mathbb{C}^n$. Let $h_K(\xi) = \sup\mathrm{Re}\,\langle\xi\cdot Z\rangle$, $\xi = (\xi_1,\ldots,\xi_n) \in \mathbb{C}^n$. Let $\psi(\xi)$ be an entire function in $\mathbb{C}^n$ with the estimate of the form: for any $\epsilon > 0$ there exists the constant C_ϵ such that

$$|\psi(\xi)| \le C_\epsilon \exp[h_k(\xi) + \epsilon|\xi|]. \tag{5.12}$$

Let $\mathfrak{D} = \mathbb{C}P^n|K$. For any $(\xi_0,\xi_1,\ldots,\xi_n) \in \mathfrak{D}^*$ there exists λ for which $\mathrm{Re}\,\xi_0\lambda > h_K(\lambda\xi)$ and, consequently, the following integral converges

$$f_0(\xi_0,\ldots,\xi_n) = \int_0^\infty \psi(\xi\cdot\lambda t)e^{\xi_0\lambda t}\lambda dt.$$

This integral does not depend on λ and it is called [36] the projective Laplace transform of the entire function ψ.

Corollary 5.8 *Let an entire function $\psi(\xi)$, $\xi \in \mathbb{C}^n$ satisfy the system of equations (5.11) and satisfy the estimate (5.12). Then the function $\psi(\xi)$ admits the exponential representation of the form: $(2\pi i)^k (k-1)! \psi(\xi)$ is the integral of*

$$= \exp(\xi \cdot W) \det[Q(W,D), \eta'(W), \bar{\partial}\eta'(W)] \frac{\partial^{k-1}}{\partial \eta_0^{k-1}} f_0(\eta(W)) dP] dW \quad (5.13)$$

over $\{W \in bK_\epsilon : P_1(W) = \ldots = P_{n-k}(W) = 0\}$.

Another consequence of Theorem 5.1 is an explicit inversion (under certain condtions) of the Abel-Radon transform (2.11) that brings, for example, to the following strengthening of Theorem 3.3.

Corollary 5.9 *Let, under the conditions of Theorem 5.1, the coefficient f_0 of the holomorphic closed 1-form $f = \sum_{j=0}^{n} f_j d\xi_j$ satisfy the system of equations (5.8) in the domain $\mathfrak{D}^*$, where $k = 1$. Then the form f is the Abel-Radon transform of the semi-algebraic current $\psi \wedge [V]$, where*

$$V = \{Z \in \mathfrak{D} : \tilde{P}_1(Z) = \ldots = \tilde{P}_{n-1}(Z) = 0\}$$

and ψ is the following holomorphic 1-form in the neighbourhood of bV

$$\psi = \frac{1}{2\pi i} \det[Q(Z,D), \eta'(Z)] f_0(\eta(Z)) \frac{dZ_n}{\det \left| \frac{\partial P_\alpha}{\partial Z_\beta} \right|_{\alpha,\beta=1}^{n-1} (Z)}, \quad (5.14)$$

where $Z \in b\mathfrak{D}_\epsilon$, and ϵ is sufficiently small.

References

[1]   Abel, N., Mémoire sur une propriété générale d'une classe très étendue de fonctions transcendantes. Présenté a l'Acad. Sc. Paris, 30 Octobre 1826, Oeuvres de N.H. Abel, vol. 1, pp. 145–211.

[2]   Abel, N., Démonstration d'une propriété générale d'une certaine classe de fonctions transcendantes, Journal fur die reine und ange-wandte Mathematik, Herausgegeben von Crelle, Dd. 4, Berlin, 1829; Oeuvres de N.H. Abel, Vol. 1, pp. 515–517.

[3]   Alexander, H., Polynomial approximation and hulls in sets of finite linear measure in $\mathbb{C}^n$, *Amer. J. of Math.* **93**, 1971, 65–74.

[4]   Andreotti, A. and Norguet, F., Cycles of algebraic manifolds and Aeppli cohomology, *Ann. Scuola Norm. Super. Pisa* **25**, 1971, 1, 59–114.

[5]   Baston, R.J. and Eastwood, M.G., The Penrose transform: its interaction with representation theory, Oxford University Press, 1989.

[6]   Berndtsson, B. and Passare, M., Integral formulas and explicit version of the Fundamental Principle, *J. Funct. Anal.* **84**, 1989, 358–372.

[7]   Bishop, E., Conditions for the analyticity of certain sets, *Michigan Math. J.* **11**, 1964, 289–304.

[8]   Blaschke, W. and Bol, G., Geometrie der gewebe, Springer, 1938.

[9]   Chern, S.S., Web Geometry, *Bull. Am. Math. Soc.* **6**, 1982, 1–8.

[10] Coleff, N.R. and Herrera, M.E., Les courants résiduels associés à une forme méromorphe, Lecture Notes in Math., vol. 633, Springer-Verlag, Berlin and New York, 1978.

[11] Darboux, L., Théorie des surfaces. I, 2e éd. Gauthiers-Villars, Paris, 1914.

[12] Dolbeault, P., General theory of multidimensional residues, Encyclopedia of Math. Sciences 7 (Several complex variables I), Springer-Verlag 1990, 215–241.

[13] Dolbeault, P. and Henkin, G., Surface de Riemann de bord donné dans $\mathbb{C}P^2$, *C.R. Ac. Sci. Paris* **316**, 1993, 27–32.

[14] Dolbeault, P. and Henkin, G., Surface de Riemann de bord donné dans $\mathbb{C}P^n$, Contribution to Complex Analysis and Analytic Geometry, ed. H. Skoda and J.-M. Trépreau, Viehweg Verlag 1994, pp.163-187.

[15] Ehrenpreis, L. Fourier analysis in several complex variables, Wiley-Interscience Publishers, 1970.

[16] Ehrenpreis, L. and Malliavin, P., Fourier analysis on nonconvex sets, Symposia Math. Inst. Nazionale de Alta Mat., vol. XII, pp. 413–419, Acad. Press, London, 1971.

[17] Fantappiè, L., L'indicatrice projettiva dei funzionali linearie i prodotti funzionnali projettivi, *Ann. di Mat. Pura ed Appl.* **21,** 1943.

[18] Folland, G.B. and Kohn, J.J., The Neumann problem for the Cauchy-Riemann complex, Princeton University Press, Princeton, 1972.

[19] Gindikin, S. and Henkin, G., Integral geometry for $\bar{\partial}$-cohomology in q-linearly concave domains in $\mathbb{C}P^n$, *Funct. Anal. Appl.* **12,** 1979, 247–261.

[20] Griffiths, P., Variations on a theorem of Abel, *Inventionnes Math.,* **35,** 1976, 321–390.

[21] Griffiths, P. and Harris, J., Principles of algebraic geometry, John Wiley and Sons, New York, 1978.

[22] Gunning, R. and Rossi, H., Analytic functions of several complex variables, Prentice-Hall, Englewood Cliffs, N.J., 1965.

[23] Harvey, R., Holomorphic chains and their boundaries, Proc. Symp. Pure Math. 30, Part 1, 1977, 309–382.

[24] Harvey, R. and Lawson, B., On boundaries of complex analytic varieties I, II, *Ann. of Math.* **102,** 1975, 233–290, **106,** 1977, 213–238.

[25] Henkin, G., La transformation de Radon pour la cohomologie de Dolbeault et un théorème d'Abel inverse, *C.R. Ac. Sci. Paris* **315,** 1992, 973–978.

[26] Henkin, G., The method of integral representation in complex analysis, Encyclopedia of Math. Sci., vol. 7 (Several complex variables I) Springer-Verlag, 1990, 19–116.

[27] Henkin, G.M. and Polyakov, P.L., Homotopy formulas for the $\bar{\partial}$-operator on $\mathbb{C}P^n$ and the Radon-Penrose transform, *Math. USSR Isvestiya* **28,** 1987, no. 3, 555–587.

[28] Herrera, M. and Lieberman, D., Residues and principal values on complex spaces, *Math. Ann.* **194,** 1971, 259–294.

[29] Jacobi, K., Theoremata nova algebraica circa systema duarum aequatuenum inter duas variables prepositorum, *Journ. für r. ang. Math.* **14,** 1835, 281–288.

[30] Kohn, J.J. and Rossi, H., On the extension of holomorphic functions from the boundary of a complex manifold, *Ann. of Math.* **81**, 1965, 451–472.

[31] Lelong, P., Intégration sur un ensemble analytique complexe, *Bull. Soc. Math. France* **85**, 1957, 239–262.

[32] Leray, J., Le calcul différentiel et intégral sur une variété analytique complexe, *Bull. Soc. Math. France* **87**, 1959, 81–180.

[33] Lie, S., Bestimmung aller Flächen die in mehzfacher Weise durch Translationsbewegung einer curve erzeugt werden, *Archiv for Mathematik* **7**, 1882, 155–176, Gesammette Abhaudlungen, Bd. 1, Abt. 1, 450–467.

[34] Little, J.B., On Lie's approach to the study of translation manifolds, *J. Diff. Geom.* **26**, 1987, 253–272.

[35] Martineau, A., Indicatrices des fonctionnelles analytiques et inversion de la transformation de Fourier-Borel par la transformation de Laplace, *C.R. Acad. Sci., Paris* **255**, 1962, 1845–1847, 2888–2890.

[36] Martineau, A., Equations différentielles d'ordre infini, *Bull. Soc. Math. France* **95**, 1967, 109–154.

[37] Ofman, S., Intégration des classes de $d'd''$-cohomologie sur les cycles analytiques. Solution du problème de l'injectivité, *Journal Math. pures et appl.* **68**, 1989, 73–94.

[38] Poincaré, H., Sur les surfaces de translation et les fonctions abéliennes, *Bull. Sc. Math. France* **29**, 1901, 61–86.

[39] Reiss, M., Mémoire sur les propriétés générales des courbes algébriques etc., *Corresp. Math. et Phys. de Quetelet* **9**, 1837, 249–308.

[40] Rothstein, W., Zur theorie der analytischen Manningfal-tigkeiten im Raume von n komplexen Veränderlichten, *Math. Ann.* **129**, 1955, 96–138.

[41] Saint-Donat, B., Variétés de translation et théorème de Torelli, *C.R. Ac. Sci. Paris* **280**, 1975, 1611–1612.

[42] Serre, J.-P., Géométrie algébrique et géométrie analytique, *Ann. Inst. Fourier* **6**, 1955–1956, 1–42.

[43] Stout, E.L., An integral formula for holomorphic functions on strictly pseudoconvex hypersurfaces, *Duke Math. J.* **42**, 1975, 347–356.

[44] Stoll, W., The growth of the area of a transcendental analytic set, *Math. Ann.* **156,** 1964, 47–78, **156**, 1964, 144–170.

[45] Stolzenberg, G., Uniform approximation on smooth curves, *Acta Math.* **115,** 1966, 185–198.

[46] Wermer, J., The hull of a curve in $\mathbb{C}^n$, *Ann. of Math.* **68,** 1958, 550–561.

[47] Wirtinger, W., Lies translationsmannigfaltigkeiten und Abelsche Integral, *Monatsh. für Math. und Physik* **46,** 1938, 384–431.

[48] Wood, J.A., A simple criterion for local hypersurfaces to be algebraic, *Duke Math. J.* **51,** 1984, no. 1, 235–237.

[49] Lawrence, M., Polynomial hulls of rectifiable curves, Am. J. of Math. **117** 1995, 405–417.

ON THE ABSENCE OF PERIODIC POINTS FOR THE RICCI CURVATURE OPERATOR ACTING ON THE SPACE OF KÄHLER METRICS

ALAN M. NADEL[1]

1. Introduction

Let M be a compact complex manifold. For any Kähler metric g on M, the Ricci curvature is given by the expression $\text{Ricci}(g) = -\partial\bar\partial \log \text{vol}_g$. The Ricci operator $\text{Ricci}(*)$ is "fully covariant" in the sense that it is well defined and does not depend on any particular choice of coordinate system or background tensor field. In fact, since the Ricci curvature of a Kähler metric coincides with the Ricci curvature of the underlying Riemannian metric, the complex structure of the manifold does not enter the picture at all, except indirectly in the assumption that the metric is Kähler . The Ricci operator is fundamental in the study of complex manifolds, but it is analytically very difficult to understand, and many of its aspects remain a mystery.

Here we study $\text{Ricci}(*)$ from a dynamical systems point of view. From this perspective, a Kähler-Einstein metric is nothing more than a fixed point of $\text{Ricci}(*)$. When $\text{Ricci}(g)$ is positive definite everywhere, it may be regarded as a Kähler metric in its own right, and its Ricci curvature $\text{Ricci}(\text{Ricci}(g))$ may be considered. More generally, we may apply $\text{Ricci}(*)$ successively n times to get $\text{Ricci}^n(g)$ provided $\text{Ricci}^k(g)$ is positive definite for each $k < n$. We may thus speak of higher-order periodic points.

We show that periodic points of orders two and three can never exist. We show, in other words, that $\text{Ricci}^n(g) = g$ implies $\text{Ricci}(g) = g$ in case $n \le 3$. Our proof is based on a form of the maximum principle. It remains unknown whether periodic points of order greater than three can ever exist. It is possible that there is a global relationship between the various order periodic points, and in particular that the existence of higher-order periodic points is related to the existence of fixed points, which depends on the complex structure. It also remains unkown whether these results hold in the Riemannian setting. I hope this note will serve as motivation for studying these further questions.

[1] Research supported by a Sloan Research Fellowship and an NSF Young Investigator Award

2. Three equivalent statements

A_n (Coupled equation form). Suppose $g_0, \ldots, g_{n-1}$ are Kähler metrics on a compact complex manifold M and suppose that $\text{Ricci}(g_0) = g_1$, $\text{Ricci}(g_1) = g_2$, $\cdots$, $\text{Ricci}(g_{n-1}) = g_0$. Then $g_0 = \ldots = g_{n-1}$ and hence these metrics are all Einstein.

B_n (Iterative form). Suppose g is a Kähler metric on a compact complex manifold M and suppose that $\text{Ricci}^n(g) = g$. [2] Then $\text{Ricci}(g) = g$.

C_n (Functional differential equation form). Suppose that g is a Kähler metric on a compact complex manifold M and that T is a biholomorphism of M into itself such that $T^n = Id$ (i.e. T has order dividing n). Suppose also that $\text{Ricci}(g) = T^* g$. Then $\text{Ricci}(g) = g = T^* g$.

Assertion. These three statements are equivalent for each n.

Proof that $A_n \Rightarrow B_n$. Assume A_n to be true, and assume the setup of B_n. We want to show that B_n is true. Set $g_k = \text{Ricci}^k(g)$ $(k = 1, \ldots, n-1)$. Then $\text{Ricci}(g_0) = g_1$, $\text{Ricci}(g_1) = g_2$, $\ldots$, $\text{Ricci}(g_{n-1}) = g_0$. By A_n we conclude that $g_0 = \ldots = g_{n-1}$. In particular, $\text{Ricci}(g) = g$, as desired.

Proof that $B_n \Rightarrow C_n$. Assume B_n to be true, and assume the setup of C_n. We want to show that C_n is true. Since the operators Ricci and T^* (on the space of Kähler metrics) commute, we have

$$\text{Ricci}^n(g) = \text{Ricci}^{n-1}(\text{Ricci}(g)) = \text{Ricci}^{n-1}(T^* g) = T^* \text{Ricci}^{n-1}(g) = \ldots$$

(continue inductively in this fashion)

$$(T^*)^n(g) = g \text{ (since } T^n = Id) .$$

Now B_n implies that $\text{Ricci}(g) = g$, as desired.

Proof that $C_n \Rightarrow A_n$. Assume C_n to be true, and assume the setup of A_n. We want to show that A_n is true. Consider the manifold M^n obtained by taking the Cartesian product of M with itself n times. Let T be the biholomorphism of this product manifold onto itself given by cyclic permutation as follows: $T(p_0, \ldots, p_{n-1}) = (p_1, \ldots, p_{n-1}, p_0)$. Note that T has order n. On this product manifold consider the Kähler metric $(g_0, \ldots, g_{n-1})$ obtained by taking the metric g_i on the $(i+1)$-th factor $(i = 0, \ldots, n-1)$. Now note that $T^*(g_0, \ldots, g_{n-1}) = (g_1, \ldots, g_{n-1}, g_0)$ and that $\text{Ricci}(g_0, \ldots, g_{n-1}) = (\text{Ricci}(g_0), \ldots, \text{Ricci}(g_{n-1}))$. The setup of A_n which we are assuming implies that

$$\text{Ricci}(g_0, \ldots, g_{n-1}) = T^*(g_0, \ldots, g_{n-1}).$$

---

[2] By this we mean that the Ricci operator is applied to g successively n times and that each time the result is positive definite.

Now C_n implies that $T^*(g_0, \ldots, g_{n-1}) = (g_0, \ldots, g_{n-1})$. In other words, $g_0 = g_1 = \cdots = g_{n-1}$, as desired.

This completes the proof that these three statements are equivalent. Note that A_1 holds true for trivial reasons. In this paper we establish A_2 and A_3. It is easy to see that if C_n holds and if m divides n then C_m is also holds. On the other hand, it does not seem possible to deduce A_6 from A_2 and A_3.

3. General maxima

Suppose that $\mathcal{C}$ is a collection of real-valued functions on a set X. A function $f \in \mathcal{C}$ is said to have a **general maximum** at a point $q \in X$ iff $f(p) \geq g(q)$ for all $g \in \mathcal{C}$ and all $q \in X$ (including, but not limited to, $g = f$ and $q = p$). In other words, a general maximum is a maximum over not only all points in our space but also all functions in our collection. Note that when we say "f has a general maximum at p" we are implicitly referring to the collection $\mathcal{C}$ even though the statement in quotes makes no explicit reference to $\mathcal{C}$. No confusion should arise.

I would like to make explicit here a very simple yet important idea that will play a crucial role in the proof of Theorem 3.2. Suppose that f has a general maximum at p and that $f(p) \leq g(p)$ for some other function g in our collection. Then g also has a general maximum at p.

Now suppose that $\mathcal{C}$ is a *finite* collection of *continuous* real-valued functions on a (nonempty) *compact* topological space X. Then a general maximum exists. That is, there exists a function $f \in \mathcal{C}$ and a point $p \in X$ such that f has a general maximum at p. This can be seen as follows. Write $\mathcal{C} = \{f_1, \ldots, f_n\}$ and pick $p_1, \ldots, p_n \in X$ such that f_i has a maximum at p_i. (Recall that any continuous real-valued function on a compact space necessarily achieves its maximum.) After re-labling if necessary, we may assume that the set $\{f_1(p_1), \ldots, f_n(p_n)\}$ has maximum $f_1(p_1)$. It is now easy to see that f_1 has a general maximum at p_1, as desired.

4. Period two and three points

Theorem 4.1 *Suppose g and h are two Kähler metrics on a compact complex manifold M such that $Ricci(g) = h$ and $Ricci(h) = g$. Then $g = h$ everywhere and hence both metrics are Kähler-Einstein .*

Proof. This proof involves a fairly simple application of the maximum principle. Consider the smooth real-valued function $\log(\text{vol}_g/\text{vol}_h)$ on the compact manifold M. Let $p \in M$ be a point at which this function achieves its maximum. At this point the function will be concave downward in the sense that its complex Hessian will be seminegative. In other words, $\partial\bar{\partial} \log(\text{vol}_g/\text{vol}_h) \leq 0$ at p. Recall that the Ricci curvature tensor of a Kähler metric is equal to minus $\partial\bar{\partial}$ of the logarithm of the volume form. Thus we can rewrite this inequality as $\text{Ricci}(h) - \text{Ricci}(g) \leq 0$ at p, and $g - h \leq 0$ at p, and $g \leq h$ at p. Now take the wedge-power of this last inequality to get $\text{vol}_g \leq \text{vol}_h$ at p, and hence $\log(\text{vol}_g/\text{vol}_h) \leq 0$ at p. In other words, when our function achieves its maximum value, that value is at most zero. It follows that our function is everywhere seminegative. By repeating the same argument with g and h reversed, we see that our function is also everywhere semipositive. Hence it is identically zero, and $\text{vol}_g = \text{vol}_h$ everywhere. Finally, take $\partial\bar{\partial}$ log of each side of this equation to get $\text{Ricci}(g) = \text{Ricci}(h)$ and $h = g$, as desired.   Q.E.D.

Theorem 4.2 *Suppose* g_0, g_1, *and* g_3 *are three Kähler metrics on a compact complex manifold* M *such that* $\text{Ricci}(g_0) = g_1$, $\text{Ricci}(g_1) = g_2$, *and* $\text{Ricci}(g_2) = g_0$. *Then* $g_0 = g_1 = g_2$ *everywhere and hence the metrics are Kähler-Einstein* .

Proof. Let vol_i denote the volume form for g_i $(i = 0, 1, 2)$. Let $\mathcal{C} = \{\text{vol}_i/\text{vol}_j\}$ $(i, j = 0, 1, 2; i \neq j)$. Thus $\mathcal{C}$ is a collection of six smooth real-valued functions on the compact manifold M.

Claim. If $\text{vol}_i/\text{vol}_j$ achieves a (local) maximum at a point $p \in M$ then $\text{vol}_{j+1}/\text{vol}_{i+1} \leq 1$ at p. (Here $i+1$ and $j+1$ are understood to be modulo 3, so that $2 + 1 = 0$.)

Proof of Claim. Assume $\text{vol}_i/\text{vol}_j$ has a maximum at p. Then by the maximum principle we have $\partial\bar{\partial} \log \text{vol}_i/\text{vol}_j \leq 0$ at p. In other words, $\text{Ricci}(g_j) - \text{Ricci}(g_i) \leq 0$ p. In other words again, $g_{j+1} - g_{i+1} \leq 0$ at p, which implies that $\text{vol}_{j+1} \leq \text{vol}_{i+1}$ at p, and the claim follows.

We know that some function in our collection achieves a general maximum at some point in our manifold. There are two (very similar) cases to consider, according to the form of the function that achieves the general maximum. Suppose first that $\text{vol}_i/\text{vol}_{i+1}$ has a general maximum at point p. Then by the claim we have $\text{vol}_{i+2}/\text{vol}_{i+1} \leq 1$ at p. A trivial algebraic manipulation now gives $\text{vol}_i/\text{vol}_{i+1} \leq \text{vol}_i/\text{vol}_{i+2}$ at p. Since the function on the left hand side of the inequality has a general maximum at p, the function on the right hand side must also have a general maximum at p. We can now apply the claim to the function on the right hand side to get $\text{vol}_{i+3}/\text{vol}_{i+1} \leq 1$ at p. In other words, $\text{vol}_i/\text{vol}_{i+1} \leq 1$ at p (since our indices are modulo three). This shows that the general maximum for our family is at most one. We conclude that every function in our family is everywhere ≤ 1 (by the definition of general maximum). Hence every function in our family is identically equal to 1 (since the reciprocal of every function in our family is also in our family). Hence all three volume forms are identically equal to each other. Hence all three metrics are indentically equal to each other (since the metrics can be recovered by taking the Ricci of the volume forms), and the theorem is proved in this case.

We now consider the other case which may occur. Suppose that the function $\text{vol}_{i+1}/\text{vol}_i$ has a general maximum at point p. The following argument is nearly identical to the argument in the first case, but we write it out anyway. The claim gives $\text{vol}_{i+1}/\text{vol}_{i+2} \leq 1$ at p. Hence $\text{vol}_{i+1}/\text{vol}_i \leq \text{vol}_{i+2}/\text{vol}_i$ at p. We conclude that the function on the right hand side also has a general maximum at p, and we apply the claim again to get $\text{vol}_{i+1}/\text{vol}_{i+3} \leq 1$ at p. As in the paragraph above, we see that every function in our family is everywhere ≤ 1, and the theorem follows. This completes the proof.   Q.E.D.

University of Southern California

THE MAXIMUM PRINCIPLE
AND RELATED TOPICS

Louis Nirenberg

This talk is a report on some joint work with Berestycki and Varadhan [BNV]. It concerns second order linear elliptic operators L acting on real functions u in a bounded domain (open connected set) Ω in $\mathbf{R}^n$. L is a general operator, not self adjoint, of the form (we use summation convention)

$$L = M + c(x)$$

where

$$M = a_{ij}(x)\frac{\partial^2}{\partial x_i\,\partial x_j} + b_i(x)\frac{\partial}{\partial x_i}$$

L is assumed to be uniformly elliptic and with real, bounded, measurable coefficients, i.e., for some constants $c_0, C_0 > 0$, $b \geq 0$

$$(1) \qquad c_0|\xi|^2 \leq a_{ij}(x)\xi_i\xi_j \leq C_0|\xi|^2 , \qquad \forall\, x \in \Omega , \ \forall\, \xi \in \mathbf{R}^n$$

and

$$(2) \qquad\qquad\qquad |b_i(x)|, |c(x)| \leq b .$$

In addition we assume $a_{ij} \in C(\Omega)$. We will state various estimates, but the constants will be independent of the modulus of continuity of the a_{ij}.

The maximum principle plays a key role in the study of second order elliptic equations. One form of it is the following

Definition. The maximum principle is said to hold for L in Ω if for any function w in Ω, the conditions

$$(3) \qquad \left.\begin{array}{l} Lw \geq 0 \quad \text{in} \quad \Omega \\ \limsup\limits_{x \to \partial\Omega} w(x) \leq 0 \end{array}\right\} \text{ imply } w \leq 0 \quad \text{in} \quad \Omega .$$

The functions to which we apply L will always be assumed to belong to $W_{loc}^{2,n}(\Omega)$, but for the purposes of this talk we may suppose they are in $C^2(\Omega)$.

The maximum principle does not, of course, always hold. Let me recall three different classical sufficient conditions:

(i) $c(x) \leq 0$;

(ii) $\exists g \in C(\overline{\Omega})$, $g > 0$ in $\overline{\Omega}$ and $Lg \leq 0$ in Ω;

(iii) Ω is narrow in, say, the x_1-direction, i.e. for some ϵ small (depending only on c_0, C_0 and b), Ω lies in $\{a < x_1 < a + \epsilon\}$, for some a.

Berestycki and I, in some work, sought some new sufficient conditions for the maximum principle to hold. In addition, we wished to treat general bounded Ω, with no regularity required of the boundary $\partial\Omega$. In this talk then, Ω is a general bounded domain in $\mathbf{R}^n$; nothing is assumed about $\partial\Omega$. Furthermore, I repeat that L is not formally self adjoint.

One of the uses of the maximum principle is to derive estimates for solutions of inhomogeneous equations:

$$(4) \qquad\qquad\qquad Lu = f .$$

Under certain boundary conditions one may sometimes estimate $|u|_{L^\infty}$ in terms of $|f|_{L^\infty}$. There is a rather different inequality which has proved very useful in recent years. It is due independently to A. D. Alexandroff, I. Ya. Bakelman, and C. Pucci (see [GT]):

Inequality (ABP): Suppose w satisfies

$$(5) \qquad \begin{cases} Lw \geq f & \text{in} \quad \Omega , \quad f \leq 0 , \ f \in L^n(\Omega) , \\ \limsup\limits_{x \to \partial\Omega} w(x) \leq 0 . \end{cases}$$

and suppose the coefficient $c(x) \leq 0$. Then

$$\sup w \leq B(n, c_0, bd)d\|f\|_{L^n}$$

Here $d = \operatorname{diam}\Omega$.

The paper [BNV] began when Varadhan pointed out to Berestycki and myself that from ABP follows a new sufficient condition for the maximum principle, namely,

(iv) The measure of Ω, denoted $|\Omega|$, is small, i.e. for a suitable positive constant ϵ, depending only on n, c_0, C_0, b and d,

$$(6) \qquad\qquad\qquad |\Omega| < \epsilon ,$$

In [BN] we made extensive use of (iv). It had in fact already been observed by Bakelman [B].

In case $\partial\Omega$ is smooth and $a_{ij} \in C(\overline{\Omega})$, there is actually a known necessary and sufficient condition for the maximum principle to hold:

$$\text{max. princ. holds for } L \text{ in } \Omega \Leftrightarrow \lambda_1(-L) > 0 .$$

Here λ_1 is the principal eigenvalue for $-L$ under Dirichlet boundary conditions; i.e. there is a positive eigenfunction ϕ_1 in Ω, satisfying

$$(L + \lambda_1)\phi_1 = 0 \quad \text{in} \quad \Omega$$
$$\phi_1 = 0 \quad \text{on} \quad \partial\Omega .$$

The eigenvalue λ_1 has algebraic multiplicity equal to 1.

For a general domain Ω, we asked if there might be an analogous principal eigenvalue, and whether its positivity is related to the maximum principle holding for L in Ω. In [DV], Donsker and Varadhan gave a min-max characterization of the principal eigenvalue λ_1 in case $\partial\Omega$ is smooth:

$$(7) \qquad \lambda_1 = - \inf_{\substack{\phi > 0 \\ \phi \in W^{2,n}_{loc}}} \sup_\Omega \frac{L\phi}{\phi} .$$

This is equivalent to

$$(8) \qquad \lambda_1 = \sup\{\lambda;\ \exists\,\phi > 0 \text{ in } \Omega \text{ satisfying } (L + \lambda)\phi \le 0\} .$$

In [BNV] we define λ_1 in any bounded domain Ω by (8). A similar definition, in a more general framework, was given by Nussbaum and Pinchover in [NP], but with equality $(L+\lambda)\phi = 0$ in place of $(L+\lambda)\phi \le 0$.

From (7) it follows rather easily that

$$(9) \qquad \lambda_1 \text{ is concave in its dependence on the function } c(\cdot) .$$

Returning to the maximum principle, we observe that the requirement in (3)

$$(10) \qquad \limsup_{x \to \partial\Omega} w(x) \le 0$$

is not reasonable in a general domain. One cannot expect to prescribe the boundary values of a solution at every point on $\partial\Omega$. For example, if $\Omega = \{|x| < 1\} \setminus \{0\}$ in $\mathbb{R}^3$, and $L = $ the Laplace operator Δ, one would not prescribe the value of a solution of the Dirichlet problem at 0, though it belongs to $\partial\Omega$. In this connection then, we introduce an important ingredient, which goes back to Stroock and Varadhan [SV]; we introduce a solution $u_0 > 0$ in Ω of

$$(11) \qquad \begin{array}{rcl} M u_0 & = -1 & \text{in} \quad \Omega \\ \text{``}u_0 & = 0\text{''} & \text{on} \quad \partial\Omega . \end{array}$$

This is constructed in the following way. Let $\{\Omega_j\}$ be a sequence of subdomains of Ω with smooth boundaries, such that $\overline{\Omega}_j \subset \Omega_{j+1}$ and

$$\cup\Omega_j = \Omega .$$

Solve

$$Mu_j = -1 \quad \text{in} \quad \Omega_j$$
$$u_j = 0 \quad \text{on} \quad \partial\Omega_j \ .$$

It is easy to see that $u_j \leq u_{j+1} \leq C$ independent of j and $u_j \nearrow u_0$, a solution of (11). The function u_0 is independent of the choice of the Ω_j, and it is not difficult to verify that u_0 extends continuously to every regular point of $\partial\Omega$ — having the value zero there. In our earlier example: $\Omega = \{|x| < 0\} \setminus \{0\}$ in $\mathbf{R}^3$, and $L = \Delta$, the function u_0 is simply the solution in $\{|x| < 1\}$ of $\Delta u_0 = -1$, $u_0 = 0$ on $|x| = 1$. The problem does not feel the origin.

For any function $u \in C(\Omega)$ we give two definitions.

Definition: For a sequence $x_j \in \Omega$, tending to $\partial\Omega$, we say

$$x_j \overset{u}{\to} \partial\Omega \quad \text{if} \quad u(x_j) \to 0 \ .$$

Definition. For a function $v \in C(\Omega)$ we say

$$v \overset{u}{=} 0 \quad \text{on} \quad \partial\Omega \ , \quad \text{or} \quad v \overset{u}{\leq} 0 \quad \text{on} \quad \partial\Omega \ ,$$

if for every sequence $x_j \overset{u}{\to} \partial\Omega$, $v(x_j) \to 0$ or $\limsup v(x_j) \leq 0$, respectively. With these notions applied with $u = u_0$, we now introduce the

Refined Maximum Principle: The refined maximum principle holds for L in Ω if for any function w bounded above,

(12)
$$\left. \begin{array}{ll} Lw \geq 0 & \text{in} \quad \Omega \\ w \overset{u_0}{\leq} 0 & \text{on} \quad \partial\Omega \end{array} \right\} \Rightarrow w \leq 0 \quad \text{in} \quad \Omega \ .$$

The requirement that w be bounded above is natural, for in the punctured ball Ω above, in $\mathbf{R}^3$, $w = |x|^{-1} - 1$ is harmonic and $w \overset{u_0}{=} 0$ on $\partial\Omega$, but w is not ≤ 0 in Ω. That the use of u_0 is not unnatural we see from our first result:

Theorem 1. *The refined maximum principle holds for L in $\Omega \Leftrightarrow \lambda_1 > 0$.*

Here are some other results from [BNV].

Theorem 2 (Refined ABP). *Assume $\lambda_1 > 0$. Suppose w is bounded above on Ω and satisfies*

$$Lw \geq f \ , \quad f \leq 0 \ , \quad f \in L^n(\Omega)$$

and

$$\limsup w(x_j) \leq \beta \quad \text{if} \quad x_j \overset{u_0}{\to} \partial\Omega \ , \quad \beta \geq 0 \ .$$

Then

$$\sup w \le A \left(\|f\|_{L^n} + \beta(\sup c^+) \cdot |\Omega|^{1/n} \right)$$

where $c^+(x) = \max\{c(x), 0\}$. *Here* A *is a constant depending only on* Ω, c_0, C_0, b *and* λ_1.

It is easy to see that $u_0 \le C$ where C depends on n, c_0, C_0, b and $d = \operatorname{diam}\Omega$. However we establish a much better estimate:

$$(13) \qquad\qquad \sup u_0 \le k|\Omega|^{2/n} \ ,$$

with k depending only on n, c_0, C_0 and $b|\Omega|^{1/n}$.

Next we turn to the existence of a principal eigenfunction $\phi_1 > 0$ in Ω associated to the eigenvalue λ_1 given by (8). We say that a function $\psi \in W_{loc}^{2,n}(\Omega)$ is an eigenfunction of $-L$ with eigenvalue λ if $\psi \not\equiv 0$, $\psi \in L^\infty$, and

$$\begin{aligned} (L + \lambda)\psi &\equiv 0 && \text{in} \quad \Omega \\ \psi &\overset{u_0}{=} 0 && \text{on} \quad \partial\Omega \ . \end{aligned}$$

Since L is not self adjoint, eigenfunctions, and eigenvalues, may be complex valued.

Theorem 3. (a) *There exists a principal eigenfunction* $\phi_1 \in W_{loc}^{2,p}(\Omega)$, $\forall p < \infty$, *satisfying*

$$(L + \lambda_1)\phi_1 = 0 \ , \quad \phi_1 > 0 \quad \text{in} \quad \Omega \ ,$$

such that, if we normalize it by

$$\phi_1(x_0) = 1 \text{ for some fixed } x_0 \in \Omega \ ,$$

then

$$\phi_1 \le Cu_0 \quad \text{in} \quad \Omega$$

and $C = C(x_0, \Omega, c_0, C_0, b)$.

(b) *In general it is not true that* $u_0 \le C\phi_1$, *however* $u_0 \overset{\phi_1}{=} 0$ *on* $\partial\Omega$.
(c) *The eigenvalue* λ_1 *is algebraically simple in the sense that (i) if* ψ *is an eigenfunction with eigenvalue* λ_1 *then* ψ *is a constant multiple of* ϕ_1, *and (ii) there is no function* ψ *in* Ω, *bounded above, and satisfying*

$$(L + \lambda_1)\psi = \phi_1 \quad \text{in} \quad \Omega \ , \quad \psi \overset{u_0}{=} 0 \quad \text{on} \quad \partial\Omega \ .$$

(d) *If* ψ *is an eigenfunction which is positive in* Ω, *with eigenvalue* λ, *then* $\lambda = \lambda_1$ *and* ψ *is a constant times* ϕ_1.
(e) *If* $\lambda \ne \lambda_1$ *is an eigenvalue of* $-L$ *then* $\operatorname{Re}\lambda > \lambda_1$.

Remarks. It seems surprising that ϕ_1 is bounded. We do not know if there are eigenvalues other than λ_1.

There is no time to describe proofs, but I remark that ϕ_1 is obtained as a limit of principal eigenfunctions $\phi_j > 0$ in our interior domains Ω_j, with principal eigenvalues λ_j, and $\lambda_j \searrow$ our λ_1 for Ω. In the proof, and at other places in [BNV], we rely on the Krylov-Safonov Harnack inequality (see [GT]), as well as on ABP. Some similar results were obtained earlier by Agmon in the interesting paper [A]; he defined

$$\lambda_1(\Omega) \equiv \inf_{\Omega_0 \subset \Omega} \lambda_1(\Omega_0)$$

— even for Ω unbounded. [A] contains many results; several are related to some in [BNV]. See also Pinchover [P] and [NP].

A few more results from [BNV].

Theorem 4. *Suppose $u > 0$ and satisfies $Lu \leq 0$ in Ω. Then either $\lambda_1 > 0$, or else $\lambda_1 = 0$ and u is a constant multiple of ϕ_1.*

Note that in this theorem, u is not assumed to be bounded, nor are boundary conditions required. The assertion (c)(ii) in Theorem 3 is a special case of

Theorem 5. *If v is bounded above and*

$$
\begin{aligned}
(L + \lambda_1)v &\underset{u_0}{\geq} 0 &&\text{in}\quad \Omega \\
v &\leq 0 &&\text{on}\quad \partial\Omega ,
\end{aligned}
$$

then v is a constant multiple if ϕ_1.

The following is an improvement of sufficient condition (iv) above for the maximum principle to hold.

Theorem 6 *(a) There exists a positive number η depending only on c_0, C_0 and b, such that the refined maximum principle holds for L in Ω if $|\Omega| < \eta$. In fact (b): for some $\tau_1 > 0$ depending only on n, c_0, C_0, and $b|\Omega|^{1/n}$,*

$$\lambda_1(L) \geq \tau_1 |\Omega|^{-2/n} - \sup c^+ .$$

Recently P. Padilla has extended Theorem 6(a) to elliptic operators on a manifold (the operators are uniformly elliptic in suitable local coordinates). The measure $|\Omega|$ is taken with respect to some Riemannian metric.

X. Cabre has obtained a significant improvement of the Inequal-ity ABP:

Theorem (Cabre). *Assume that the coefficient $c(x) \leq 0$. Suppose w is bounded above in Ω and satisfies*

$$
\begin{aligned}
Lw &\underset{u_0}{\geq} f &&\text{in}\quad \Omega , \quad f \leq 0, \quad f \in L^n(\Omega) \\
w &\leq 0 &&\text{on}\quad \partial\Omega .
\end{aligned}
$$

Then
$$\sup w \leq \widetilde{B}(n, c_0, C_0, b|\Omega|^{1/n})|\Omega|^{1/n}\|f\|_{L^n} \ .$$

Acknowledgment. The work was supported by NSF grant CMS-8806-731.

References

[A] S. Agmon, On positivity and decay of solutions of second order elliptic equations on Riemannian manifolds, Methods of functional analysis and theory of elliptic equations, ed. D. Greco, Liguori Ed. Napoli (1983) 19–52.

[B] I. Ya Bakelman, Convex functions and elliptic equations, Springer-Verlag.

[BN] H. Berestycki, L. Nirenberg, S. R. S. Varadhan, The principal eigenvalue and maximum principle for second order elliptic operators in general domains, Comm. Pure Appl. Math., Math. 47 (1994) 47-92.

[DV] M. Donsker, S. R. S. Varadhan, On the principal eigenvalue of second-order elliptic differential operators, Comm. Pure Appl. Math. 29 (1976) 595–621.

[GT] D. Gilbarg, N. S. Trudinger, Elliptic partial differential equations of second order, Grundlehren der math. Wiss. Springer Verlag, 2nd edition, 1983.

[NP] R. D. Nussbaum, Y. Pinchover, On variational principles for the generalized principal eigenvalue of second order elliptic operators and some applications, J. Analyse Math. 59 (1992) 161-177.

[P] Y. Pinchover, Large scale properties of multiparameter oscillator problems, Comm. Partial Diff'l. Eqns. 15 (1990) 647–673.

[SV] D. Stroock, S. R. S. Varadhan, On degenerate elliptic-parabolic operators of second order and their associated diffusions, Comm. Pure Appl. Math. 25 (1972) 651–713.

VERY AMPLENESS CRITERION OF DOUBLE ADJOINTS OF AMPLE LINE BUNDLES

YUM-TONG SIU[1]

Introduction

In this paper we give a numerical criterion for $L+2K_X$ to be very ample for an ample line bundle over a compact complex manifold X. Such a very ampleness criterion is motivated by the following conjecture of Fujita [Fu87]. For an ample line bundle L on a compact complex manifold X of complex dimension n, $(n+1)L+K_X$ is free (*i.e.* generated by global holomorphic sections) and $(n+2)L+K_X$ is very ample. The case $n=1$ is well-known and the case $n=2$ was proved by Reider [Re88]. Ein-Lazarsfeld [EL93] proved the freeness part for $n=3$. For a general n, Demailly [D93] used the complex Monge-Ampère equation and the Lelong numbers of closed positive currents to prove the very ampleness of $12n^n L+2K_X$ and the generation of r-jets by global holomorphic sections of $6(n+r)^n L+2K_X$. By using purely algebraic geometric methods Kollar [Kol93] proved the freeness of $2(n+1)((n+2)!)(L+K_X)$ and the very ampleness of $2(n+1)(n+3)((n+2)!)(L+K_X)+K_X$. In [Siu94] a simple algebraic method using the lower bound of the polynomial from the theorem of Riemann-Roch was introduced to prove the generation of r-jets by global holomorphic sections for $mL+2K_X$ with m at least $2(n+2+n\binom{3n+2r-1}{n}))$, which Demailly [D94] improved to $2+\binom{3n+2r-1}{n}$. The bound for m from this simple algebraic method is of the order $(3e)^n$ which is better than the order n^n in [D93]. Ein-Lazarsfeld-Nakamaye [ELN94] introduced another algebraic mehtod which gives the following numerical criterion for the very ampleness of $L+2K_X$. Let X be a smooth projective variety of dimension n. Let L be an ample line bundle over X such that $L+K_X$ is ample. If $(L+K_X)^{n-d}\cdot W > ((2n+1)d+n)^d (n+1)^n$ for any subvariety W of codimension d in X for $0 \le d \le n-1$, then $L+2K_X$ is very ample. For comparison the condition which Demailly [D94, Remark (11.20)] obtained, by using the analytic method of the Monge-Ampère equation, for the global holomorphic sections of $L+2K_X$ to generate r-jets, is $(L^\nu\cdot W)^{1/\nu} \ge 6(n+6(n+r)^n)^n$ for any subvariety W of dimension ν in X. In particular, the result of Ein-Lazarsfeld-Nakamaye [ELN94] gives the very ampleness of $mL+2K_X$ for $m \ge ((2n+1)(n-1)+n)^{n-1}(n+1)^n$

[1] Partially supported by a grant from the National Science Foundation

which is of the order n^{2n} and it is weaker than the order n^n by the method of [D93] and the order $(3e)^n$ by the method of [Siu94]. However, the method [Siu94] of using the lower bound of the polynomial from the theorem of Riemann-Roch cannot give the very ampleness of $mL + 2K_X$ with $m = 1$. In this paper we introduce a technique to replace the use of the polynomial from the theorem of Riemann-Roch. This technique gives a numerical criterion for the very ampleness of $L + 2K_X$ which corresponds to the very ampleness of $mL + 2K_X$ with an order for m better than those obtained in [Siu94] and [D94] by a factor of n. The main result is the following.

Main Theorem. *Let L be an ample line bundle over a compact complex manifold X of complex dimension $n \geq 2$. Let $P_1, \cdots, P_J$ be distinct points in X and $q_1, \cdots, q_J$ be nonnegative integers. Let κ_d be the supremum of $(L^d \cdot W)^{-1/d}$ as W ranges over all the subvarieties of dimension d in X. Let $N = \sum_{j=1}^{J} \binom{3n+2q_j-3}{n}$. Suppose $N\nu\kappa_\nu + \sum_{\mu=\nu+1}^{n} \mu\kappa_\mu + Jn\kappa_n < \frac{1}{2}$ for $1 \leq \nu \leq n$. Then global holomorphic sections of $L+2K_X$ over X generate simultaneously the q_j-jets at P_j for $1 \leq j \leq J$ (in the sense that the restriction map $\Gamma(X, L+2K_X) \to \oplus_{j=1}^{J} \mathcal{O}_X/\mathfrak{m}_{P_j}^{q_j}$ is surjective, where $\mathfrak{m}_{P_j}$ is the maximum ideal at P_j). More generally, if $0 < \alpha \leq \frac{1}{2}$ and $(\alpha + \frac{1}{2})L$ is a line bundle, then global holomorphic sections of $(\alpha + \frac{1}{2})L + 2K_X$ over X generate simultaneously the q_j-jets at P_j for $1 \leq j \leq J$.*

Corollary (0.1). *Let L be an ample line bundle over a compact complex manifold X of complex dimension $n \geq 2$. Let $P_1, \cdots, P_J$ be distinct points in X and $q_1, \cdots, q_J$ be nonnegative integers. Suppose $\epsilon > 0$ and $(L^d \cdot W)^{1/d} \geq 2n \sum_{j=1}^{J} \binom{3n+2q_j-3}{n} + 2Jn + \epsilon$ for any subvariety W of dimension d in X. Then global holomorphic sections of $L+2K_X$ over X generate simultaneously the q_j-jets at P_j for $1 \leq j \leq J$.*

Corollary (0.2). *Let L be an ample line bundle over a compact complex manifold X of complex dimension $n \geq 2$. Let $P_1, \cdots, P_J$ be distinct points in X and $q_1, \cdots, q_J$ be nonnegative integers. Then for any integer $m > 2n \sum_{j=1}^{J} \binom{3n+2q_j-3}{n} + 2Jn$, global holomorphic sections of $mL+2K_X$ over X generate simultaneously the q_j-jets at P_j for $1 \leq j \leq J$.*

In Corollary (0.2) the lower bound for m can be improved when it is proved separately instead of as a corollary to the Main Theorem (see Proposition (5.1) below). The consequence of the Main Theorem which corresponds to the formulation of [ELN94] is the following.

Corollary (0.3). *Let L be an ample line bundle over a compact complex manifold X of complex dimension $n \geq 2$ such that $L + K_X$ is ample. Let $P_1, \cdots, P_J$ be distinct points in X and $q_1, \cdots, q_J$ be nonnegative integers. Let N be the maximum of $\sum_{j=1}^{J} \binom{4n+3q_j-1}{n}$, $\binom{3n-1}{n}$, and $2\binom{3n-3}{n}$. For any $1 \leq d \leq n$ let κ_d be the supremum of $((L + K_X)^d \cdot W)^{-1/d}$ as W ranges over all the subvarieties of dimension d in X. Suppose $N\nu\kappa_\nu + \sum_{\mu=\nu+1}^{n} \mu\kappa_\mu + 2n\kappa_n < \frac{1}{2}$ for $1 \leq \nu \leq n$. Then global holomorphic* *sections of* $L + 2K_X$ *over* X *generate simultaneously the* q_j-*jets at* P_j *for* $1 \leq j \leq J$.

Remark. The numerical bounds given in the results in this paper can be somewhat improved by keeping better track of the arithmetics in the steps where non-optimal estimates are used so that the expressions in the final results can assume a simplier form. However, such improvements do not contribute to any substantial difference in the final results.

We sketch here the techniques used in this paper. First let us recall the technique introduced in [Siu94] of using the polynomial from the theorem of Riemann-Roch for a subvariety. Consider the metric of $mL + K_X$ defined by multivalued holomorphic sections which have a high vanishing order at a prescribed point. Consider the subvariety defined by the local non-integrability of the metric. One uses the pigeon hole principle to rule out the low nonnegative values of the polynomial from the theorem of Riemann-Roch and then uses the vanishing theorem to give a lower bound of the dimension of the space of holomorphic sections of $m'L + 2K_X$ over the subvariety. From the lower bound we conclude that there exists a non-identically-zero holomorphic section of $m'L + 2K_X$ over the subvariety which can be extended to a holomorphic section of $m'L + 2K_X$ over the ambient space with high vanishing order at the prescribed point. We use such holomorphic sections over the ambient space to modify the metric to eventually get a metric $m''L + K_X$ with isolated high singularity at the prescribed point. Then the results of very ampleness and simultaneous generation of jets by global holomorphic sections follow from Nadel's vanishing theorem.

When we use only $L + 2K_X$ instead of $mL + 2K_X$, we can no longer use the polynomial in m from the theorem of Riemann-Roch applied to the line bundle $mL + 2K_X$ on a subvariety. So we have to use a new method to get a lower bound for the dimension of the space of holomorphic sections of $L + 2K_X$ over a subvariety. The new method is to use a modification of Shukorov's technique for his nonvanishing theorem [Sh85]. We now sketch this new method with some small sacrifice on the accuracy of some minor points. In our sketch all the metrics of **Q**-bundles are constructed from multivalued holomorphic sections. We start out with a singular metric h of some numerically effective **Q**-bundle B over X defined by multivalued holomorphic sections with a high vanishing order at a prescribed point P of X. Suppose $B + \gamma L$ is a **Z**-bundle for some $0 < \gamma < 1$. We take a subvariety V_1 of X which is a branch of the subvariety of local non-integrability of h. We use a resolution of singularieties to get a complex manifold X_1 with a map $f_1 : X_1 \to X$ such that $f_1(X_1) = V_1$. The idea is to use the theorem of Riemann-Roch to get non-identically-zero multivalued holomorphic sections of some $\alpha_1 f_1^* L$ with $0 < \alpha_1 < \gamma$ so that we can construct from h a singular metric h_1 of $f_1^*(B + \alpha_1 L + K_X) -$

$K_{X_1} + Z_1$, where Z_1 is a **Z**-divisor of X_1 with the property that the support $\text{Supp} Z_1$ of Z_1 is mapped by f_1 to a proper subvariety of V_1 and every holomorphic section of $f_1^*(B + \gamma L + K_X) - K_{X_1} + Z_1$ comes from a holomorphic section of $B + \gamma L + K_X$ over X (as a consequence of Nadel's vanishing theorem for multiplier ideal sheaves). The key point is that we require that a branch V_2 of the subvariety of local non-integrability of h_1 is not contained $(f_1)^{-1}(\text{Supp} Z_1)$ and the dimension of $f_1(V_2)$ is less than that of V_1. We repeat the above argument with X (respectively h and V_1) replaced by X_1 (respectively h_1 and V_2) to construct by resolution of singularites a complex manifold X_2 (instead of X_1) with a map $g_2 :$ $X_2 \to X_1$ such that $g_2(X_2) = V_2$. Let $f_2 = f_1 \circ g_2$. By continuing the process, finally we end up with a sequence of manifolds X, X_1, X_2, et cetera so that the dimension of the image of X_ν in X under the map $f_\nu : X_\nu \to X$ is strictly decreasing. We assume that the degree of L on every subvariety of positive dimension is large that we can choose α_ν with $\sum_{\nu=1}^{\lambda-1} \alpha_\nu L < \gamma$. In every step we keep the key requirement that V_ν is not contained in the union of $(f_\nu)^{-1}(f_\mu(\text{Supp}(Z_\mu))$ for any $\mu < \nu$. At some step the dimension of $f_\lambda(X_\lambda)$ becomes zero and we stop and we set Z_λ to be the zero divisor. In every step the spaces X_1, X_2, et cetera are all connected, because V_1, V_2, et cetera are all irreducible. When we use $\alpha_{\lambda-1} L$ to construct a singular metric for the **Q**-bundle $f_{\lambda-1}^*(B + \sum_{\nu=1}^{\lambda-1} \alpha_\nu L) + f_{\lambda-1}^* K_X - K_{X_{\lambda-1}} + Z_{\lambda-1}$, we can modify the construction so that the last complex manifold X_λ is not connected and the number N of components of X_λ is large (and the number of points in $f_\lambda(V_\lambda)$ is N). For the modification we need a large lower bound for the degree of L on every subvariety of positive dimension. We still keep the key requirement that every branch of V_λ is not contained in the union of $(f_\lambda)^{-1}(f_\mu(\text{Supp}(Z_\mu))$ for any $\mu < \lambda$. Since every element of $\Gamma(X_\lambda, f_\lambda^*(B + \gamma L + K_X))$ comes from an element of $\Gamma(X_\nu, f_\nu^*(B + \gamma L + K_X) + Z_\nu)$, it follows that the dimension of $\Gamma(X_1, f_1^*(B + \gamma L + K_X) + Z_1)$ is at least N. This argument replaces the use of the polynomial from the theorem of Riemann-Roch. The final result of this argument means that we are able to find a holomorphic section s of $B + \gamma L + K_X$ over X which is not identically zero on V_1 and yet vanishes to high order at the prescribed point P. In our situation B will be $\frac{1}{2} L + K_X$ and γ will be $\frac{1}{2}$. We will then use $s^{1/2}$ to modify the metric h of B to eventually get a singular metric of $\frac{1}{2} L + K_X$ with high singularity at P but locally integrable at every point of some deleted neighborhood of P in X. Then the results of very ampleness and simultaneous generation of jets by global holomorphic sections follow from Nadel's vanishing theorem.

For the purpose of cutting down the dimension of the subvariety of local non-integra-bility, though it is sufficient to use multivalued holomorphic sections not vanishing identically on any branch of the subvariety, it is also possible to allow low order vanishing for the multivalued holomorphic sections. The technique of using such low order vanishing is discussed in the section below on background singular metrics of Lelong mumber one (§3). We get better results by using multivalued holomorphic sections with such low order vanishing. Such low order vanishing can also be used in the criterion for the very ampleness of $mL + 2K_X$ with some improvement on the bound for m (see Proposition (5.1) below).

The result in this paper deals with the double adjoint $L + 2K_X$ instead of the more desirable adjoint $L + K_X$ because of the limitation of the methods involved. For freeness there is the following recent result of Angehrn and myself ([AS94], [A94]) for the adjoint bundle $L + K_X$. If L is an ample line bundle L on a compact complex manifold X of complex dimension n and if $\epsilon > 0$ and $(L^d \cdot W)^{1/d} \geq \frac{n}{2}(n + 2J - 1) + \epsilon$ for any irreducible subvariety W of dimension d in X and $1 \leq d \leq n$, then the global holomorphic sections of $L + K_X$ over X separate any set of J distinct points $P_1, \cdots, P_J$ of X in the sense that the restriction map $\Gamma(X, L + K_X) \to \oplus_{\nu=1}^{J} \mathcal{O}_X/\mathbf{m}_{P_\nu}$ is surjective, where $\mathbf{m}_{P_\nu}$ is the maximum ideal at P_ν.

Another kind of result related to those in this paper is the effective Matsusaka big theorem ([Si93], [D94]).

1. Multiplier Ideal Sheaves and Almost Ample Line Bundles

A **Q**-*bundle* means that some positive multiple of it is a holomorphic line bundle. We also call a holomorphic line bundle a **Z**-*bundle*. Let L be a holomorphic line bundle over a compact complex manifold X of complex dimension n. Let γ be a positive rational number. We say that the **Q**-bundle γL is *ample* (respectively *numerically effective*) if L is ample (respectively numerically effective). By a *multivalued holomorphic section* u of the **Q**-bundle γL we simply mean that there exists a positive integer p such that $\gamma p L$ is a line bundle and u^p is a holomorphic section of $p\gamma L$. We say that the multivalued holomorphic section u vanishes at a point P to order at least q if u^p vanishes to order at least pq. We will consider metrics of the form $(\sum_{\nu=1}^{\ell} |u_\nu|^2)^{-\alpha}$ for the **Q**-bundle $\alpha\gamma L$ with $u_1, \cdots, u_\ell$ being multivalued holomorphic sections of γL. The *multiplier ideal sheaf* for the metric $(\sum_{\nu=1}^{\ell} |u_\nu|^2)^{-\alpha}$ is defined as the the sheaf of the germs of all holomorphic functions f with $|f|^2(\sum_{\nu=1}^{\ell} |u_\nu|^2)^{-\alpha}$ locally integrable. The multiplier ideal sheaf is coherent [N89]. It is a consequence of the methods of the L^2 estimates of $\bar{\partial}$ and also can be derived by using desingularization by monoidal transformations. In particular, the *set of local non-integrability* (which is the set of points of X where

$(\sum_{\nu=1}^{\ell} |u_\nu|^2)^{-\alpha}$ is not locally integrable) is a subvariety of X. We will not use the *curvature current* of the metric $(\sum_{\nu=1}^{\ell} |u_\nu|^2)^{-\alpha}$ but will mention the *Lelong number* of the curvature current of the metric $(\sum_{\nu=1}^{\ell} |u_\nu|^2)^{-\alpha}$. Such a Lelong number at a point P of X can be defined as the minimum of the vanishing order of $(u_\nu)^\alpha|C$ when C ranges over all local smooth complex curves passing through P and ν ranges over $1, \cdots, \ell$ [Si74]. If the Lelong number of the curvature current of the metric $(\sum_{\nu=1}^{\ell} |u_\nu|^2)^{-\alpha}$ is less than 1 at P, then $(\sum_{\nu=1}^{\ell} |u_\nu|^2)^{-\alpha}$ is locally integrable at P [Sk72]. If the Lelong number of the curvature current of the metric $(\sum_{\nu=1}^{\ell} |u_\nu|^2)^{-\alpha}$ at P is $\geq n + q$ for some nonnegative integer q, then $|f|^2(\sum_{\nu=1}^{\ell} |u_\nu|^2)^{-\alpha}$ is not locally integrable at P for any holomorphic function germ f at P whose vanishing order at P is less than q [Sk72].

We will use Nadel's vanishing theorem for multiplier ideal sheaves [N89, D93] in the following form. If A is an ample **Q**-bundle and B is a **Q**-bundle over a compact complex manifold X such that $A + B$ is a **Z**-bundle and if $\mathcal{I}$ is the multiplier ideal sheaf of the metric $(\sum_{\nu=1}^{\ell} |u_\nu|^2)^{-\alpha}$ of B defined by multivalued holomorphic sections $u_1, \cdots, u_\ell$ of the **Q**-bundle $\frac{1}{\alpha} B$ for some positive rational number α, then $H^\nu(X, \mathcal{O}_X(A + B + K_X) \otimes \mathcal{I})$ vanishes for $\nu \geq 1$. We would like to remark that the notion of multiplier ideal sheaves was first introduced by J. J. Kohn [Koh79] in his work on subellipticity of the $\bar{\partial}$-Neumann problems on pseudo-convex domains. One special form of Nadel's vanishing theorem is the vanishing theorem of Kawamata-Viehweg ([Ka82], [V82]). We will use the vanishing theorem of Kawamata-Viehweg expressed in terms of almost ample line bundles which are defined as follows.

Definition. Let X be a compact complex manifold and F be a **Q**-bundle over X. We say that F is *almost ample* if there exist a family of nonsingular divisors $\{Z_j\}$ with normal crossing in X so that $G := F - \sum_j e_j Z_j$ is ample for some rational numbers $0 < e_j < 1$.

The vanishing theorem of Kawamata-Viehweg states that, if F is an almost ample line bundle over a compact complex manifold X, then $H^\nu(X, F + K_X)$ vanishes for $\nu \geq 1$. Nadel's vanishing theorem implies the vanishing theorem of Kawamata-Viehweg when one sets $A = G$ and $B = \sum_j e_j Z_j$ with the metric $|u|^{-2}$, where u is the multivalued holomorphic section of B whose divisor is $\sum_j e_j Z_j$.

The multiplier ideal sheaf can be described as a direct image sheaf under a desingularization map in the following way. Consider the metric $(\sum_{\nu=1}^{\ell} |u_\nu|^2)^{-\alpha}$ of a **Q**-bundle B over a compact complex manifold X, where $(u_\nu)^p$ is a holomorphic section of the **Z**-bundle $\frac{p}{\alpha} B$ for some positive integer p. Let $\mathcal{J} \subset \mathcal{O}_X$ be the ideal sheaf on X generated locally by $(u_\nu)^p$ $(1 \leq \nu \leq \ell)$. We use a blowup map $f : Y \to X$ given by successive monoidal transformations with nonsingular centers so that there exist a finite number of nonsingular divisors E_μ in Y in normal crossing with the property that $K_Y - f^* K_X = \sum_\mu r_\mu E_\mu$ for some nonnegative integers r_μ and that the pullback of the ideal sheaf $\mathcal{J}$ by f is equal to the ideal sheaf of the divisor $\sum_\mu e_\mu^* E_\mu$ for some nonnegative integers e_μ^*. Let $e_\mu = \alpha e_\mu^* / p$. For a real number a we denote by $\lfloor a \rfloor$ the round-down of a which means the largest integer not exceeding a and denote by $\lceil a \rceil$ the round-up of a which means the smallest integer not smaller than a. Let I be the set of indices μ with $e_\mu - r_\mu \geq 1$. Then the zeroth direct image under f of the ideal sheaf of $\sum_{\mu \in I} \lfloor e_\mu - r_\mu \rfloor E_\mu$ is the multiplier ideal sheaf $\mathcal{I}$ of the metric $(\sum_{\nu=1}^\ell |u_\nu|^2)^{-\alpha}$.

We give here two lemmas which we will need later.

Lemma (1.1). *Let A and B be $\mathbf{Q}$-bundles over a compact complex manifold X such that A is ample and B is numerically effective and $A + B$ is a $\mathbf{Z}$-bundle. Let $u_1, \cdots, u_\ell$ be non-identically-zero multivalued holomorphic sections of $\frac{1}{\alpha} B$ over X for some positive rational number α. Let $\mathcal{I}$ be the multiplier ideal sheaf of the metric $(\sum_{\nu=1}^\ell |u_\nu|^2)^{-\alpha}$ of B. Let V be a branch of the zero set of $\mathcal{I}$. Then there exist two coherent ideal sheaves $\mathcal{F} \subset \mathcal{G}$ with the following properties:*
(i) The zero-set of $\mathcal{F}$ is V.
(ii) The zero-set of $\mathcal{G}$ is a proper subvariety of V.
(iii) The product of $\mathcal{G}$ and the ideal sheaf of V is contained in $\mathcal{F}$.
(iv) For any numerically effective holomorphic line bundle L over X, both $H^\nu(X, \mathcal{F} \otimes (L + A + B + K_X))$ and $H^\nu(X, \mathcal{G} \otimes (L + A + B + K_X))$ vanish for $\nu \geq 1$.

Proof: We have a blowup map $f : Y \to X$ given by successive monoidal transformations with nonsingular centers so that there exist a finite number of nonsingular divisors E_μ in Y in normal crossing with the property that

(i) $K_Y - f^* K_X = \sum_\mu r_\mu E_\mu$ for some nonnegative integers r_μ,
(ii) for a suitable positive integer p the pullback by f of the ideal sheaf generated locally by the holomorphic sections $(u_\nu)^p$ ($1 \leq j \leq \ell$) of the $\mathbf{Z}$-bundle $\frac{p}{\alpha} B$ is equal to the ideal sheaf of the divisor $\frac{p}{\alpha} \sum_\mu e_\mu E_\mu$ for some nonnegative rational numbers e_μ.

We choose $0 < \delta_\mu < 1$ so small that the $\mathbf{Q}$-bundle $\frac{1}{2} f^* A - \sum_\mu \delta_\mu E_\mu$ on Y is ample and $e_\mu - r_\mu + \delta_\mu \neq e_\lambda - r_\lambda + \delta_\lambda$ for $\mu \neq \lambda$. There exists some index μ such that $f(E_\mu)$ is equal to V. Then $e_\mu - r_\mu \geq 1$. Let $\beta = (1 + r_\mu - \delta_\mu)/e_\mu$. Then $0 < \beta < 1$. Choose $0 < \epsilon < \beta$ so small that $\beta - \epsilon > (1 + r_\lambda - \delta_\lambda)/e_\lambda$ for any λ with $\beta > (1 + r_\lambda - \delta_\lambda)/e_\lambda$. Let I be the set of all indices λ such that $\beta e_\lambda - r_\lambda + \delta_\lambda \geq 1$. Let $\mathcal{F}$ be the direct image under f of the ideal sheaf $\sum_{\lambda \in I} \lfloor \beta e_\lambda - r_\lambda + \delta_\lambda \rfloor E_\lambda$. Let J be the set of all indices λ such that $(\beta - \epsilon) e_\lambda - r_\lambda + \delta_\lambda \geq 1$. Let $\mathcal{G}$ be the direct image under f of the ideal sheaf $\sum_{\lambda \in J} \lfloor (\beta - \epsilon)e_\lambda - r_\lambda + \delta_\lambda \rfloor E_\lambda$. Both $\mathcal{F}$ and $\mathcal{G}$ are ideal sheaves on X and $\mathcal{F} \subset \mathcal{G}$. Let $v_1', \cdots, v_q'$ be multivalued holomorphic sections of $\frac{1}{2}f^*(A + (1 - \beta)B) - \sum_\mu \delta_\mu E_\mu$ over Y without common zeroes. Let $w'_1, \cdots, w'_t$ be multivalued holomorphic sections of $\frac{1}{2}f^*(A + (1 - \beta + \epsilon)B) - \sum_\mu \delta_\mu E_\mu$ over Y without common zeroes. Let v be the multivalued holomorphic section of $\sum_\mu \delta_\mu E_\mu$ over Y whose divisor is precisely $\sum_\mu \delta_\mu E_\mu$. Let v_j be the multivalued holomorphic section of $\frac{1}{2}A + (1 - \beta)B$ which is the direct image under f of vv'_j. Let w_j be the multivalued holomorphic section of $\frac{1}{2}A + (1 - \beta + \epsilon)B$ which is the direct image under f of vw'_j. The sheaf $\mathcal{F}$ is the multiplier ideal sheaf of the metric $(\sum_{j=1}^q |v_j|^2)^{-1}(\sum_{\nu=1}^\ell |u_\nu|^2)^{-\alpha\beta}$ of $\frac{1}{2}A + B$ and the sheaf $\mathcal{G}$ is the multiplier ideal sheaf of the metric $(\sum_{j=1}^t |w_j|^2)^{-1}(\sum_{\nu=1}^\ell |u_\nu|^2)^{-\alpha(\beta-\epsilon)}$ of $\frac{1}{2}A + B$. The zero-set of $\mathcal{F}$ is precisely V and the zero-set of $\mathcal{G}$ is a proper subvariety of V. The product of $\mathcal{G}$ and the ideal sheaf of V is contained in $\mathcal{F}$. By Nadel's vanishing theorem, for any numerically effective line bundle L over X both $H^\nu(X, \mathcal{F} \otimes (L + A + B + K_X))$ and $H^\nu(X, \mathcal{G} \otimes (L + A + B + K_X))$ vanish for $\nu \geq 1$.

Lemma (1.2). *Let B be an ample $\mathbf{Q}$-bundles over a reduced irreducible compact complex space X of complex dimension n and P be a regular point of X and q be a positive rational number. If $B^n \cdot X > q$, then there exists a non-identically-zero multivalued holomorphic section of B over X whose vanishing order at P is at least q.*

Proof: Take a sufficiently large integer p such that pB is a line bundle. By applying the theorem of Riemann-Roch to pB over X, we can find a non-identically-zero holomorphc section s of pB over X so that s vanishes to order at least pq at P. The required multivalued holomorphic section of B is then obtained by taking the p-th root of s.

2. Lower Bound Argument for the Dimension of the Space of Holomorphic Sections

Proposition (2.1). Let X be a compact complex manifold of complex dimension $n \geq 2$. Let B be a numerically effective $\mathbf{Q}$-bundle over X and let $s_1, \cdots, s_\ell$ be non-identically-zero multivalued holomorphic sections of B over X. Let Σ be the set of points of X where $(\sum_{j=1}^\ell |s_j|^2)^{-1}$ is not locally integrable. Let Σ' be a positive dimensional branch of Σ. Let L be an ample line bundle over X. For any $1 \leq \nu \leq n$ let κ_ν be the supremum of $(L^\nu \cdot W)^{-1/\nu}$ as W ranges over all the subvarieties of complex dimension ν in X. Let $P_1, \cdots, P_J$ be a finite number of distinct points in X and $q_1, \cdots, q_J$ be nonnegative integers. Let c be the complex dimension of Σ'. Let N be $\sum_{j=1}^J \binom{q_j + n - 1}{n}$. Let γ be a rational number such that $\gamma > N\nu\kappa_\nu + \sum_{\mu=\nu+1}^c \mu\kappa_\mu$ for any $1 \leq \nu \leq c$ and $\gamma L + B$ is a $\mathbf{Z}$-bundle. Then there exists a holomorphic section of $\gamma L + B + K_X$ over X which vanishes at P_j to order at least q_j for $1 \leq j \leq J$ but is not identically zero on Σ'.

Proof of Proposition (2.1): Choose $\gamma_\nu > \nu \kappa_\nu$ for $1 \leq \nu < c$ and choose $\gamma' > 0$ such that $\gamma' < \gamma - N\gamma_\nu + \sum_{\mu=\nu+1}^{c} \gamma_\mu$ for any $1 \leq \nu \leq c$. For the proof we will construct by induction a collection of objects satisfying certain requirements. We give below (I) a list of the objects to be constructed, (II) a list of other objects defined from the constructed objects, (III) a list of the properties of the constructed and defined objects, (IV) the construction procedure, and (V) the verification of the properties.

(I) *A list of the objects to be constructed.*
(I.1) a positive integer Λ,
(I.2) a sequence of irreducible positive dimensional subvarieties of X :

$$\tilde{\Sigma}^{(0)} \supset \Sigma^{(0)} \supset \tilde{\Sigma}^{(1)} \supset \Sigma^{(1)} \supset \cdots \supset \tilde{\Sigma}^{(\lambda)} \supset \Sigma^{(\lambda)} \supset \cdots \supset \tilde{\Sigma}^{(\Lambda)} \supset \Sigma^{(\Lambda)}$$

such that $\Sigma^{(\lambda)}$ is a branch of $\tilde{\Sigma}^{(\lambda)}$ and $\tilde{\Sigma}^{(0)} = \Sigma^{(0)} = X$ and $\tilde{\Sigma}^{(1)} = \Sigma$ and $\Sigma^{(1)} = \Sigma'$,
(I.3) a proper subvariety $\Sigma_1^{(\lambda)}$ of $\Sigma^{(\lambda)}$ ($1 \leq \lambda \leq \Lambda$) with $\Sigma_1^{(0)}$ being the empty set,
(I.4) for $1 \leq \lambda \leq \Lambda$ a sequence of holomorphic maps of compact complex manifolds $\overline{f}^{(\lambda)} : Y^{(\lambda)} \to X^{(\lambda-1)}$ given by successive monoidal transformations with nonsingular centers such that there exist a collection of nonsingular divisors $\{E_j^{(\lambda)}\}$ in $Y^{(\lambda)}$ with normal crossing and an index j_λ with $X^{(\lambda)} = E_{j_\lambda}^{(\lambda)}$ and $X^{(0)} = X$,
(I.5) for $1 \leq \lambda \leq \Lambda$ nonnegative divisors $P^{(\lambda)} = \sum_{\mu \neq j_\lambda} p_\mu^{(\lambda)} E_\mu^{(\lambda)}$ and $N^{(\lambda)} = \sum_{\mu \neq j_\lambda} n_\mu^{(\lambda)} E_\mu^{(\lambda)}$ in $Y^{(\lambda)}$.

(II) *A list of other objects defined from the constructed objects.*
(II.1) d_λ is the dimension of $\Sigma^{(\lambda)}$ ($0 \leq \lambda \leq \Lambda$).
(II.2) $h^{(\lambda)}$ is the restriction of $f^{(\lambda)} : Y^{(\lambda)} \to X^{(\lambda-1)}$ to $X^{(\lambda)} = E_{j_\lambda}^{(\lambda)}$ ($1 \leq \lambda \leq \Lambda$).
(II.3) $g^{(\lambda)} = h^{(1)} \circ h^{(2)} \circ \cdots \circ h^{(\lambda)}$ for $1 \leq \lambda \leq \Lambda$ and $g^{(0)}$ is the identity map of $X^{(0)}$.
(II.4) $A^{(\lambda)} = (g^{(\lambda)})^* (\gamma - \gamma' - \sum_{\nu=1}^{\lambda} \gamma_{d_\nu} L)$ for $0 \leq \lambda \leq \Lambda$.
(II.5) $B^{(\lambda)} = (g^{(\lambda)})^* (\gamma_{d_\lambda} L)$ ($1 \leq \lambda \leq \Lambda$) with $B^{(0)} = B$.
(II.6) Let $F^{(0)} = \gamma' L$. Inductively define $F^{(\lambda)}$ on $X^{(\lambda)}$ as the restriction of $(f^{(\lambda)})^* (B^{(\lambda-1)} + F^{(\lambda-1)} + K_{X^{(\lambda-1)}}) + P^{(\lambda)} - N^{(\lambda)} - K_{E_{j_\lambda}^{(\lambda)}}$ to $X^{(\lambda)}$.

(III) *A list of the properties of the constructed and defined objects.*
(III.1) $g^{(\lambda)}(X^{(\lambda)}) = \Sigma^{(\lambda)}$.
(III.2) $\Sigma_1^{(\lambda)}$ contains $\Sigma_1^{(\lambda-1)} \cap \Sigma^{(\lambda)}$ and $\Sigma^{(\lambda)} \cap \text{Sing}(\tilde{\Sigma}^{(\lambda)})$ and contains $g^{(\lambda)}(x)$ for any x in $X^{(\lambda)} - (g^{(\lambda)})^{-1}(\Sigma_1^{(\lambda)})$ where the rank of $g^{(\lambda)}$ at x is

less than the complex dimension d_λ of $\Sigma^{(\lambda)}$. Here $\mathrm{Sing}(\tilde\Sigma^{(\lambda)})$ means the singular set of $\tilde\Sigma^{(\lambda)}$.

(III.3) The intersection of $\Sigma^{(\lambda-1)} - \Sigma_1^{(\lambda-1)}$ and $\tilde\Sigma^{(\lambda)}$ is precisely the set of points of $\Sigma^{(\lambda-1)} - \Sigma_1^{(\lambda-1)}$ where $|\sigma^{(\lambda)}|^{-2}$ is not locally integrable for some multivalued holomorphic section $\sigma^{(\lambda)}$ of $\gamma_{d_\lambda} L$.

(III.4) $\tilde\Sigma^{(\lambda)} - \Sigma_1^{(\lambda-1)}$ is dense in $\tilde\Sigma^{(\lambda)}$.

(III.5) Every exceptional divisor in the map $f^{(\lambda)} : Y^{(\lambda)} \to X^{(\lambda-1)}$ is contained in the collection $\{E_j^{(\lambda)}\}$.

(III.6) Every component of the support of $P^{(\lambda)}$ is an exceptional divisor for the map $f^{(\lambda)} : Y^{(\lambda)} \to X^{(\lambda-1)}$.

(III.7) For $1 \leq \lambda \leq \Lambda$ the $\mathbf{Q}$-bundle $(f^{(\lambda)})^*(B^{(\lambda-1)} + F^{(\lambda-1)} + K_{X^{(\lambda-1)}}) + P^{(\lambda)} - N^{(\lambda)} - K_{Y^{(\lambda)}} - E_{j_\lambda}^{(\lambda)}$ is almost ample on $Y^{(\lambda)}$ and is of the form $G^{(\lambda)} + \sum_{j \neq j_\lambda} e_j^{(\lambda)} E_j^{(\lambda)}$ on $Y^{(\lambda)}$ so that $G^{(\lambda)}$ is ample on $Y^{(\lambda)}$ and $0 \leq e_j^{(\lambda)} < 1$. When $\lambda = 0$ the collection $\{E_j^{(0)}\}$ is the empty collection and we set $\sum_{j \neq j_0} e_j^{(0)} E_j^{(0)} = 0$.

(III.8) $(g^{(\lambda-1)} \circ f^{(\lambda)})(E_j^{(\lambda)})$ is contained in $\Sigma_1^{(\lambda)}$ for $j \neq j_\lambda$. In particular, $(g^{(\lambda-1)} \circ f^{(\lambda)})(\mathrm{Supp} P^{(\lambda)})$ and $(g^{(\lambda-1)} \circ f^{(\lambda)})(\mathrm{Supp} N^{(\lambda)})$ are both contained in $\Sigma_1^{(\lambda)}$.

(III.9) For every x in $\Sigma^{(\Lambda)} - \Sigma_1^{(\Lambda)}$, we can find a non-identically-zero multivalued holomorphic section $\sigma_x^{(\Lambda)}$ of $\gamma_\Lambda L$ on $\Sigma^{(\Lambda)}$ which vanishes at $g^{(\Lambda)}(x)$ to order at least equal to d_Λ so that x belongs to the subvariety $S_x^{(\Lambda)}$ of $\Sigma^{(\Lambda)} - \Sigma_1^{(\Lambda)}$ where $|\sigma_x^{(\Lambda)}|^{-2}$ is not locally integrable and, moreover, the dimension of $S_x^{(\Lambda)}$ is zero.

(III.10) The restriction map

$$\Gamma(Y^{(\lambda+1)}, (f^{(\lambda+1)})^*(A^{(\lambda)} + B^{(\lambda)} + F^{(\lambda)} + K_{X^{(\lambda)}}) + P^{(\lambda+1)} - N^{(\lambda+1)}) \to$$

$$\Gamma(E_{j_{\lambda+1}}^{(\lambda+1)}, ((f^{(\lambda+1)})^*(A^{(\lambda)} + B^{(\lambda)} + F^{(\lambda)} + K_{X^{(\lambda)}}) + P^{(\lambda+1)} - N^{(\lambda+1)})|E_{j_{\lambda+1}}^{(\lambda+1)})$$

is surjective for $0 \leq \lambda \leq \Lambda - 1$.

(III.11) The bundle $F^{(\lambda+1)} + K_{X^{(\lambda+1)}}$ is equal to $(g^{(\lambda)})^*((\sum_{\nu=1}^{\lambda} \gamma_{d_\nu}) L + B + F^{(0)} + K_X)$ on

$$X^{(\lambda+1)} - \mathrm{Supp} P^{(\lambda+1)} \cup \mathrm{Supp} N^{(\lambda+1)}$$
$$- \cup_{\nu=1}^{\lambda} (h^{(\nu+1)} \circ \cdots \circ h^{(\lambda+1)})^*(\mathrm{Supp} P^{(\nu)} \cup \mathrm{Supp} N^{(\nu)}).$$

(IV) *The Construction Procedure.*

Suppose we have constructed the objects on the list indexed by λ up to the index λ and we would like to construct the objects on the list indexed by $\lambda + 1$.

We distinguish between the cases of $\lambda = 0$ and $\lambda \geq 1$.

Case 1. $\lambda = 0$.

Let p be a positive integer such that pB is a **Z**-bundle and $(s_1)^p, \cdots, (s_\ell)^p$ are holomorphic sections of pB. Let $\mathcal{J} \subset \mathcal{O}_X$ be the ideal sheaf which is locally generated by $(s_\mu)^p$ $(1 \leq \mu \leq \ell)$. Let $f^{(1)} : Y^{(1)} \rightarrow X^{(0)}$ be a blowup map of $X^{(0)} = X$ by successive monoidal transformations with nonsingular centers such that there exist a collection $\{E_j^{(1)}\}$ of nonsingular divisors with normal crossing in $Y^{(1)}$ with the two properties given below. We can require that at every stage the nonsingular center of the monoidal transformation is contained in the singular set or in the codimensional ≥ 2 branch of the inverse image of the union of the zero-sets of $s_1, \cdots, s_\ell$.

(i) $(f^{(0)})^* \mathcal{J}$ is equal to the ideal sheaf of $\Sigma_j r_j^* E_j^{(1)}$ for some nonnegative integers r_j^*.

(ii) $K_{Y^{(1)}} - (f^{(1)})^* K_{X^{(0)}} = \sum_j b_j^{(1)} E_j^{(1)}$ for some nonnegative integers $b_j^{(1)}$.

We can assume without loss of generality that the collection $\{E_j^{(1)}\}$ is minimal with respect the above two properties. Let $r_j^{(1)} = r_j^*/p$. We set $\tilde{\Sigma}^{(1)} = \Sigma$ and $\Sigma^{(1)} = \Sigma'$ and set $a_j^{(1)} = 0$.

Case 2. $\lambda \geq 1$.

For every x in $\Sigma^{(\lambda)} - \Sigma_1^{(\lambda)}$, by Lemma (1.2) we can find a non-identically-zero multivalued holomorphic section $\sigma_x^{(\lambda)}$ of $\gamma_{d_\lambda} L$ on $\Sigma^{(\lambda)}$ which vanishes at $g^{(\lambda)}(x)$ to order at least equal to d_λ, because $(\gamma_{d_\lambda} L)^{d_\lambda} \cdot W > d_\lambda^{d_\lambda}$ for any subvariety W of complex dimension d_λ and in particular for $W = \Sigma^{(\lambda)}$. If for every x in $\Sigma^{(\lambda)} - \Sigma_1^{(\lambda)}$ the dimension of the subvariety of $\Sigma^{(\Lambda)} - \Sigma_1^{(\Lambda)}$ where $|\sigma_x^{(\lambda)}|^{-2}$ is not locally integrable is zero, then we set $\Lambda = \lambda$ and the procedure is finished. So we assume that there exists some x in $\Sigma^{(\lambda)} - \Sigma_1^{(\lambda)}$ such that one branch of the subvariety of $\Sigma^{(\lambda)} - \Sigma_1^{(\lambda)}$ where $|\sigma_x^{(\lambda)}|^{-2}$ is not locally integrable has positive dimension. We denote that particular $\sigma_x^{(\lambda)}$ by $\sigma^{(\lambda)}$. Let $\tilde{\Sigma}^{(\lambda+1)}$ be the topological closure in $\Sigma^{(\lambda)}$ of the subvariety of $\Sigma^{(\lambda)} - \Sigma_1^{(\lambda)}$ where $|\sigma^{(\lambda)}|^{-2}$ is not locally integrable. Let $\Sigma^{(\lambda+1)}$ be a branch of $\tilde{\Sigma}^{(\lambda+1)}$ which has positive dimension $d_{\lambda+1}$ and is not entirely contained in $\Sigma_1^{(\lambda)}$.

Let $\tilde{\sigma}^{(\lambda)} = (g^{(\lambda)})^*(\sigma^{(\lambda)})$. We apply a finite number of successive monoidal transformations to $X^{(\lambda)}$ to get $f^{(\lambda+1)} : Y^{(\lambda+1)} \rightarrow X^{(\lambda)}$ with nonsingular centers so that there exist a collection of nonsingular divisors $\{E_j^{(\lambda+1)}\}$ in $Y^{(\lambda+1)}$ in normal crossing with the three properties given below. We can require that at every stage the nonsingular center of the monoidal transformation is contained in the singular set of the inverse image of $(g^{(\lambda)})^{-1}(\Sigma_1^{(\lambda)} \cup \tilde{\Sigma}^{(\lambda+1)})$ or a branch of codimension ≥ 2 of the inverse image of $(g^{(\lambda)})^{-1}(\Sigma_1^{(\lambda)} \cup \tilde{\Sigma}^{(\lambda+1)})$.

(i) The pullback $(f^{(\lambda+1)})^* \mathrm{div}(\tilde{\sigma}^{(\lambda)})$ of the divisor $\mathrm{div}(\tilde{\sigma}^{(\lambda)})$ of $\tilde{\sigma}^{(\lambda)}$ is of the form $\sum_j r_j^{(\lambda+1)} E_j^{(\lambda+1)}$ for some nonnegative rational numbers $r_j^{(\lambda+1)}$.

(ii) $K_{Y^{(\lambda+1)}} - (f^{(\lambda+1)})^* K_{X^{(\lambda)}} = \sum_j b_j^{(\lambda+1)} E_j^{(\lambda+1)}$ for some nonnegative integers $b_j^{(\lambda+1)}$.

(iii) $(f^{(\lambda+1)})^* \sum_{j \neq j_\lambda} e_j^{(\lambda)} (E_j^{(\lambda)} \cap X^{(\lambda)}) = \sum_j a_j^{(\lambda+1)} E_j^{(\lambda+1)}$.

We can assume without loss of generality that the collection $\{E_j^{(\lambda+1)}\}$ is minimal with respect the above three properties. We now treat both cases at the same time and does not distinguish between them.

There exists some $j_{\lambda+1}$ such that $r_{j_{\lambda+1}}^{(\lambda+1)} + a_{j_{\lambda+1}}^{(\lambda+1)} - b_{j_{\lambda+1}}^{(\lambda+1)} \geq 1$ and $(g^{(\lambda)} \circ f^{(\lambda+1)})(E_{j_{\lambda+1}}^{(\lambda+1)})$ is $\Sigma^{(\lambda+1)}$. Let $\Sigma_1^{(\lambda+1)}$ be the subvariety of $\Sigma^{(\lambda+1)}$ consisting of all the points z which either belong to $\tilde{\Sigma}_1^{(\lambda)} \cup \mathrm{Sing}(\tilde{\Sigma}^{(\lambda)})$ or satisfy the property that the rank of $g^{(\lambda)} \circ f^{(\lambda+1)}$ at some point of $(g^{(\lambda)} \circ f^{(\lambda+1)})^{-1}(z)$ is less than $d_{\lambda+1}$. By the restriction on the centers for the monoidal transformations used in the construction of $f^{(\lambda+1)}$: $Y^{(\lambda+1)} \to X^{(\lambda)}$, we know that $(g^{(\lambda)} \circ f^{(\lambda+1)})(E_j^{(\lambda+1)})$ is contained in $\Sigma_1^{(\lambda+1)}$ for $j \neq j_{\lambda+1}$.

We choose $\delta_j^{(\lambda+1)} > 0$ sufficiently small and relabel the divisors $E_j^{(\lambda+1)}$ so that

$$(f^{(\lambda+1)})^* G^{(\lambda)} - \sum_j \delta_j^{(\lambda+1)} E_j^{(\lambda+1)}$$

is ample on $Y^{(\lambda+1)}$ and

$$\frac{1 + b_j^{(\lambda+1)} - a_j^{(\lambda+1)} - \delta_j^{(\lambda+1)}}{r_j^{(\lambda+1)}} < \frac{1 + b_k^{(\lambda+1)} - a_k^{(\lambda+1)} - \delta_k^{(\lambda+1)}}{r_k^{(\lambda+1)}}$$

for $j < k$ when $r_j^{(\lambda+1)}$ and $r_k^{(\lambda+1)}$ are both nonzero. We let

$$\alpha^{(\lambda+1)} = \frac{1 + b_{j_{\lambda+1}}^{(\lambda+1)} - a_{j_{\lambda+1}}^{(\lambda+1)} - \delta_{j_{\lambda+1}}^{(\lambda+1)}}{r_{j_{\lambda+1}}^{(\lambda+1)}}.$$

We let

$$n_\nu^{(\lambda+1)} = \lfloor \alpha^{(\lambda+1)} r_\nu^{(\lambda+1)} + a_\nu^{(\lambda+1)} - b_\nu^{(\lambda+1)} + \delta_\nu^{(\lambda+1)} \rfloor$$

for $\nu < j_{\lambda+1}$ and $p_\nu^{(\lambda+1)} = -\lfloor \alpha^{(\lambda+1)} r_\nu^{(\lambda+1)} + a_\nu^{(\lambda+1)} - b_\nu^{(\lambda+1)} + \delta_\nu^{(\lambda+1)} \rfloor$ for $\nu > j_{\lambda+1}$. Let

$$P^{(\lambda+1)} = \sum_{\nu > j_{\lambda+1}} p_\nu^{(\lambda+1)} E_\nu^{(\lambda+1)}$$

and

$$N^{(\lambda+1)} = \sum_{\nu < j_{\lambda+1}} n_\nu^{(\lambda+1)} E_\nu^{(\lambda+1)}$$

The divisors $P^{(\lambda+1)}$ and $N^{(\lambda+1)}$ are both nonnegative.

(V) *Verification of the Properties.*
Most of the properties are immediate consequences of the construction. We now verify the ones that are not immediate consequences of the construction. First we verify Property (III.6). If $p_\nu^{(\lambda)}$ is positive, then $b_\nu^{(\lambda)}$ must be positive and $E_\nu^{(\lambda)}$ must be an exceptional divisor for the map $f^{(\lambda)} : Y^{(\lambda)} \to X^{(\lambda-1)}$. Hence every component of $P^{(\lambda)}$ is an exceptional divisor for the map $f^{(\lambda)} : Y^{(\lambda)} \to X^{(\lambda-1)}$ and Property (III.6) is verified.

Now we verify Property (III.7). We have to verify that the **Q**-bundle

$$(f^{(\lambda+1)})^*(B^{(\lambda)} + F^{(\lambda)} + K_{X^{(\lambda)}}) + P^{(\lambda+1)} - N^{(\lambda+1)} - K_{Y^{(\lambda+1)}} - E_{j_\lambda+1}^{(\lambda+1)}$$

is almost ample on $Y^{(\lambda+1)}$ and is of the form

$$G^{(\lambda+1)} + \sum_{j \neq j_{\lambda+1}} e_j^{(\lambda+1)} E_j^{(\lambda+1)}$$

on $Y^{(\lambda+1)}$ so that $G^{(\lambda+1)}$ is ample on $Y^{(\lambda+1)}$ and $0 \leq e_j^{(\lambda+1)} < 1$. Since we have not defined $G^{(0)}$, $e_j^{(0)}$, j_0, and $E_j^{(0)}$, in the following argument we use for notational simplicity the convention that $G^{(0)} = F^{(0)}$ and $(f^{(1)})^*(\sum_{j \neq j_0} e_j^{(0)}(E_j^{(0)} \cap X^{(0)})) = 0$. The **Q**-bundle $(f^{(\lambda+1)})^*(B^{(\lambda)} + F^{(\lambda)} + K_{X^{(\lambda)}}) + P^{(\lambda+1)} - N^{(\lambda+1)} - K_{Y^{(\lambda+1)}} - E_{j_\lambda+1}^{(\lambda+1)}$ is equal to

$$(f^{(\lambda+1)})^*((1 - \alpha^{(\lambda+1)})B^{(\lambda)} + F^{(\lambda)})$$
$$- \sum_j b_j^{(\lambda+1)} E_j^{(\lambda+1)} + (f^{(\lambda+1)})^*(\alpha^{(\lambda+1)} B^{(\lambda)}) + P^{(\lambda+1)} - N^{(\lambda+1)} - E_{j_\lambda+1}^{(\lambda+1)}$$

$$= (f^{(\lambda+1)})^*((1 - \alpha^{(\lambda+1)})B^{(\lambda)}) + (f^{(\lambda+1)})^*(G^{(\lambda)}) +$$
$$\quad (f^{(\lambda+1)})^*(\sum_{j \neq j_\lambda} e_j^{(\lambda)}(E_j^{(\lambda)} \cap X^{(\lambda)})) - \sum_j b_j^{(\lambda+1)} E_j^{(\lambda+1)}$$
$$\quad + (f^{(\lambda+1)})^*(\alpha^{(\lambda+1)} B^{(\lambda)}) + P^{(\lambda+1)} - N^{(\lambda+1)} - E_{j_\lambda+1}^{(\lambda+1)}$$

$$= (f^{(\lambda+1)})^*((1 - \alpha^{(\lambda+1)})B^{(\lambda)}) + (f^{(\lambda+1)})^*(G^{(\lambda)}) + \sum_j a_j^{(\lambda+1)} E_j^{(\lambda+1)}$$

$$- \sum_j b_j^{(\lambda+1)} E_j^{(\lambda+1)} + (f^{(\lambda+1)})^*(\alpha^{(\lambda+1)} B^{(\lambda)}) + P^{(\lambda+1)} - N^{(\lambda+1)} - E_{j_\lambda+1}^{(\lambda+1)}$$

$$= (f^{(\lambda+1)})^*((1 - \alpha^{(\lambda+1)})B^{(\lambda)}) + (f^{(\lambda+1)})^*(G^{(\lambda)}) - \sum_j \delta_j^{(\lambda+1)} E_j^{(\lambda+1)}$$

$$+ \sum_j \delta_j^{(\lambda+1)} E_j^{(\lambda+1)} + \sum_j a_j^{(\lambda+1)} E_j^{(\lambda+1)} - \sum_j b_j^{(\lambda+1)} E_j^{(\lambda+1)}$$

$$+ \sum_j \alpha^{(\lambda+1)} r_j^{(\lambda+1)} E_j^{(\lambda+1)} + P^{(\lambda+1)} - N^{(\lambda+1)} - E_{j_{\lambda+1}}^{(\lambda+1)}$$

$$= (f^{(\lambda+1)})^* ((1 - \alpha^{(\lambda+1)}) B^{(\lambda)}) + ((f^{(\lambda+1)})^* (G^{(\lambda)}) - \sum_j \delta_j^{(\lambda+1)} E_j^{(\lambda+1)})$$

$$+ \sum_j (\alpha^{(\lambda+1)} r_j^{(\lambda+1)} + a_j^{(\lambda+1)} - b_j^{(\lambda+1)} + \delta_j^{(\lambda+1)}) E_j^{(\lambda+1)}$$

$$+ P^{(\lambda+1)} - N^{(\lambda+1)} - E_{j_{\lambda+1}}^{(\lambda+1)}$$

$$= (f^{(\lambda+1)})^* ((1 - \alpha^{(\lambda+1)}) B^{(\lambda)}) + ((f^{(\lambda+1)})^* (G^{(\lambda)})$$

$$- \sum_j \delta_j^{(\lambda+1)} E_j^{(\lambda+1)}) + \sum_{j \neq j_{\lambda+1}} e_j^{(\lambda+1)} E_j^{(\lambda+1)},$$

where

$$e_j^{(\lambda+1)} = (\alpha^{(\lambda+1)} r_j^{(\lambda+1)} + a_j^{(\lambda+1)} - b_j^{(\lambda+1)} + \delta_j^{(\lambda+1)})$$

$$- \lfloor \alpha^{(\lambda+1)} r_\nu^{(\lambda+1)} + a_\nu^{(\lambda+1)} - b_\nu^{(\lambda+1)} + \delta_\nu^{(\lambda+1)} \rfloor$$

is in the interval $[0,1)$. Let $G^{(\lambda+1)}$ be the ample bundle

$$(f^{(\lambda+1)})^* ((1 - \alpha^{(\lambda+1)}) B^{(\lambda)}) + ((f^{(\lambda+1)})^* (G^{(\lambda)}) - \sum_j \delta_j^{(\lambda+1)} E_j^{(\lambda+1)})$$

on $Y^{(\lambda+1)}$. Then

$$(f^{(\lambda+1)})^* (B^{(\lambda)} + F^{(\lambda)} + K_{X^{(\lambda)}}) + P^{(\lambda+1)} - N^{(\lambda+1)} - K_{Y^{(\lambda+1)}} - E_{j_{\lambda+1}}^{(\lambda+1)}$$

which is equal to $G^{(\lambda+1)} + \sum_{j \neq j_{\lambda+1}} e_j^{(\lambda+1)} E_j^{(\lambda+1)}$ is almost ample on $Y^{(\lambda+1)}$. This finishes the verification of Property (III.7).

We now verify Property (III.10). Since by (III.7) the **Q**-bundle

$$(f^{(\lambda+1)})^* (B^{(\lambda)} + F^{(\lambda)} + K_{X^{(\lambda)}}) + P^{(\lambda+1)} - N^{(\lambda+1)} - K_{Y^{(\lambda+1)}} - E_{j_{\lambda+1}}^{(\lambda+1)}$$

is almost ample on $Y^{(\lambda+1)}$, it follows that

$$H^1 (Y^{(\lambda+1)}, (f^{(\lambda+1)})^* (A^{(\lambda)} + B^{(\lambda)} + F^{(\lambda)} + K_{X^{(\lambda)}}) + P^{(\lambda+1)} - N^{(\lambda+1)} - E_{j_{\lambda+1}}^{(\lambda+1)})$$

vanishes. Thus the restriction map

$$\Gamma(Y^{(\lambda+1)}, (f^{(\lambda+1)})^* (A^{(\lambda)} + B^{(\lambda)} + F^{(\lambda)} + K_{X^{(\lambda)}}) + P^{(\lambda+1)} - N^{(\lambda+1)}) \rightarrow$$

$$\Gamma(E_{j_{\lambda+1}}^{(\lambda+1)}, ((f^{(\lambda+1)})^* (A^{(\lambda)} + B^{(\lambda)} + F^{(\lambda)} + K_{X^{(\lambda)}}) + P^{(\lambda+1)} - N^{(\lambda+1)}) | E_{j_{\lambda+1}}^{(\lambda+1)})$$

is surjective and Property (III.10) is verified.

We now verify Property (III.11). Inductively we have the following definition for $F^{(\lambda+1)}$. The bundle $F^{(\lambda+1)}$ is equal to the restriction of

$$(f^{(\lambda+1)})^*(B^{(\lambda)} + F^{(\lambda)} + K_{X^{(\lambda)}}) + P^{(\lambda+1)} - N^{(\lambda+1)} - K_{Y^{(\lambda+1)}} - E_{j_{\lambda+1}}^{(\lambda+1)}$$

to $X^{(\lambda+1)}$. This means that $F^{(\lambda+1)} + K_{X^{(\lambda+1)}}$ is equal to

$$(f^{(\lambda+1)})^*(B^{(\lambda)} + F^{(\lambda)} + K_{X^{(\lambda)}}) + P^{(\lambda+1)} - N^{(\lambda+1)}$$

on $X^{(\lambda+1)}$. In particular, $F^{(\lambda+1)} + K_{X^{(\lambda+1)}}$ is equal to $(f^{(\lambda+1)})^*(B^{(\lambda)} + F^{(\lambda)} + K_{X^{(\lambda)}})$ on $X^{(\lambda+1)} - \operatorname{Supp} P^{(\lambda+1)} - \operatorname{Supp} N^{(\lambda+1)}$. Inductively on

$$X^{(\lambda+1)} - \operatorname{Supp} P^{(\lambda+1)} \cup \operatorname{Supp} N^{(\lambda+1)}$$
$$- \cup_{\nu=1}^{\lambda} (h^{(\nu+1)} \circ \cdots \circ h^{(\lambda+1)})^*(\operatorname{Supp} P^{(\nu)} \cup \operatorname{Supp} N^{(\nu)})$$

the bundle $F^{(\lambda+1)} + K_{X^{(\lambda+1)}}$ is equal to $(g^{(\lambda)})^*((\sum_{\nu=1}^{\lambda} \gamma_{d_\nu})L + B + F^{(0)} + K_X)$ and Property (III.11) is verified. This finishes the verification of all the listed properties.

Now we are going to get a subspace of dimension N in $\Gamma(X, \gamma L + B + K_X)$ which is mapped injectively into $\Gamma(\Sigma', (\gamma L + B + K_X)|\Sigma')$. By (III.10) we have the surjectivity of the restriction map

$$\Gamma(Y^{(\lambda+1)}, (f^{(\lambda+1)})^*(A^{(\lambda)} + B^{(\lambda)} + F^{(\lambda)} + K_{X^{(\lambda)}}) + P^{(\lambda+1)} - N^{(\lambda+1)}) \rightarrow$$
$$\Gamma(E_{j_{\lambda+1}}^{(\lambda+1)}, ((f^{(\lambda+1)})^*(A^{(\lambda)} + B^{(\lambda)} + F^{(\lambda)} + K_{X^{(\lambda)}}) + P^{(\lambda+1)} - N^{(\lambda+1)})|E_{j_{\lambda+1}}^{(\lambda+1)}).$$

From (II.4) and (II.5) we have $(f^{(\lambda+1)})^*(A^{(\lambda)}) = A^{(\lambda+1)} + B^{(\lambda+1)}$. From (II.6) we have

$$(f^{(\lambda+1)})^*(A^{(\lambda)} + B^{(\lambda)} + F^{(\lambda)} + K_{X^{(\lambda)}}) + P^{(\lambda+1)} - N^{(\lambda+1)})|E_{j_{\lambda+1}}^{(\lambda+1)}$$
$$= A^{(\lambda+1)} + B^{(\lambda+1)} + F^{(\lambda+1)} + K_{X^{(\lambda+1)}}$$

Since by (III.6) every component of the support of $P^{(\lambda+1)}$ is an exceptional divisor for the map $f^{(\lambda+1)} : Y^{(\lambda+1)} \rightarrow X^{(\lambda)}$, it follows that every element of $\Gamma(Y^{(\lambda+1)}, (f^{(\lambda)})^*(A^{(\lambda)} + B^{(\lambda)} + F^{(\lambda)} + K_{X^{(\lambda)}}) + P^{(\lambda+1)} - N^{(\lambda+1)})$ descends to an element of $\Gamma(X^{(\lambda)}, A^{(\lambda)} + B^{(\lambda)} + F^{(\lambda)} + K_{X^{(\lambda)}})$. This implies that

$$(2.1.1) \quad \dim_{\mathbf{C}} \Gamma(X^{(\lambda)}, A^{(\lambda)} + B^{(\lambda)} + F^{(\lambda)} + K_{X^{(\lambda)}})$$
$$\geq \dim_{\mathbf{C}} \Gamma(X^{(\lambda+1)}, A^{(\lambda+1)} + B^{(\lambda+1)} + F^{(\lambda+1)} + K_{X^{(\lambda+1)}}).$$

By Property (III.8) we know that $(g^{(\lambda-1)} \circ f^{(\lambda)})(E_j^{(\lambda)})$ is contained in $\Sigma_1^{(\lambda)}$ for $j \neq j_\lambda$ and $1 \leq \lambda \leq \Lambda$ and as a consequence $(g^{(\lambda-1)} \circ f^{(\lambda)})(\operatorname{Supp} P^{(\lambda)})$ and $(g^{(\lambda-1)} \circ f^{(\lambda)})(\operatorname{Supp} P^{(\lambda)})$ are both contained in $\Sigma_1^{(\lambda)}$.

By Property (III.2) we know that $\Sigma_1^{(\lambda)} \cap \Sigma^{(\Lambda)} \subset \Sigma_1^{(\Lambda)}$ for $1 \leq \lambda \leq \Lambda$. Since by Property (III.11) on

$$X^{(\Lambda)} - \operatorname{Supp} P^{(\Lambda)} \cup \operatorname{Supp} N^{(\Lambda)} -$$

$$\cup_{\lambda=1}^{\Lambda-1} (h^{(\Lambda+1)} \circ \cdots \circ h^{(\lambda+1)})^*(\operatorname{Supp} P^{(\lambda)} \cup \operatorname{Supp} N^{(\lambda)})$$

the bundle $F^{(\Lambda)} + K_{X^{(\Lambda)}}$ is equal to $(g^{(\Lambda)})^*((\sum_{\nu=1}^{\Lambda-1} \gamma_{d_\nu})L + B + F^{(0)} + K_X)$, we conclude that

$$(2.1.2) \qquad A^{(\Lambda)} + B^{(\Lambda)} + F^{(\Lambda)} + K_{X^{(\Lambda)}}$$

$$= (g^{(\Lambda)})^*(\gamma L + B + K_X) \quad \text{on} \quad X^{(\Lambda)} - (g^{(\Lambda)})^{-1}(\Sigma_1^{(\Lambda)}).$$

By Property (III.9), for every x in $\Sigma^{(\Lambda)} - \Sigma_1^{(\Lambda)}$, we can find a non-identically-zero multivalued holomorphic section $\sigma_x^{(\Lambda)}$ of $\gamma_\Lambda L$ on $\Sigma^{(\Lambda)}$ which vanishes at $g^{(\Lambda)}(x)$ to order at least equal to d_Λ such that x belongs to the subvariety $S_x^{(\Lambda)}$ of $\Sigma^{(\Lambda)} - \Sigma_1^{(\Lambda)}$ where $|\sigma_x^{(\Lambda)}|^{-2}$ is not locally integrable and, moreover, the dimension of $S_x^{(\Lambda)}$ is zero. We now select N distinct points $x_1, \cdots, x_N$ in $\Sigma^{(\Lambda)} - \Sigma_1^{(\Lambda)}$ inductively in the following manner. The point x_1 is chosen arbitrarily in $\Sigma^{(\Lambda)} - \Sigma_1^{(\Lambda)}$. After $x_1, \cdots, x_j$ have been chosen for some $1 \leq j < N$, we choose x_{j+1} in $\Sigma^{(\Lambda)} - \Sigma_1^{(\Lambda)}$ so that $\prod_{\nu=1}^j \sigma_{x_\nu}^{(\Lambda)}$ is locally integrable at x_{j+1}. We are going to prove by induction on $1 \leq j \leq N$ the following statement

$$(2.1.3)_j \quad \Gamma(X^{(\Lambda)}, A^{(\Lambda)} + B^{(\Lambda)} + F^{(\Lambda)} + K_{X^{(\lambda+1)}})$$

$$\to \oplus_{\nu=1}^j (g^{(\Lambda)})^*((\mathcal{O}_X/\mathbf{m}_{x_\nu})(\gamma L + B + K_X))$$

is surjective, where the map is defined by using (2.1.2) which gives the equality of $A^{(\Lambda)} + B^{(\Lambda)} + F^{(\Lambda)} + K_{X^{(\Lambda)}}$ and $(g^{(\Lambda)})^*(\gamma L + B + K_X)$ on $X^{(\Lambda)} - (g^{(\Lambda)})^{-1}(\Sigma_1^{(\Lambda)})$. We use the convention that the statement $(2.1.3)_0$ is vaccuous. Suppose the statement $(2.1.3)_j$ has been proved for some $0 \leq j < N$. To prove the statement $(2.1.3)_{j+1}$ it suffices to show that the subset of elements $\Gamma(X^{(\Lambda)}, A^{(\Lambda)} + B^{(\Lambda)} + F^{(\Lambda)} + K_{X^{(\lambda+1)}})$ which vanish on $(g^{(\Lambda)})^{-1}(x_\nu)$ for $1 \leq \nu \leq j$ is mapped surjectively onto $(g^{(\Lambda)})^*((\mathcal{O}_X/\mathbf{m}_{x_{j+1}})(\gamma L + B + K_X))$. Let σ be the pullback of $\prod_{\nu=1}^{j+1} \sigma_{x_\nu}^{(\Lambda)}$ under $g^{(\Lambda)}$. Then σ is a multivalued holomorphic section of $(g^{(\Lambda)})^*((j+1)\gamma_\Lambda L)$ over $X^{(\Lambda)}$. Since by (III.2) the rank of $g^{(\Lambda)}$ is d_Λ at every point of $\Sigma^{(\Lambda)} - \Sigma_1^{(\Lambda)}$ and since $\prod_{\nu=1}^j \sigma_{x_\nu}^{(\Lambda)}$ is locally integrable at x_{j+1}, it follows that for some open neighborhood U of $(g^{(\Lambda)})^{-1}(x_{j+1})$ the metric $|\sigma|^{-2}$ is locally integrable at every point of $U - (g^{(\Lambda)})^{-1}(x_{j+1})$. Moreover, $|\sigma|^{-2}$ is not locally integrable at any point of $(g^{(\Lambda)})^{-1}(x_\nu)$ for $1 \leq \nu \leq j+1$. Let $\tilde{\sigma}$ be the canonical multivalued section of the **Q**-bundle $F^{(\Lambda)} - G^{(\Lambda)} = \sum_{j \neq j_\Lambda} e_j^{(\Lambda)}(E_j^{(\Lambda)} \cap E_{j_\Lambda}^{(\Lambda)})$ over $X^{(\Lambda)}$ so that the divisor of $\tilde{\sigma}$ is $\sum_{j \neq j_\Lambda} e_j^{(\Lambda)}(E_j^{(\Lambda)} \cap E_{j_\Lambda}^{(\Lambda)})$. We know that $E_j^{(\Lambda)} \cap E_{j_\Lambda}^{(\Lambda)}$ is contained in $(g^{(\Lambda)})^{-1}(\Sigma_1^{(\Lambda)})$ for $j \neq j_\Lambda$. Thus $\tilde{\sigma}$ does not have any zero on $X^{(\Lambda)} - (g^{(\Lambda)})^{-1}(\Sigma_1^{(\Lambda)})$. Let $\mathcal{I} \subset \mathcal{O}_{X^{(\Lambda)}}$ be the multiplier ideal sheaf of the singular metric $|\tilde{\sigma}\sigma|^{-2}$ of $(g^{(\Lambda)})^*((j+1)\gamma_\Lambda L) + (F^{(\Lambda)} - G^{(\Lambda)})$. Let $\mathcal{J} \subset \mathcal{O}_{X^{(\Lambda)}}$ be the ideal sheaf which agrees with $\mathcal{O}_{X^{(\Lambda)}}$ on U and agrees with $\mathcal{I}$ on $X^{(\Lambda)} - (g^{(\Lambda)})^{-1}(x_{j+1})$. By (II.4) and (II.5)

$$A^{(\Lambda)} + B^{(\Lambda)} - (g^{(\Lambda)})^*((j+1)\gamma_\Lambda L) = (g^{(\Lambda)})^*((\gamma - \gamma' - (j+1)\gamma_\Lambda - \sum_{\nu=1}^{\Lambda-1} \gamma_\mu)L)$$

which by $\gamma - \gamma' - (j+1)\gamma_\Lambda - \sum_{\nu=1}^{\Lambda-1} \gamma_\mu > 0$ is numerically effective. Since the $\mathbf{Q}$-bundle $A^{(\Lambda)} + B^{(\Lambda)} + F^{(\Lambda)}$ is equal to the sum of the $\mathbf{Q}$-bundle $(g^{(\Lambda)})^*((j+1)\gamma_\Lambda L) + (F^{(\Lambda)} - G^{(\Lambda)})$ and the ample $\mathbf{Q}$-bundle $G^{(\Lambda)} + (A^{(\Lambda)} + B^{(\Lambda)} - (g^{(\Lambda)})^*((j+1)\gamma_\Lambda L))$ and since $\mathcal{I}$ is the multiplier ideal sheaf of the metric $|\tilde{\sigma}\sigma|^{-2}$ of the $\mathbf{Q}$-bundle $(g^{(\Lambda)})^*((j+1)\gamma_\Lambda L) + (F^{(\Lambda)} - G^{(\Lambda)})$, it follows from Nadel's vanishing theorem that

$$H^1(X^{(\Lambda)}, \mathcal{I}(A^{(\Lambda)} + B^{(\Lambda)} + F^{(\Lambda)} + K_{X^{(\Lambda)}}))$$

vanishes. From the exact sequence

$$0 \to \mathcal{I} \to \mathcal{J} \to \mathcal{J}/\mathcal{I} \to 0$$

it follows that

$$\Gamma(X^{(\Lambda)}, \mathcal{J}(A^{(\Lambda)} + B^{(\Lambda)} + F^{(\Lambda)} + K_{X^{(\Lambda)}}))$$
$$\to \Gamma(X^{(\Lambda)}, (\mathcal{J}/\mathcal{I})(A^{(\Lambda)} + B^{(\Lambda)} + F^{(\Lambda)} + K_{X^{(\Lambda)}})).$$

is surjective. Since by (2.1.2) the bundle $A^{(\Lambda)} + B^{(\Lambda)} + F^{(\Lambda)} + K_{X^{(\Lambda)}}$ is equal to $(g^{(\Lambda)})^*(\gamma L + B + K_X)$ on $X^{(\Lambda)} - (g^{(\Lambda)})^{-1}(\Sigma_1^{(\Lambda)})$, we conclude that the subset of elements $\Gamma(X^{(\Lambda)}, A^{(\Lambda)} + B^{(\Lambda)} + F^{(\Lambda)} + K_{X^{(\Lambda+1)}})$ which vanish on $(g^{(\Lambda)})^{-1}(x_\nu)$ for $1 \leq \nu \leq j$ is mapped surjectively onto $(g^{(\Lambda)})^*((\mathcal{O}_X/\mathbf{m}_{x_{j+1}})(\gamma L + B + K_X))$. Thus we have the statement $(2.1.3)_{j+1}$.

The statement $(2.1.3)_N$ implies that the dimension of $\Gamma(X^{(\Lambda)}, A^{(\Lambda)} + B^{(\Lambda)} + F^{(\Lambda)} + K_{X^{(\Lambda+1)}})$ over $\mathbf{C}$ is at least N. By (2.1.1) we have the final conclusion that $\dim_{\mathbf{C}} \Gamma(X^{(1)}, A^{(1)} + B^{(1)} + F^{(1)} + K_{X^{(1)}})$ is at least N. By Property (III.10) We know that every element of $\Gamma(X^{(1)}, A^{(1)} + B^{(1)} + F^{(1)} + K_{X^{(1)}})$ can be extended to an element of $\Gamma(Y^{(1)}, (f^{(1)})^*(A^{(0)} + B^{(0)} + F^{(0)} + K_{X^{(0)}}) + P^{(1)} - N^{(1)})$ which descends to an element of $\Gamma(X, \gamma L + B + K_X)$. Since $f^{(1)}(X^{(1)}) = \Sigma'$, it follows that there is a linear subspace V of dimension N in $\Gamma(X, \gamma L + B + K_X)$ which is mapped injectively into $\Gamma(\Sigma', (\gamma L + B + K_X)|\Sigma')$ by restriction. Since

$N = \sum_{j=1}^{J} \binom{q_j+n-1}{n}$, it follows that there exists a nonzero element of V which vanishes at P_j to order at least q_j for $1 \leq j \leq J$. This concludes the proof of the Proposition.

Remark (2.2). Proposition (2.1) holds also when B is the trivial line bundle and $\Sigma = \Sigma' = X$.

3. Background Singular Metrics of Lelong Number One

Proposition (3.1). *Let X be a compact complex manifold of complex dimension $n \geq 2$ and L be an ample line bundle over X such that $L+2K_X$ is numerically effective. Let $P_1, \cdots, P_J$ be distinct points in X and $q_1, \cdots, q_J$ be nonnegative integers. For any $1 \leq \nu \leq n$ let κ_ν be the supremum of $(L^\nu \cdot W)^{-1/\nu}$ as W ranges over all the subvarieties of complex dimension ν in X. Let N be $\sum_{j=1}^{J} \binom{3n+2q_j-1}{n}$. Let $0 < \eta < \frac{1}{2}$ and $1 \leq \lambda \leq J$ and U_λ be an open neighborhood of P_λ in X. Let $\tau_1, \cdots, \tau_k$ be a finite number of non-identically-zero multivalued holomorphic sections of ηL over X vanishing to order at least 1 at P_ν ($1 \leq \nu \leq J$) such that $(\sum_{\nu=1}^{k} |\tau_\nu|^2)^{-1}$ is locally integrable on $U_\lambda - P_\lambda$. Suppose $N\nu\kappa_\nu + \sum_{\mu=\nu+1}^{n} \mu\kappa_\mu < \frac{1}{2} - \eta$ for $1 \leq \nu \leq n$. Then there exist a finite number of holomorphic sections s_μ ($1 \leq \mu \leq \ell$) of $L + 2K_X$ over X vanishing to order at least $2(n+q_\nu)$ at P_ν ($1 \leq \nu \leq J$) such that the set of points of U_λ where the singular metric $(\sum_{\nu=1}^{k} |\tau_\nu|^2)^{-1} (\sum_{\mu=1}^{\ell} |s_\mu|)^{-1}$ of $(\eta + \frac{1}{2})L + K_X$ is not locally integrable is of dimension zero.*

Proof: We are going to prove by decending induction on $0 \leq d \leq n-1$ the following statement.

$(3.1.1)_d$ There exist a finite number of holomorphic sections $s_\mu^{(d)}$ ($1 \leq \mu \leq k_d$) of $L + 2K_X$ over X vanishing to order at least $2n + 2q_\nu$ at P_ν ($1 \leq \nu \leq J$) such that the set of points of U_λ where the metric $(\sum_{\nu=1}^{k} |\tau_\nu|^2)^{-1} (\sum_{\mu=1}^{k_d} |s_\mu^{(d)}|)^{-1}$ of $(\eta + \frac{1}{2})L + K_X$ is not locally integrable is of dimension at most d.

Take $0 \leq d \leq n$. Suppose we have already proved $(3.1.1)_d$ if $d < n$ (we do not start with any assumption if $d = n$). We are going to prove $(3.1.1)_{d-1}$. If $d = n$, we let $\Sigma = X$, $p = 1$, $V_1 = X$ and let $\{s_\mu^{(d)} \mid 1 \leq \mu \leq k_d\}$ be the vacuous set and $k_d = 0$. If $d < n$, we let Σ be the set of points where the singular metric $(\sum_{\nu=1}^{k} |\tau_\nu|^2)^{-1} (\sum_{\mu=1}^{k_d} |s_\mu^{(d)}|)^{-1}$ of $(\eta+\frac{1}{2})L+K_X$ is not locally integrable and let $V_1, \cdots, V_p$ be the set of all d-dimensional branches of Σ. Without loss of generality we assume that $p \geq 1$, otherwise Σ contains no d-dimensional branch, and we can simply use $\{s_\mu^{(d)} \mid 1 \leq \mu \leq k_d\}$ as $\{s_\mu^{(d-1)} \mid 1 \leq \mu \leq k_{d-1}\}$ to get the statement $(3.1.1)_{d-1}$. Fix arbitrarily $1 \leq \lambda \leq p$. Let $B = (\frac{1}{2}+\eta)L + K_X$ and $\gamma = \frac{1}{2} - \eta$. Let $\ell = k_d$

and $s_\nu = s_\nu^{(d)}$ $(1 \le \nu \le \ell)$. We now apply Proposition (2.1) and Remark (2.2) with $\Sigma' = V_\lambda$. Since $N\nu\kappa_\nu + \sum_{\mu=\nu+1}^n \mu\kappa_\mu < \frac{1}{2} - \eta$ for $1 \le \nu \le n$ with $N = \sum_{j=1}^J \binom{3n+2q_j-1}{n}$, there exists a multivalued holomorphic section σ_λ of $L + 2K_X = \gamma L + B + K_X$ over X which vanishes to order at least $2(n + q_\nu)$ at P_ν $(1 \le \nu \le J)$ but does not vanish identically on V_λ. Let $k_{d-1} = k_d + p$ and $s_\nu^{(d-1)} = s_\nu^{(d)}$ $(1 \le \nu \le k_d)$ and $s_{k_d+\lambda}^{(d-1)} = \sigma_\lambda$ $(1 \le \lambda \le p)$. We are going to verify the following statement.

(3.1.2) The set of points of $U_\lambda - P_\lambda$ where

$$\left(\sum_{\nu=1}^k |\tau_\nu|^2\right)^{-1} \left(\sum_{\mu=1}^{k_{d-1}} |s_\mu^{(d-1)}|\right)^{-1}$$

is not locally integrable is contained in the intersection of Σ and the common zero-set of $\sigma_1, \cdots, \sigma_p$.

Take a point x in $U_\lambda - P_\lambda$. Suppose x is not in Σ. Then by the definition of Σ we know that the metric $(\sum_{\nu=1}^k |\tau_\nu|^2)^{-1}(\sum_{\mu=1}^{k_d} |s_\mu^{(d)}|)^{-1}$ of $(\eta + \frac{1}{2})L + K_X$ is locally integrable at x. Since

$$\left(\sum_{\nu=1}^k |\tau_\nu|^2\right)^{-1} \left(\sum_{\mu=1}^{k_{d-1}} |s_\mu^{(d-1)}|\right)^{-1} \le \left(\sum_{\nu=1}^k |\tau_\nu|^2\right)^{-1} \left(\sum_{\mu=1}^{k_d} |s_\mu^{(d)}|\right)^{-1},$$

it follows that the metric $(\sum_{\nu=1}^k |\tau_\nu|^2)^{-1}(\sum_{\mu=1}^{k_{d-1}} |s_\mu^{(d-1)}|)^{-1}$ of $(\eta + \frac{1}{2})L + K_X$ is locally integrable at x. Assume that x is not in the common zero set of $\sigma_1, \cdots, \sigma_p$. Then $\sum_{\mu=1}^p |\sigma_\mu| \ge \theta$ on some open neighborhood U of x for some positive number θ and

$$\left(\sum_{\nu=1}^k |\tau_\nu|^2\right)^{-1} \left(\sum_{\mu=1}^{k_{d-1}} |s_\mu^{(d-1)}|\right)^{-1} \le$$

$$\left(\sum_{\nu=1}^k |\tau_\nu|^2\right)^{-1} \left(\sum_{\mu=1}^p |\sigma_\mu|\right)^{-1} \le \left(\sum_{\nu=1}^k |\tau_\nu|^2\right)^{-1} \theta^{-1}$$

is locally integrable on $U \cap (U_\lambda - P_\lambda)$. Thus the set of points of $U_\lambda - P_\lambda$ where the metric $(\sum_{\nu=1}^k |\tau_\nu|^2)^{-1}(\sum_{\mu=1}^{k_{d-1}} |s_\mu^{(d-1)}|)^{-1}$ of $(\eta + \frac{1}{2})L + K_X$ is not locally integrable is contained in the intersection of Σ and the common zero-set of $\sigma_1, \cdots, \sigma_p$ and (3.1.2) is verified. The statement $(3.1.1)_{d-1}$ follows from (3.1.2). This concludes the proof of the Proposition.

Lemma (3.2). *Let X be a compact complex manifold of complex dimension $n \ge 2$ and L be an ample line bundle over X. Let $P_1, \cdots, P_J$ be distinct points in X and let η be a positive rational number such that $\eta > (L^n)^{-1/n}$. Then there exist a positive number δ and a finite number of non-identically-zero multivalued holomorphic sections $\tau_1, \cdots, \tau_k$ of*

$J\eta L$ over X such that the Lelong number of the curvature current of $(\sum_{\nu=1}^{k} |\tau_{\nu}|^2)^{-1}$ is at least $1+\delta$ at each P_{ν} $(1 \leq \nu \leq J)$ and is less than $1-\delta$ on $U_{\lambda} - P_{\lambda}$ for some $1 \leq \lambda \leq J$ and some open neighborhood U_{λ} of P_{λ} in X.

Proof: Choose $\eta > \eta' > (L^n)^{-1/n}$. Since $(\eta'L)^n > 1$, by Lemma (1.2) there exists, for $1 \leq \nu \leq J$, some non-identically-zero multivalued holomorphic section t_{ν} of $\eta'L$ over X whose vanishing order is at least 1 at P_{ν}. Let $t = \prod_{\nu=1}^{J} t_{\nu}$. Choose $1 \leq \lambda \leq J$ so that the vanishing order γ of t at P_{λ} is no more than the vanishing order of t at P_{ν} for $1 \leq \nu \leq J$. Since L is ample, we can find a finite number of multivalued holomorphic sections $u_1, \cdots, u_k$ of $(\eta - \eta')L$ without common zero on $X - P_{\lambda}$ so that each u_{ν} vanishes to order $> \epsilon$ at each of the points $P_1, \cdots, P_q$ for some positive number $\epsilon < 1$. Let $\tau_{\nu} = t^{(1-\epsilon)/\gamma}(u_{\nu})^{\theta}$, where $\theta = \frac{J}{\eta-\eta'}(\eta - \eta'\frac{(1-\epsilon)}{\gamma}) > J$ so that τ_{ν} is a multivalued holomorphic section of $J\eta L$. Then the Lelong number of the curvature current of $(\sum_{\nu=1}^{k} |\tau_{\nu}|^2)^{-1}$ is greater than $1 - \epsilon + J\epsilon \geq 1$ at each P_{ν} $(1 \leq \nu \leq J)$ and is no more than $1 - \epsilon$ on a deleted neighborhood of P_{λ}.

Proposition (3.3). *Let X be a compact complex manifold of complex dimension $n \geq 2$ and L be an ample line bundle over X such that $L + 2K_X$ is numerically effective. For any $1 \leq \nu \leq n$ let κ_{ν} be the supremum of $(L^{\nu} \cdot W)^{-1/\nu}$ as W ranges over all the subvarieties of dimension ν in X. Let $P_1, \cdots, P_J$ be distinct points in X and $q_1, \cdots, q_J$ be nonnegative integers. Let N be $\sum_{j=1}^{J} \binom{3n+2q_j-3}{n}$. Suppose $\eta > 0$ such that $Jn\kappa_n < \eta$ and $N\kappa_{\nu} + \sum_{\mu=\nu+1}^{n} \mu\kappa_{\mu} < \frac{1}{2} - \eta$ for $1 \leq \nu \leq n$. Let β be a rational number with $0 \leq \beta \leq \frac{1}{2} - \eta$ such that $(\beta + \eta + \frac{1}{2})L$ is a $\mathbf{Z}$-bundle. Then global holomorphic sections of $(\beta + \eta + \frac{1}{2})L + 2K_X$ over X generate simultaneously the q_{ν}-jets at P_{ν} for $1 \leq \nu \leq J$. In particular, global holomorphic sections of $(\beta + \eta + \frac{1}{2})L + 2K_X$ over X generate simultaneously the q_{ν}-jets at P_{ν} for $1 \leq \nu \leq J$.*

Proof: Since $Jn\kappa_n < \eta$, by Lemma (3.2) there exist a finite number of non-identically-zero multivalued holomorphic sections $\tau_1, \cdots, \tau_k$ of ηL over X such that the Lelong number of the curvature current of $(\sum_{\nu=1}^{k} |\tau_{\nu}|^2)^{-1}$ is at least 1 at each P_{ν} $(1 \leq \nu \leq J)$ and is locally integrable on $U_{\lambda} - P_{\lambda}$ for some $1 \leq \lambda \leq J$ and some open neighborhood U_{λ} of P_{λ}.

By Proposition (3.1) there exist a finite number of non-identically-zero holomorphic sections s_{μ} $(1 \leq \mu \leq \ell)$ of $L + 2K_X$ over X vanishing to order at least $2(n + q_{\nu} - 1)$ at P_{ν} $(1 \leq \nu \leq J)$ such that the set of points of U_{λ} where $(\sum_{\nu=1}^{k} |\tau_{\nu}|^2)^{-1}(\sum_{\mu=1}^{\ell} |s_{\mu}|)^{-1}$ is not locally integrable is of dimension zero.

We are going to use induction on J to prove the Proposition. Suppose we are given some q_{ν}-jet g_{ν} at P_{ν} for $1 \leq \nu \leq J$. By induction hypothesis there exists some holomorphic section v of $(\beta + \eta + \frac{1}{2})L + 2K_X$ over X whose q_{ν}-jet at P_{ν} is equal to g_{ν} for $1 \leq \nu \leq J$ and $\nu \neq \lambda$. Let $\mathcal{I}$ be the multiplier ideal sheaf of the metric $(\sum_{\nu=1}^{k}|\tau_\nu|^2)^{-1}(\sum_{\mu=1}^{\ell}|s_\mu|)^{-1}$ of $(\eta+\frac{1}{2})L+K_X$. Then for some neighborhood U of P_λ in U_λ the zero-set of the sheaf $\mathcal{I}|U$ is the single point P_λ and $\mathcal{I} \subset (\mathbf{m}_{P_\lambda})^{q_\lambda}$ on U, where $\mathbf{m}_{P_\lambda}$ is the maximum ideal at P_λ. Thus we can write $\mathcal{I}$ as the product of two coherent sheaves $\mathcal{J}$ and $\mathcal{F}$ such that $\mathcal{J}$ is equal to $\mathcal{O}_X$ on U and $\mathcal{F}$ is equal to $\mathcal{O}_X$ on $X - P_\mu$ and $\mathcal{F} \subset (\mathbf{m}_{P_\lambda})^{q_\lambda}$ on U. Since L is ample and β is positive, we conclude from Nadel's vanishing theorem that $H^1(X, \mathcal{I}((\beta+\eta+\frac{1}{2})L+2K_X))$ is zero. From the short exact sequence

$$0 \to \mathcal{I} \to \mathcal{J} \to \mathcal{J}/\mathcal{I} \to 0$$

it follows that the map

$$\Gamma(X, \mathcal{J}((\beta+\eta+\frac{1}{2})L+2K_X)) \to \Gamma(X,(\mathcal{J}/\mathcal{I})((\beta+\eta+\frac{1}{2})L+2K_X))$$

is surjective. Since $\Gamma(X,(\mathcal{J}/\mathcal{I})((\beta+\eta+\frac{1}{2})L+2K_X))$ is equal to the stalk of $(\mathcal{O}_X/\mathcal{F})((\beta+\eta+\frac{1}{2})L+2K_X)$ at P_μ and since $\mathcal{F} \subset (\mathbf{m}_{P_\lambda})^{q_\lambda}$ on U, it follows that we can find an element v' of $\Gamma(X, \mathcal{J}((\beta+\eta+\frac{1}{2})L+2K_X))$ whose q_λ-jet at P_λ is equal to the q_λ-jet defined by $g_\lambda - v$ at P_λ. Since the Lelong number of the curvature current of the metric $(\sum_{\nu=1}^{k}|\tau_\nu|^2)^{-1}(\sum_{\mu=1}^{\ell}|s_\mu|)^{-1}$ is at least $n+q_\nu$ at P_ν for $1 \le \nu \le J$, any element of $\Gamma(X, \mathcal{J}((\beta+\eta+\frac{1}{2})L+2K_X))$ is a holomorphic section of $(\beta+\eta+\frac{1}{2})L+2K_X$ which vanishes to order at least q_ν at P_ν for $1 \le \nu \le J$ and $\nu \ne \lambda$. Thus the holomorphic section $v'+v$ of $(\beta+\eta+\frac{1}{2})L+2K_X$ over X has the property that its q_ν-jet at P_ν is equal to g_ν for $1 \le \nu \le J$. This finishes the proof of the Proposition.

Corollary (3.4). *Let X be a compact complex manifold of complex dimension $n \ge 2$ and L be an ample line bundle over X such that $L+2K_X$ is numerically effective. Let P_1 be a point in X. Let N be $\binom{3n-3}{n}$. For any $1 \le \nu \le n$ let κ_ν be the supremum of $(L^\nu \cdot W)^{-1/\nu}$ as W ranges over all the subvarieties of complex dimension ν in X. Suppose $N\kappa_\nu + \sum_{\mu=\nu+1}^{n}\mu\kappa_\mu < \frac{1}{2} - \eta$ for $1 \le \nu \le n$ and $n\kappa_n < \eta$ for some rational number η. Then there exists some singular metric of $(\eta+\frac{1}{2})L+K_X$ so that the Lelong number of its curvature current is at least n at P_1 and is less than 1 in some deleted neighborhood of P_1.*

Proof: Consider the case $J = 1$ and $q_1 = 0$ in Proposition (3.3). The metric

$$\left(\sum_{\nu=1}^{k}|\tau_\nu|^2\right)^{-1}\left(\sum_{\mu=1}^{\ell}|s_\mu|\right)^{-1}$$

of $(\eta+\frac{1}{2})L+K_X$ in the proof of Proposition (3.3) satisfies the requirement.

4. Verification of Numerical Positivity of Double

Adjoint and Proof of Main Theorem

Now we would like to get rid of the additional assumption that $L + 2K_X$ is numerically effective.

Lemma (4.1). *Let X be a compact complex manifold of complex dimension $n \geq 2$ and L be an ample line bundle over X. Let N be the maximum of $\binom{3n-1}{n}$ and $2\binom{3n-3}{n}$. For any $1 \leq \nu \leq n$ let κ_ν be the supremum of $(L^\nu \cdot W)^{-1/\nu}$ as W ranges over all the subvarieties of complex dimension ν in X. Suppose $N\kappa_\nu + \sum_{\mu=\nu+1}^{n} \mu\kappa_\mu < \frac{1}{2} - \eta$ for $1 \leq \nu \leq n$ and $2n\kappa_n < \eta$. Then the line bundle $L + 2K_X$ is numerically effective.*

Proof: Let k be the smallest positive integer such that $kL + 2K_X$ is numerically effective. Since L is ample, we know that such a positive integer k exists. By applying Proposition (3.3) to the line bundle kL instead of L and to the case of $J = 2$ and $q_1 = q_2 = 0$ and to the case $J = 1$ and $q_1 = 1$, we conclude that $kL + 2K_X$ is very ample. Let P_1 be any point of X. By Corollary (3.4), for some $0 < \gamma < 1$ there exists a singular metric h of $\gamma\frac{k}{2}L + K_X$ so that the Lelong number of its curvature current is at least n at P_1 and is less than 1 in some deleted neighborhood of P_1. By applying Nadel's vanishing theorem to the multiplier ideal sheaf of that singular metric h of $\gamma\frac{k}{2}L + K_X$, we conclude that there exists some holomorphic section of $\lceil k/2 \rceil L + 2K_X$ over X which is nonzero at P_1. Thus $\lceil k/2 \rceil L + 2K_X$ is numerially effective. By the definition of k we know that $\lceil k/2 \rceil = k$, which implies that $k = 1$.

Proof of the Main Theorem: The Main Theorem follows from Proposition (3.3) and Lemma (4.1).

Proof of Corollary (0.3): By Lemma (4.1) applied to $L + K_X$ instead of L, we conclude that $L + 3K_X$ is numerically effective. By applying the proof of Proposition (3.1) to $L + K_X$ instead of L, we conclude that there exist a finite number of non-identically-zero holomorphic sections s_μ ($1 \leq \mu \leq \ell$) of $L + 3K_X$ over X vanishing to order at least $3(n+q_\nu)$ at P_ν ($1 \leq \nu \leq J$) such that the set of points where the metric $(\sum_{\mu=1}^{\ell} |s_\mu|)^{-1}$ of $\frac{1}{2}(L+3K_X)$ is not locally integrable is of dimension zero. Thus the set of points where the metric $(\sum_{\mu=1}^{\ell} |s_\mu|^2)^{-1/3}$ of $\frac{1}{3}L + K_X$ is not locally integrable is of dimension zero. The Lelong number of the curvature current of the metric $(\sum_{\mu=1}^{\ell} |s_\mu|^2)^{-1/3}$ of $\frac{1}{3}L + K_X$ is at least $n + q_\nu$ at P_ν ($1 \leq \nu \leq J$). It follows from applying Nadel's vanishing theorem to the multiplier ideal sheaf of the metric $(\sum_{\mu=1}^{\ell} |s_\mu|^2)^{-1/3}$ of $\frac{1}{3}L + K_X$ that global holomorphic sections of $L + 2K_X$ over X generate simultaneously the q_ν-jets at P_ν for $1 \leq \nu \leq J$.

Remark (4.2) In the proof of Corollary (0.3) we do not use the method of §3 on the background singular metrics of Lelong mumber one, because we do not want to include the condition on the upper bound of $(L^n)^{-1/n}$ as an assumption in the statement of Corollary (0.3).

5. Improvement of Bounds by Using Polynomials from the Theorem of Riemann-Roch

When we consider $mL + 2K_X$ instead of $L + 2K_X$, we can use the polynomial from the theorem of Riemann-Roch. The method of §3 on the background singular metrics of Lelong number 1 can improve the bound of m for the very ampleness of $mL + 2K_X$ in the method of [Si94] and we now carry out this improvement.

If a polynomial $P(m)$ of degree k with leading coefficient $\alpha/(k!)$ assumes nonnegative values for every integer on the interval $[m_0, m_0 + k\delta]$ for some nonnegative integer m_0 and some positive integer δ, then there exists some integer $m_0 \leq m \leq m_0 + k\delta$ such that $P(m) \geq \alpha\delta^k/(2^{k-1})$ (see [D94]). So if $\alpha \geq 1$, for $P(m) \geq N$ it suffices to set $k\delta = k\lceil 2(\frac{N}{2})^{1/k}\rceil$. Demailly obtained this result by using Newton's formula for iterated differences $(\Delta^k P)(m) = (\Delta^{k-1}P)(m+1) - (\Delta^{k-1}P)(m)$ with $(\Delta^0 P)(m) = P(m)$. If one simply uses the factorization $|P(m)| = (\alpha/(k!)) \prod_{j=1}^{k} |m - c_j|$ and argues that an interval long enough must contain an integral point which is at least of a certain distance from each root c_j, then one could only get the weaker result that there exists some integer $m_0 \leq m \leq m_0 + 2k\delta + k + 1$ with $P(m) \geq (\alpha\delta^k)/(k!)$ when $P(m)$ assumes nonnegative values for every integer on the interval $[m_0, m_0 + 2k\delta + k + 1]$.

Proposition (5.1) *Let L be an ample line bundle over a compact complex manifold X of complex dimension $n \geq 2$. Let $P_1, \cdots, P_J$ be distinct points in X and $q_1, \cdots, q_J$ be nonnegative integers. Let $N = \sum_{j=1}^{J} \binom{3n+2q_j-3}{n}$ and $N_1 = \sum_{j=1}^{J} \binom{2n+q_j-2}{n}$. Then for any integer*

$$m \geq \frac{1}{2^{n-2}}\left(\sum_{j=1}^{J}(n + q_j - 1) + (n-1)\lceil 2(N_1/2)^{1/(n-1)}\rceil\right)$$

$$+ \sum_{\mu=0}^{n-2} \frac{J+1}{2^\mu} + \sum_{\mu=1}^{n-1} \frac{\mu}{2^\mu}\lceil 2(\frac{N}{2})^{1/\mu}\rceil.$$

the global holomorphic sections of $mL + 2K_X$ over X generate simultaneously the q_ν-jets at P_ν for $1 \leq \nu \leq J$.

Remark (5.2) In the required lower bound of m in Proposition (5.1) the dominant term is of the order $\frac{N}{2}$ which is equal to $\frac{1}{2}\sum_{j=1}^{J}\binom{3n+2q_j-3}{n}$.

Proof of Proposition (5.1): By Lemma (3.2) for some $0 < \epsilon < 1$ there exist a finite number of non-identically-zero multivalued holomorphic sections $\tau_1, \cdots, \tau_\ell$ of $(J + \frac{1}{4})L$ over X such that the Lelong number of the curvature current of $(\sum_{\nu=1}^{\ell} |\tau_\nu|^2)^{-1}$ is at least $1+\epsilon$ at each P_ν $(1 \leq \nu \leq J)$ and is less than $1-\epsilon$ on $U_\lambda - P_\lambda$ for some $1 \leq \lambda \leq J$ and some open neighborhood U_λ of P_λ. Take a basis $\sigma_1, \cdots, \sigma_r$ of $\Gamma(X, p_0 L)$ for some positive integer p_0 with $p_0 L$ very ample. Let $\hat{h}$ be the metric $(\sum_{\nu=1}^{r} |\sigma_\nu|^2)^{-1/p_0}$ of L. Define the metric h of $(J + \frac{1}{2})L$ by $h = (\sum_{\nu=1}^{\ell} |\tau_\nu|^2)^{-1} \hat{h}^{1/2}$.

Let $m_n = \sum_{j=1}^{J} (n + q_j - 1)$. Choose a non-identically-zero multivalued holomorphic section $s^{(n)}$ of $m_n L$ that vanishes to order $n + q_j - 1 - \epsilon$ at P_j. Denote by h_n the metric $|s^{(n)}|^{-2}$ of $m_n L$.

Let $m_{n-1} = J + 1 + m_n + (n-1)\lceil 2(N_1/2)^{1/(n-1)} \rceil$. For $1 \le d \le n-2$ inductively we let $m_d = \frac{1}{2}(\lceil m_{d+1} + J + \frac{1}{2} \rceil + d\lceil 2(N/2)^{1/d} \rceil)$. We claim that $m_d \ge m_{d+1}$ for $1 \le d \le n-1$. The case $d = n-1$ is clear and when $n \ge 3$ the case $d = n-2$ follows from $(n-2)\lceil 2(N/2)^{1/(n-2)} \rceil \ge \lceil m_{n-1} + J + \frac{1}{2} \rceil$. When $n \ge 4$ the case $1 \le d \le n-3$ follows by induction on d and from the inequality

(5.2.1) $$d\lceil 2(N/2)^{1/d} \rceil \ge (d+1)\lceil 2(N/2)^{1/(d+1)} \rceil.$$

The inequality (5.2.1) is verified by using

$$(N/2)^{1/d} = (N/2)^{1/(d+1)}(N/2)^{1/(d(d+1))},$$

$$\left(\frac{d+1}{d}\right)^{d(d+1)} \le e^{(d+1)},$$

$$\lceil 2(N/2)^{1/(d+1)} \rceil \le 2(N/2)^{1/(d+1)} + 1,$$

$$\left(1 + (2(N/2)^{1/(d+1)})^{-1}\right)^{d(d+1)} \le \exp(d(d+1)/(2(N/2)^{1/(d+1)})) \le e,$$

and $\frac{N}{2} \ge e^{d+2}$, where the last inequality is checked directly for small n and by using $\binom{q}{n} = \frac{q}{n}\binom{q-1}{n-1}$ for large n.

We are going to construct, by descending induction on $1 \le d \le n-1$, a metric h_d of $m_d L + K_X$ so that the Lelong number of its curvature current is at least $n + q_j - 1$ at P_j ($1 \le j \le J$) and the subvariety of local non-integrability of $h_d h$ is of dimension at most $d - 1$ at every point of U_λ.

For the first step of the induction we consider the metric $h|s^{(n)}|^{-2}$ of $(m_n + J + \frac{1}{2})L$ and let $V^{(n)}$ be the subvariety of local non-integrability of $h|s^{(n)}|^{-2}$. For every $(n-1)$-dimensional branch $V_\nu^{(n)}$ of the subvariety $V^{(n)}$ with $V_\nu^{(n)}$ intersecting U_λ, by Lemma (1.1) there exist two coherent ideal sheaves $\mathcal{I}_\nu^{(n)} \subset \mathcal{J}_\nu^{(n)}$ on X such that
(i) the zero-set of $\mathcal{I}_\nu^{(n)}$ is precisely $V_\nu^{(n)}$,
(ii) the zero-set of $\mathcal{J}_\nu^{(n)}$ is a proper subvariety of $V_\nu^{(n)}$,
(iii) the product of $\mathcal{J}_\nu^{(n)}$ and the ideal sheaf of $V_\nu^{(n)}$ is contained in $\mathcal{I}_\nu^{(n)}$,
(iii) $H^p(X, \mathcal{I}_\nu^{(n)} \otimes (mL + K_X)) = 0$ for $p \ge 1$ and $m \ge J + 1 + m_n$,
(iv) $H^p(X, \mathcal{J}_\nu^{(n)} \otimes (mL + K_X)) = 0$ for $p \ge 1$ and $m \ge J + 1 + m_n$.

Since $H^p(X, (\mathcal{J}_\nu^{(n)}/\mathcal{I}_\nu^{(n)}) \otimes (mL + K_X)) = 0$ for $p \ge 1$ and $m \ge J + 1 + m_n$, by considering the polynomial $\dim_{\mathbb{C}} \Gamma(X, (\mathcal{J}_\nu^{(n)}/\mathcal{I}_\nu^{(n)}) \otimes (mL + K_X))$

of m we conclude that there exists some $J + 1 + m_n \leq m_\nu^{(n)} \leq J + 1 + m_n + (n-1)\lceil 2(N_1/2)^{1/(n-1)} \rceil$ such that

$$\dim_{\mathbf{C}} \Gamma(X, (\mathcal{J}_\nu^{(n)}/\mathcal{I}_\nu^{(n)}) \otimes (m_\nu^{(n)} L + K_X))$$

is at least N_1. From the surjectivity of

$$\Gamma(X, \mathcal{J}_\nu^{(n)} \otimes (m_\nu^{(n)} L + K_X)) \to$$
$$\Gamma(X, (\mathcal{J}_\nu^{(n)}/\mathcal{I}_\nu^{(n)}) \otimes (m_\nu^{(n)} L + K_X))$$

we conclude that there exists some element $u_\nu^{(n)}$ of $\Gamma(X, m_\nu^{(n)} L + K_X)$ whose restriction to $V_\nu^{(n)}$ is not identically zero and which vanishes to order $n + q_j - 1$ at P_j for $1 \leq j \leq J$. Let $\rho_{n-1,\nu} = m_{n-1} - m_\nu^{(n)}$. Define the metric h_{n-1} of $m_{n-1}L$ by $(h_{n-1})^{-1} = \sum_\nu (\hat{h}^{-\rho_{n-1,\nu}} |u_\nu^{(n)}|^2)$. The Lelong number of the curvature current of h_{n-1} is at least $n + q_j - 1$ at P_j ($1 \leq j \leq J$) and the subvariety of local non-integrability of $h_{n-1}h$ is of dimension at most $n-2$ at every point of U_λ. The reason is that at a point of $U_\lambda - P_\lambda$ where some $u_\nu^{(n)}$ is nonzero the local integrability of $h_{n-1}h$ follows from the local integrability of h on $U_\lambda - P_\lambda$. Thus the subvariety of local non-integrability of $h_{n-1}h$ is contained in the intersection of $V^{(n)}$ and the common zero-set of the multivalued holomorphic sections $u_\nu^{(n)}$ for all ν and is of dimension at most $n - 2$.

Now suppose we have constructed the metric h_{d+1} for some $1 \leq d \leq n - 2$ and we are going to construct the metric h_d. Let $V^{(d+1)}$ be the subvariety of local non-integrability of $h_{d+1}h$. For every d-dimensional branch $V_\nu^{(d+1)}$ of $V^{(d+1)}$ with $V_\nu^{(d+1)}$ intersecting U_λ, by Lemma (1.1) there exist two coherent ideal sheaves $\mathcal{I}_\nu^{(d+1)} \subset \mathcal{J}_\nu^{(d+1)}$ on X such that
(i) the zero-set of $\mathcal{I}_\nu^{(d+1)}$ is precisely $V_\nu^{(d+1)}$,
(ii) the zero-set of $\mathcal{J}_\nu^{(d+1)}$ is a proper subvariety of $V_\nu^{(d+1)}$,
(iii) the product of $\mathcal{J}_\nu^{(d+1)}$ and the ideal sheaf of $V_\nu^{(d+1)}$ is contained in $\mathcal{I}_\nu^{(d+1)}$,
(iii) $H^p(X, \mathcal{I}_\nu^{(d+1)} \otimes (mL + 2K_X)) = 0$ for $p \geq 1$ and $m \geq m_{d+1} + J + \frac{1}{2}$,
(iv) $H^p(X, \mathcal{J}_\nu^{(d+1)} \otimes (mL + 2K_X)) = 0$ for $p \geq 1$ and $m \geq m_{d+1} + J + \frac{1}{2}$
.

Since $H^p(X, (\mathcal{J}_\nu^{(d+1)}/\mathcal{I}_\nu^{(d+1)}) \otimes (mL + 2K_X)) = 0$ for $p \geq 1$ and $m \geq m_{d+1}+J+\frac{1}{2}$, by considering the polynomial $\dim_{\mathbf{C}} \Gamma(X, (\mathcal{J}_\nu^{(d+1)}/\mathcal{I}_\nu^{(d+1)}) \otimes (mL + 2K_X))$ of m we conclude that there exists some integer

$$m_{d+1} + J + \frac{1}{2} \leq m_\nu^{(d+1)} \leq \lceil m_{d+1} + J + \frac{1}{2} \rceil + d \lceil 2(N/2)^{1/d} \rceil$$

such that $\dim_{\mathbf{C}} \Gamma(X, (\mathcal{J}_\nu^{(d+1)}/\mathcal{I}_\nu^{(d+1)}) \otimes (m_\nu^{(d+1)} L + 2K_X))$ is at least N. From the surjectivity of

$$\Gamma(X, \mathcal{J}_\nu^{(d+1)} \otimes (m_\nu^{(d+1)} L + 2K_X)) \to$$

$$\Gamma(X, (\mathcal{J}_\nu^{(d+1)}/\mathcal{I}_\nu^{(d+1)}) \otimes (m_\nu^{(d+1)} L + 2K_X))$$

we conclude that there exists some element $u_\nu^{(d+1)}$ of $\Gamma(X, m_\nu^{(d+1)} L + 2K_X)$ whose restriction to $V_\nu^{(d+1)}$ is not identically zero and which vanishes to order at least $2(n + q_j) - 1$ at P_j for $1 \le j \le J$. Let $\rho_{d,\nu} = m_d - \frac{1}{2} m_\nu^{(d+1)}$. We define a metric h_d of $m_d L + K_X$ as follows.

$$(h_d)^{-1} = (\sum_\nu |u_\nu^{(d+1)}|(\hat{h})^{-\rho_{d,\nu}}) + (h_{d+1})^{-1}(\hat{h})^{-(m_d - m_{d+1})}.$$

The Lelong number of the curvature current of h_d is at least $n + q_j - 1$ at P_j $(1 \le j \le J)$ and the subvariety of local non-integrability of $h_d h$ is of dimension at most $d-1$ at every point of U_λ. The reason is that at a point of $U_\lambda - P_\lambda$ where some $u_\nu^{(d+1)}$ is nonzero the local integrability of $h_d h$ follows from the local integrability of h on $U_\lambda - P_\lambda$. Thus the subvariety of local non-integrability of $h_d h$ is contained in the intersection of $V^{(d+1)}$ and the common zero-set of the multivalued holomorphic sections $u_\nu^{(d+1)}$ for all ν and is of dimension at most $d-1$. This concludes the construction of the metric h_d of $m_d L + K_X$ by descending induction on $1 \le d \le n-1$.

Since the Lelong number of the curvature current of the metric $h_1 h$ of $(m_1 + J + \frac{1}{2})L + K_X$ is at least $n + q_j$ at P_j $(1 \le j \le J)$ and the subvariety of local non-integrability of $h_1 h$ is of dimension 0 at every point of U_λ, it follows from Nadel's vanishing theorem that global holomorphic sections of $\lceil m_1 + J + \frac{1}{2} \rceil L + 2K_X$ generate simultaneously the q_j-jets at P_j $(1 \le j \le J)$.

Define $m_d' = \frac{1}{2}(m_{d+1} + J + 1 + d\lceil 2(N/2)^{1/d} \rceil)$ for $1 \le d \le n - 2$. Since m_n and m_{n-1} are integers and m_d is an integer or a half-integer for $1 \le d \le n-2$, it follows that $m_d' \ge m_d$ for $1 \le d \le n - 2$. From $m_d' \le \frac{1}{2}(m_{d+1}' + J + 1 + d\lceil 2(N/2)^{1/d} \rceil)$ it follows that

$$m_{n-\nu}' \le \frac{1}{2^{\nu-1}} m_{n-1} + \sum_{\mu=1}^{\nu-1} \frac{J+1}{2^\mu} + \sum_{\mu=1}^{\nu-1} \frac{n - \nu + \mu - 1}{2^\mu} \lceil 2(\frac{N}{2})^{1/(n-\nu+\mu-1)} \rceil.$$

Thus

$$m_1' \le \frac{1}{2^{n-2}}(\sum_{j=1}^{J}(n + q_j - 1) + J + 1 + \lceil 2(n-1)(N_1/2)^{1/(n-1)} \rceil)$$

$$+ \sum_{\mu=1}^{n-2} \frac{J+1}{2^\mu} + \sum_{\mu=1}^{n-2} \frac{\mu}{2^\mu} \lceil 2(\frac{N}{2})^{1/\mu} \rceil.$$

It follows that the global holomorphic sections of $mL + 2K_X$ over X generate simultaneously the q_ν-jets at P_ν for $1 \le \nu \le J$ for m no less than

$$\frac{1}{2^{n-2}}(\sum_{j=1}^{J}(n+q_j-1)+(n-1)\lceil 2(N_1/2)^{1/(n-1)}\rceil)+\sum_{\mu=0}^{n-2}\frac{J+1}{2^\mu}+$$

$$\sum_{\mu=1}^{n-1}\frac{\mu}{2^\mu}\lceil 2(\frac{N}{2})^{1/\mu}\rceil.$$

References

[A94] U. Angehrn, An Effective Polynomial Bound or Base Point Freeness and Point Separation of Adjoint Bundles, Harvard thesis 1994.

[AS94] U. Angehrn and Y.-T. Siu, Effective freeness and separation of points for adjoint bundles, Invent. Math., to appear.

[D93] J.-P. Demailly. A numerical criterion for very ample line bundles, J. Diff. Geom., 37 (1993), 323-374.

[D94] J.-P. Demailly. L^2 vanishing theorems for positive line bundles and adjunction theory, Preprint, 1994.

[EL93] L. Ein and R. Lazarsfeld. Global generation of pluricanonical and adjoint linear series on smooth projective threefolds, J. of the A.M.S., 6 (1993), 875-903.

[ELN94] L. Ein, R. Lazarsfeld and M. Nakamaye. Preprint 1994.

[Fu87] T. Fujita. On polarized manifolds whose adjoint bundles are not semipositive. In: Algebraic Geometry, Sendai, Advanced Studies in Pure Math., 10 (1987), 167-178.

[Ka82] Y. Kawamata. A generalization of Kodaira-Ramanujam's vanishing theorem, Math. Ann., 261 (1982), 43-46.

[Koh79] J. J. Kohn. Subellipticity of the $\bar{\partial}$-Neumann problems on pseudoconvex domains: sufficient conditions, Acta Math. 142 (1979), 79-122.

[Kol93] J. Kollar. Effective base point freeness, Math. Ann., 296 (1993), 595-605.

[N89] N. Nadel. Multiplier ideal sheaves and the existence of Kähler-Einstein metrics of positive scalar curvature, Proc. Natl. Acad. Sci. USA, 86 (1989), 7299-7300, and Ann. of Math., 132 (1989), 549-596.

[Re88] I. Reider. Vector bundles of rank 2 and linear systems on algebraic surfaces, Ann. of Math., 127 (1988), 309-316.

[Sh85] V. Shukorov. The non-vanishing theorem. Math. U.S.S.R. Izv. 19 (1985), 591-607.

[Si74] Y.-T. Siu. Analyticity of sets associated to Lelong numbers and the extension of closed positive currents. Invent. Math. 27 (1974), 53-156.

[Si93] Y.-T. Siu. An effective Matsusaka big theorem. Ann. Inst. Fourier (Grenoble) 43 (1993), 1387-1405.

[Si94] Y.-T. Siu. Effective very ampleness. Invent. Math., to appear.

[Sk72] H. Skoda. Sous-ensembles analytiques d'order fini ou infini dans C^n. Bull. Soc. Math. France 100 (1972), 353-408.

[V82] E. Viehweg. Vanishing theorems, J. reine und angew. Math., 335 (1982), 1-8.

Department of Mathematics, Harvard University, Cambridge, MA 02138

INTEGRABILITY OF ELLIPTIC OVERDETERMINED SYSTEMS OF NONLINEAR FIRST-ORDER COMPLEX PDE

François Treves[1]

1. General systems of complex first-order PDE

Our aim is to relate the *local solvability* of an over-determined system of complex *first-order* PDE,

$$(1) \qquad p_j(x, w_x) = 0, \quad j = 1, \cdots, n,$$

to the integrability of the associated Hamiltonian system,

$$(2) \quad H_{p_j} = \sum_{k=1}^{N} (\partial p_j / \partial \xi_k) \partial / \partial x_k - (\partial p_j / \partial x_k) \partial / \partial \xi_k, \quad j = 1, \cdots, n.$$

As the equations (1) are complex, in general the solution w will not be real. One must therefore hypothesize that the functions $p_j(x, \xi)$ are holomorphic with respect to the "fibre variables" ξ_k, which vary in $\mathbf{C}$ - and which, for this very reason, will be called ζ_j throughout the remainder of this note. The functions $p_j(x, \zeta)$ are of class C^∞ with respect to the pair (x, ζ). We could look at PDE that also involve the unknown function u and not only its gradient u_x; but an elementary trick reduces this apparently more general case to systems (1).

Actually we are interested in the *stable solvability* of (1) at a point (x_0, ζ_0) of $\mathbf{R}^N \times \mathbf{C}^N$. This means that there is an open neighborhood $U \times \mathcal{O}$ of (x_0, ζ_0) with the following property: there is a C^∞ solution u of (1) in U such that $u_x(x_0) = \zeta$ for every $\zeta \in \mathcal{O}$.

In order to prove that stable solvability follows from suitable hypotheses one can try to extend to this situation the classical method of characteristics. The difficulty lies in the fact that the base manifold (which we call $\mathcal{M}$ in the sequel, though for all practical purposes it is an open subset of $\mathbf{R}^N$) in which x varies is real, and the functions p_j are merely C^∞, not analytic; and therefore, that in general characteristics will not

[1] Work partially supported by NSF grant DMS-8903007. AMS Classification No **35.** Key words and sentences: *Symplectic, Hamiltonian, holomorphic, integrable structure, characteristic set, Levi form*

exist. The symplectic geometry foundation must be adapted to a set-up
in which part of the variables (x) are real and part (ζ) are complex and
all functions in $CT^*\mathcal{M}$, *i.e.*, the complex cotangent spaces $CT_x^*\mathcal{M}$, are
holomorphic with respect to ζ. The (complex) symplectic structure is not
carried by the tangent spaces to phase-space (as it is in the real situation)
but by their quotient modulo the subspace of vectors tangent to the fibres
$CT_x^*\mathcal{M}$ that are of type $(0,1)$, *i.e.* modulo the linear combinations of the
Cauchy-Riemann operators $\partial/\partial\bar{\zeta}_i$. The sections of the resulting vector
bundle are not true vector fields; but they act as true vector fields on
functions $f(x,\zeta)$ that are holomorphic with respect to ζ. This is how the
Hamiltonian fields (2) must be interpreted.

This symplectic geometry, partly real and partly complex, only calls
into play submanifolds of $CT^*\mathcal{M}$ whose intersections with any fibre
$CT_x^*\mathcal{M}$ are complex analytic submanifolds of $CT_x^*\mathcal{M}$ (of a fixed complex
dimension). It makes sense to say that such a submanifold is involutive
(or isotropic, or Lagrangian).

Our basic hypothesis is that the zero-set Σ of the functions $p_1, \cdots, p_n$
is an involutive submanifold of $CT^*\mathcal{M}$ on which the rank of the base
projection π is equal to N (and for the sake of simplicity, π maps Σ
onto $\mathcal{M}$). It is the datum of an involutive submanifold such as Σ that
we call *a system of differential equations* (in short, *of DE*). The rank
condition on $\pi|_\Sigma$ means that we may assume the fibre-differentials $\partial_\zeta p_j$
to be linearly independent on Σ. Below we use the notation $m = N - n =
\dim_{\mathbf{C}}(\Sigma \cap CT_{x_0}^*\mathcal{M})$. Note that $\dim_{\mathbf{R}}\Sigma = N + 2m = 3m + n$. We
affix the adjective "fibre-holomorphic" to the various ingredients of this
symplectic geometry, *e.g.*, the fibre-holomorphic Poisson bracket $\{f, g\}$
of two C^1 functions in $CT^*\mathcal{M}$ whose restrictions to the fibres $CT_x^*\mathcal{M}$ are
holomorphic. Also, if $\mathcal{O}$ is an open subset of $CT^*\mathcal{M}$, or of Σ, we denote
by $Hol_f(\mathcal{O})$ the space of continuous functions in $\mathcal{O}$ whose restrictions to
each intersection $\Sigma \cap CT_{x_0}^*\mathcal{M}$ are holomorphic.

In this set-up the Darboux theorem is not generally true. As a matter
of fact, the following holds:

Theorem 1. *For the system of DE Σ to be stably solvable at (x_0, ζ_0)
it is necessary and sufficient that the following equivalent conditions be
satisfied:*

*(i) There exist an open neighborhood $\tilde{\mathcal{O}}$ of (x_0, ζ_0) in Σ and $2m$ solutions
of the homogeneous Hamiltonian equations in $\tilde{\mathcal{O}}$, $u_1, \cdots, u_m, v_1, \cdots, v_m \in
C^\infty(\tilde{\mathcal{O}})$ such that*

$$(3) \qquad \{u_j, u_k\} \equiv \{v_j, v_k\} \equiv \{u_j, v_k\} - \delta_{jk} \equiv 0 \quad (j, k = 1, \cdots, m).$$

(ii) There exist an open neighborhood $\tilde{O}$ of (x_0, ζ_0) in Σ and m solutions of the homogeneous Hamiltonian equations in $\tilde{O}$, $u_1, \cdots, u_m \in C^\infty(\tilde{O})$, such that $\{u_j, u_k\} \equiv 0$ $(j, k = 1, \cdots, m)$ and that

(4) the map $\zeta \to u(x, \zeta)$ $(u = (u_1, \cdots, u_m))$ is a biholomorphism of $O \cap CT_x^* \mathcal{M}$ onto an open subset of $\mathbf{C}^m$ independent of $x \in \pi(\tilde{O})$.

For a proof see TREVES [1]. The proof of the entailment $(ii) => (i)$ shows that the solutions u_i and v_j of the homogeneous Hamiltonian equations, in Property (i), can be selected to satisfy the following requirement, in addition to (3):

(5) The Jacobian matrix of $u_1, \cdots, u_m, v_1, \cdots, v_m$ with respect to $\zeta_1, \cdots, \zeta_m, x_1, \cdots, x_m$ is equal to I_{2m} at (x_0, ζ_0).

The proof also provides local representations of Σ that are especially convenient. They suggest that we change the notation for the coordinates: we write t_k instead of x_{m+k} and τ_k instead of ζ_{m+k} $(1 \le k \le n)$. Now ζ stands for $(\zeta_1, \cdots, \zeta_m)$ and the functions p_k in (1) have the expressions

$$(6) \qquad p_k = \tau_k - q_k(x, t, \zeta) \quad (1 \le k \le n).$$

We have

$$(7) \qquad H_{p_k} = \partial/\partial t_k - \sum_{i=1}^{m}\{(\partial q_k/\partial \zeta_i)\partial/\partial x_i - (\partial q_k/\partial x_i)\partial/\partial \zeta_i\}+$$

$$\sum_{\ell=1}^{n}(\partial q_k/\partial t_\ell)\partial/\partial x_\ell,$$

whence

$$\{p_k, p_\ell\} = \partial q_k/\partial t_\ell - \partial q_\ell/\partial t_k - \sum_{i=1}^{m}\{(\partial q_\ell/\partial \zeta_i)\partial q_k/\partial x_i - (\partial q_\ell/\partial x_i)\partial q_k \partial \zeta_i\}.$$

This shows that the Poisson brackets $\{p_i, p_j\}$ are independent of $\tau = (\tau_1, \cdots, \tau_m)$. Since they vanish when $\tau = q(x, t, \zeta)$ $(q = (q_1, \cdots, q_n))$, we conclude that, *in Ω,*

$$(8) \qquad \{p_k, p_\ell\} \equiv 0, \forall k, \ell = 1, \cdots, n;$$

(8) entails that $[H_{p_k}, H_{p_\ell}] \equiv 0$ in Ω.

Let U be an open subset of $\pi(\Omega)$. That $w \in C^1(\Omega)$ is a solution of the system Σ means that, in U,

$$(9) \qquad \partial w/\partial t_j = q_j(x, t, w_x), \quad j = 1, \cdots, n.$$

The central point, which lies in U, will now be called (x_0, t_0). We reason about a point $(x_0, t_0, \zeta_0, \tau_0)$ of Σ ($\tau_0 = q(x_0, t_0, \zeta_0)$). We may coordinatize $\Sigma \cap \Omega$ by means of x_i, t_j, ζ_k ($i, k = 1, \cdots, m$, $j = 1, \cdots, n$) and identify a point $\theta \in \Sigma \cap CT^*_{x_0} \mathcal{M}$ to a point in $\mathbf{C}^m$ by means of its ζ-coordinates, $\theta_1, \cdots, \theta_m$. The solution w is required to satisfy

$$(10) \qquad\qquad w_x(x_0, \theta) = \theta.$$

Next, going to Condition (i) we may extend the functions u_i and v_j to Ω (possibly contracted about $(x_0, t_0, \zeta_0, \tau_0)$) as functions independent of τ; such extensions are obviously fibre-holomorphic. By virtue of (7) it follows that all the Poisson brackets of u_i, v_j, p_k are independent of τ. But then (i) entails

$$(11) \quad \{u_i, u_j\} = \{v_i, v_j\} = \{u_i, v_j\} - \delta_{ij} = \{u_i, p_k\} = \{u_j, p_k\} = 0,$$
$$\forall i, j = 1, \cdots, m, k = 1, \cdots, n.$$

Finally we observe that, in Ω,

$$(12) \qquad \{u_i, t_\ell\} = \{v_j, t_\ell\} = \{p_k, t_\ell\} - \delta_{k\ell} = \{t_k, t_\ell\} = 0,$$
$$\forall i, j = 1, \cdots, m, k = 1, \cdots, n.$$

In other words, Property (i) states that we may find functions u_i, v_j, p_k, $t_\ell \in Hol_f(\Omega) \cap C^\infty(\Omega)$ such that, in Ω, the symplectic form ω has the expression

$$(13) \qquad\qquad \omega = \sum_{i=1}^{m} du_i \wedge dv_i + \sum_{k=1}^{n} dp_k \wedge dt_k.$$

This can be regarded as a fibre-holomorphic variant of Darboux's theorem.

The next remark serves as a preparation to the next section. Call $T_f^{1,0}(\Sigma)$ the vector subbundle of $CT\Sigma$ whose sections are the vector fields tangent to the intersections $\Sigma \cap CT^*_x \mathcal{M}$ of type $(0,1)$. Call then $\mathbf{E}_\Sigma$ the vector subbundle of $CT\Sigma/T_f^{1,0}(\Sigma)$ whose sections are orthogonal, for the symplectic form ω, to those of the whole bundle $CT\Sigma/T_f^{1,0}(\Sigma)$; $\mathbf{E}_\Sigma$ is generated by the holomorphic Hamiltonians of defining functions of Σ (such as the p_j). We shall denote by $\mathcal{V}_\Sigma$ the preimage of $\mathbf{E}_\Sigma$ under the quotient map $CT\Sigma \to CT\Sigma/T_f^{1,0}(\Sigma)$. In the local parametrization (x, t, ζ) of Σ, sections of $\mathcal{V}_\Sigma$ are linear combinations of

$$(14) \qquad\qquad H_{p_1}, \cdots, H_{p_n}, \partial/\partial \bar{\zeta}_1, \cdots, \partial/\partial \bar{\zeta}_m.$$

Such linear combinations are true tangent vectors to Σ.

2. Elliptic Systems

Let $\mathcal{X}$ be a C^∞ manifold. In accordance with classical terminology one says that a vector subbundle T' of $\mathbf{C}T^*\mathcal{X}$ defines an *elliptic structure* on $\mathcal{X}$ if T' is formally integrable (*i.e.*, if the differential of any smooth section of T' is a section of the ideal in the exterior algebra $\wedge \mathbf{C}T^*\mathcal{X}$ generated by T') and if, moreover, $T' \cap \overline{T}' = 0$. If we introduce the orthogonal $\mathcal{V} \subset \mathbf{C}T\mathcal{X}$ of T' for the duality between complex tangent and cotangent vectors, ellipticity of the structure defined by T' (or by $\mathcal{V}$) is equivalent to the fact that $\mathcal{V}$ satisfies the Frobenius condition (*i.e.*, the commutation bracket of two smooth sections of $\mathcal{V}$ is a section of $\mathcal{V}$) and that the characteristic set of $\mathcal{V}$ (*i.e.*, the intersection of T' with the real cotangent bundle of $T^*\mathcal{X}$) is the zero-section.

It is traditional terminology to say that the system of DE Σ is *elliptic at a point* $\gamma \in \Sigma$ if the linearized system at γ is elliptic, *i.e.*, if the pushdown of $(\mathcal{V}_\Sigma)_\gamma$ under the base projection is elliptic, in the sense just defined, *i.e.*, in the sense that

$$(15) \qquad \pi_*((\mathcal{V}_\Sigma)_\gamma) + \overline{\pi_*((\mathcal{V}_\Sigma)_\gamma)} = \mathbf{C}T_{\pi(\gamma)}\mathcal{M}.$$

Ellipticity is a stable property: if valid at γ it is valid at all nearby points. In our local chart (x, t, ζ) Property (15) is equivalent to the fact that the constant coefficients operators

$$(16) \qquad \partial/\partial t_k - \sum_{i=1}^{n}(\partial q_k/\partial \zeta_i)(x_0, t_0, \zeta_0)\partial/\partial t_i \quad (k = 1, \cdots, n)$$

form an elliptic system [(16) are the linearlizations of the p_j at $(x_0, t_0, \zeta_0, \tau_0)$]. The ellipticity of the system (16) is equivalent to that of the system (14). Thus *for Σ to be elliptic at γ it is necessary and sufficient that $\mathcal{V}_\Sigma$ be elliptic in some neighborhood of γ in Σ*.

We shall apply the Newlander-Nirenberg theorem (see NIRENBERG [1], also TREVES [2], Ch.VI): *every elliptic structure on a smooth manifold is locally integrable*. Recall that $\dim_{\mathbf{R}} \Sigma = n + 3m$ and that the fibre dimension of $\mathcal{V}_\Sigma$, *i.e.* the number of vector fields (14), is equal to $m + n$. The local integrability of $\mathcal{V}_\Sigma$ requires that, in any sufficiently small open subset Ω of $\mathbf{C}T^*\mathcal{M}$ that intersects Σ there be 2m C^∞ functions $f_1, \cdots, f_{2m} \in Hol_f(\Sigma \cap \Omega)$ such that, at every point of $\Sigma \cap \Omega$,

$$(17) \qquad H_{p_k}f_j = 0, \quad j = 1, \cdots, 2m, \quad k = 1, \cdots, n;$$

$$(18) \qquad df_1 \wedge \cdots \wedge df_{2m} \neq 0.$$

Theorem 2. *Suppose V_Σ defines an elliptic structure on Σ. Then Σ is stably integrable at every one of its points.*

Proof. Let $f_1, \cdots, f_{2m} \in Hol_f(\Sigma \cap \Omega) \cap C^\infty(\Sigma \cap \Omega)$ satisfy (17) and (18). By virtue of the Jacobi identity we have

$$H_{p_k}\{f_i, f_j\} \equiv 0, \, k = 1, \cdots, n.$$

We further exploit the Newlander-Nirenberg theorem: it states that an elliptic structure on a smooth manifold (of dimension N) is locally isomorphic to the structure defined on $\mathbf{C}^\nu \times \mathbf{R}^{n-\nu}$ ($n + \nu = N$) defined by the vector fields $\partial/\partial\bar{z}_1, \cdots, \partial/\partial\bar{z}_\nu, \partial/\partial t_1, \cdots, \partial/\partial t_{n-\nu}$. Any distribution annihilated by these vector fields is a holomorphic function of z, independent of t. As a consequence, there is a holomorphic function A_{ij} in an open neighborhood $\tilde{\mathcal{O}}$ of $f(\Sigma \cap \Omega)$ in $\mathbf{C}^{2m}$ ($f = (f_1, \cdots, f_{2m})$) such that, in $\Sigma \cap \Omega$, $\{f_i, f_j\} = A_{ij}(f)$. Condition (18) implies $\det(A_{ij}) \neq 0$. Call (B_{ij}) the inverse of the skew-symmetric matrix (A_{ij}); the two-form

$$\tilde{\omega} = \sum_{i,j=1}^{2m} B_{ij}(z)dz_i \wedge dz_j$$

defines a complex symplectic structure on $\tilde{\mathcal{O}}$. The holomorphic Darboux theorem entails the existence of complex symplectic coordinates, *i.e.*, holomorphic functions $\tilde{u}_i, \tilde{v}_j$ ($1 \leq i, j \leq m$) in $\tilde{\mathcal{O}}$ (possibly after further contractions of Ω and of $\tilde{\mathcal{O}}$) such that

$$\sum_{i,j=1}^{2m} A_{ij}(z)((\partial\tilde{u}_k/\partial z_i)\partial\tilde{v}_\ell/\partial z_j - (\partial\tilde{u}_k/\partial z_j)\partial\tilde{v}_\ell/\partial z_i) = \delta_{i\ell} \, (1 \leq k, \ell \leq m).$$

If we take $u_i = \tilde{u}_i(f)$, $v_j = \tilde{v}_j(f)$ all requirements in (i) are satisfied. $\square$

3. Hypocomplex systems

Inspection of the proof of Theorem 2 shows that its conclusion can be extended to a class of systems of DE larger than the elliptic ones. Indeed, we have only used two consequences of ellipticity; the exitence of the "first integral" $f_1, \cdots, f_{2m}$ (*i.e.*, the local integrability of V_Σ); and the fact that any distribution annihilated by the sections of V_Σ is a holomorphic function of the first integrals. These two properties define *hypocomplex structures* (TREVES [2], Sect.III.5): Let $\mathcal{X}$ be a C^∞ manifold, $\dim \mathcal{X} = p+q$. A smooth vector subbundle T' of $\mathbf{C}T^*\mathcal{X}$ of rank p is said to be define a hypocomplex structure on $\mathcal{X}$ if $\mathcal{X}$ can be covered with open sets U in which there are p C^∞ functions $Z_1, \cdots, Z_p$ whose differentials span T' over U and that have the following property: *if the differential dh of a C^1 function h in an open neighborhood $U' \subset U$ of some point $x_0 \in U$ is a section of T' over U', then there is a holomorphic function $\tilde{h}$ in an open neighborhood of $Z(x_0)$ in $\mathbf{C}^p$ $(Z = (Z_1, \cdots, Z_p))$ such that $h = \tilde{h} \circ Z$ in a neighborhood of x_0 in U'.* It is one possible reading of the Newlander-Nirenberg theorem that *every elliptic structure is hypocomplex* (see TREVES [22], Ch.VI; in Ch.III of the same book the reader will find nonelliptic examples of hypcomplex structures).

By the *characteristic set of the system* of DE Σ we mean the characteristic set of the formally integrable structure on Σ, $\mathcal{V}_\Sigma$; we shall denote it by $Char\Sigma$. By the *Levi form of the system* Σ at a point $\hat{\gamma} \in Char\Sigma$ we shall mean the Levi form of $\mathcal{V}_\Sigma$ at $\hat{\gamma}$.

Theorem 3. *Suppose the structure on Σ defined by $\mathcal{V}_\Sigma$ is locally integrable and that, at any point $\hat{\gamma} \in Char\Sigma$, the Levi matrix of Σ has at least one eigenvalue < 0. Then Σ is stably solvable at every one of its points.*

Proof. It is a consequence of Th.6.1 of Ch.II in BAOUENDI-CHANG-TREVES [1] that if the Levi matrix of the structure on Σ defined by $\mathcal{V}_\Sigma$ has at least one eigenvalue < 0 at every characteristic point of $\mathcal{V}_\Sigma$ then the structure is hypocomplex. It suffices then to repeat the proof of Theorem 2. $\square$

If the Levi matrix of the structure on Σ defined by $\mathcal{V}_\Sigma$ has at least one eigenvalue < 0 at every characteristic point of $\mathcal{V}_\Sigma$, then by antipodality, it also has one eigenvalue > 0 at every such point.

To take advantage of Theorem 3 results of local integrability are needed. In the case of CR structures of the hypersurface type, with precisely one eigenvalue < 0 and all the remaining ones > 0, there are counterexamples to local integrability (see JACOBOWITZ-TREVES [1]). For the same structures, provided there are at least three eigenvalues < 0, one might be able to apply the local embedding results of D. Catlin.

References

M.S. Baouendi, C.H. Chang, and F. Treves. — [1] *Microlocal hypoanalyticity and extension of CR functions*, J. Diff. Geom. 18 (1983), 331-391.

D. Catlin. — [1] *Sufficient conditions for the extension of CR structures*, to appear.

H. Jacobowitz and F. Treves. — [1] *Aberrant CR structures*, Hokkaido Math. J. 22 (1983), 276-292.

L. Nirenberg. — [1] *A complex Frobenius theorem*, Seminar on analytic functions I, Princeton 1957, 172-189.

Treves, F. — [1] *Remarks on the integrability of first-order complex PDE*, J. Functional Analysis 106 (1992), 329-352.
[2] Hypo-Analytic Structures. Local Theory, Princeton University Press, Princeton, N.J. 1992.

Department of Mathematics, Rutgers University, New Brunswick, N.J. 08903

THE HOLOMORPHIC CONTACT GEOMETRY OF A REAL HYPERSURFACE

S. M. WEBSTER[1]

Introduction

A primary goal of this work is a better understanding of the biholo-morphic invariants of a real hypersurface in complex space. At a point where the Levi form is non-degenerate the work of Chern and Moser provides a systematic approach. Up to now a comparable result is lacking in the degenerate case, despite a great deal of study. The approach here is to pass from the pseudogroup of biholomorphic point transformations to the larger one of holomorphic contact transformations. The number of invariants is greatly reduced at this level, the Chern-Moser invariants completely disappearing, in fact. The main idea is that what invariants remain at a degenerate point should be the most accessible ones.

Another goal here is to find what role contact transformations may play in complex analysis. They have long been important in classical geometry and mechanics, and are now central to some aspects of modern PDE theory. We recall some of the basic formalism in section 1 which would suffice for the study of complex submanifolds. For real submani-folds, however, a new concept, that of real contact, is needed.

For a point x in a real submanifold M of $\mathbf{C}^n$, a complex hyper-plane $V^{n-1} \subset T_x(\mathbf{C}^n)$ intersects the tangent plane $T_x(M)$ in a real space generically of codimension 2. If this codimension drops to 1 or 0, then we say that V has *real contact* with M at x. For M a real hypersurface this just means that V is the the holomorphic tangent space $H_x(M)$. For M a generic real submanifold, the set of all such V forms a real $(2n-1)$-dimensional submanifold $\tilde{M}$ of $\mathbf{C}^n \times \mathbf{P}_{n-1}^*$, the space of all complex contact elements. The real and imaginery parts of the holomorphic contact 1-form are linearly dependent upon restriction to $\tilde{M}$. Such *real-contact Lagrangians*, under holomorphic contact transformation, are the basic objects of study.

In section 2 we recall from [7] the correspondence between Levi-null vectors of a real hypersurface M and complex tangent vectors to $\tilde{M}$. We also show how to reverse the sign of a non-zero Levi eigenvalue, and prove

[1] Partially supported by NSF Grant DMS-8913354

the local equivalence of all non-degenerate analytic real hypersurfaces, in contact geometry. For generic real submanifolds M of higher codimension, Levi degeneracies are also tied to complex tangents on $\tilde{M}$. This gives rise to a new reflection in section 3. In section 4 we consider briefly a case in which M itself has a CR singularity. $\tilde{M}$ then has an actual (topological) singularity.

In section 5 we introduce the *cubic form* of a real hypersurface. It is a third-order relative point-transformation invariant along the Levi degeneracy set. In section 6 we give a formal partial contact normalization which is based only on the Levi form rank at a point. Under it the cubic form largely disappears. In section 7 we give a complete contact normalization at a point of generic Levi degeneracy for a real hypersurface $M^3 \subset \mathbf{C}^2$. It shows that all such M^3 are formally equivalent to the real hypersurface $\operatorname{Im} w = (\operatorname{Re} z)^3$.

1. Holomorphic contact structure

We recall some of the basic formalism of contact geometry. A clear view of this very classical theory will help to guide us in the next section. One may consult S. Lie [5], Goursat [4], or Arnold [1] for further details. For coordinates of a point x in $\mathbf{C}^n$ we take

$$x = (z, w), \quad z = (z^1, \cdots, z^{n-1}), \quad w = z^n.$$

A complex contact element is a pair

$$(x, V) \in \mathbf{C}^n \times \mathbf{P}^*_{n-1} \equiv P, \quad V : dw = p_\alpha dz^\alpha \equiv p \cdot dz,$$

consisting of a point x and a tangent complex hyperplane $V^{n-1} \subset T_x \mathbf{C}^n$, the representation being valid so long as V is not vertical. (As above we shall often use either a vector dot-product notation or the summation convention with α, β, γ running from 1 to n-1). We denote by $\pi : P \to \mathbf{C}^n$ the projection $\pi(x, V) = x$. The contact form and sub-bundle are

$$(1.1) \qquad\qquad \theta = dw - p \cdot dz, \quad G : \theta = 0,$$

the p now being local fiber coordinates. θ is only determined up to $\theta \to \lambda\theta, \lambda \neq 0$, as only G has invariant meaning.

A non-singular complex analytic hypersurface

$$M^{n-1} \subset \mathbf{C}^n : \varphi(z, w) = 0, \quad \varphi_w \neq 0,$$

generates the manifold of its tangent planes,

$$\tilde{M}^{n-1} \subset P : \varphi = 0, \quad \varphi_z + p\varphi_w = 0,$$

$$\varphi_z = (\varphi_\alpha) \equiv (\partial\varphi/\partial z^\alpha), \quad \varphi_w = \partial\varphi/\partial w.$$

$\tilde{M}$ may be described as the particular integral submanifold of the system

$$(1.2) \qquad\qquad \varphi = 0, \quad \theta = \sigma d\varphi,$$

gotten by equating the coefficients of the coordinate differentials in the second equation and eliminating σ. Hence, it is an integral submanifold of $\theta = 0$, i. e. isotropic, and maximal dimensional among such. We shall use the term contact Lagrangian (Legendrian in [1]) for such manifolds in this work.

Next consider an analytic submanifold of codimension l

$$M^{n-l} \subset \mathbf{C}^n : \varphi^j = 0, \quad 1 \le j \le l, \quad d\varphi^1 \wedge \cdots \wedge d\varphi^l \ne 0.$$

Then

$$\tilde{M} = \{(x, V) \in P : x \in M, \ V \supset T_x M\}.$$

If $(\sigma_1, \cdots, \sigma_l)$ are parameters, not all zero, then $\Sigma\sigma_j\varphi^j = 0$ gives a hypersurface M_σ containing M. Taking $V = T_x M_\sigma$ and varying σ, and $x\epsilon\, M$, gives all points of $\tilde{M}$, which has dimension $n-1$. It is the integral submanifold of

$$(1.3) \qquad\qquad \varphi^j = 0, \quad \theta = \Sigma\sigma_j d\varphi^j,$$

which again is gotten by equating coefficients and eliminating the σ_j. Any contact Lagrangian projecting to a complex submanifold is of this form.

A (local) holomorphic contact transformation $\Phi : P \to P$ is characterized by the condition $\Phi^*\theta = \lambda\theta, \ \lambda \ne 0$. In the above coordinates $x^* = \Phi(x)$ and

$$(1.4) \qquad\qquad dw^* - p^* \cdot dz^* - \lambda(dw - p \cdot dz) = 0.$$

This equation suggests at least one functional relation among the variables (z, w, z^*, w^*). Thus, to construct Φ we may seek its graph as an integral submanifold of the system

$$(1.5) \qquad\qquad Q(z, w, z^*, w^*) = 0, \quad \theta^* - \lambda\theta = \sigma dQ,$$

for suitable functions Q, λ, σ. Comparing coefficients and eliminating λ, σ gives the 2n-1 equations

$$(1.6) \qquad\qquad Q = 0, \ Q_z + pQ_w = 0, \ Q_{z^*} + p^*Q_{w^*} = 0.$$

If they can be solved for (z^*, w^*, p^*), the resulting map will be a contact transformation.

Notice that if we take λ as a new independent variable, then the left hand side of (1.4) is a contact form in $4n - 1$ variables, and (1.5), as a special case of (1.2), defines a contact Lagrangian. Its projection along the λ -axis is the graph of Φ.

For the Legendre transform we take $Q = w^* - w + z \cdot z^*$,

(1.7) $L : z^* = p, \ p^* = -z, \ w^* = w - z \cdot p.$

The Ampere (or partial Legendre) transform

(1.8) $z_1^* = p_1, \ p_1^* = -z_1, \ w^* = w - z_1 p_1,$

$$z_\mu^* = z_\mu, \ p_\mu^* = p_\mu, \ 2 \le \mu \le n - 1,$$

is, in analogy with (1.3), generated by $n - 1$ relations, $Q^1 = \ldots = Q^{n-1} = 0$, among the variables (z, w, z^*, w^*), where

$$Q^1 = w^* - w + z_1 z_1^*, \ Q^\mu = z_\mu^* - z_\mu, \ 2 \le \mu \le n - 1.$$

(The differential of) a point transformation is generated by n independent such relations.

2. Real hypersurfaces

We consider a real hypersurface

(2.1) $M^{2n-1} \subset \mathbf{C}^n : r(z, w) = 0, \ r = \bar{r}, \ r_w \ne 0.$

For each $x \in M$, $T_x M$ contains the unique complex hyperplane $V = H_x M$. The set of all such $(x, H_x M)$ is the $2n - 1$-dimensional manifold

(2.2) $\tilde{M}^{2n-1} \subset P : r = 0, \ R \equiv r_z + p r_w = 0.$

From the definition $\tilde{M}$ satisfies

$$r = 0, \ \theta = \sigma \partial r, \ (\sigma = 1/r_w).$$

Since $\partial r = -\bar{\partial} r$ on M, it follows that $\tilde{M}$ is an integral submanifold of

(2.3) $\theta \wedge \bar{\theta} = 0, \ \theta \ne 0.$

Any integral submanifold of (2.3) will be called *real isotropic*. Those of the maximal dimension, which turns out to be $2n - 1$, will be called *real-contact Lagrangians*. (Strictly speaking, we should drop the inequality in (2.3). However, it will be automatic in all cases considered, except in section 4, and will eliminate technical difficulties in the present section.) If $K \subset P$ is any real-isotropic submanifold which projects diffeomorphically to M under π, then $K = \tilde{M}$. To see this note that K has the form $r(z, w) = 0, p = F(z, w)$, and

$$\theta|_K = \pi^*[(dw - F \cdot dz)|_M] = \pi^*[(\partial r - (F + r_z/r_w) \cdot dz)|_M].$$

Since $\partial r, dz, d\bar{z}$ are a basis of 1-forms on M, $\theta|_K$ and $\bar{\theta}|_K$ are dependent if and only if $F = -r_z/r_w$.

To transform a real hypersurface M by a holomorphic contact transformation Φ, consider $K' \equiv \Phi(\tilde{M})$, which is a real-contact Lagrangian. The projection $M' \equiv \pi(K')$ could be singular, or lower dimensional, but generically will be a real hypersurface. To find its defining function, we need only find a real function r' independent of p vanishing on K', by the above remark.

It's clear that $H(M) = \pi_*(T(\tilde{M}) \cap G)$, where, however, $T(\tilde{M}) \cap G$ is only a real vector bundle. The complex tangents to $\tilde{M}$ are precisely those vectors which project to null vectors of the Levi form of M [7]. To see this note that

$$H(\tilde{M}) : \partial r = 0, \ \partial R = 0, \ \partial \bar{R} = 0.$$

The first equation says that $\pi_* H(\tilde{M}) \subset H(M)$, while the second gives dp. Since
(2.4) $$\partial R_{\bar\alpha} = \partial(r_{\bar\alpha} + p_{\bar\alpha} r_{\bar w}) \equiv g_{\beta\bar\alpha} dz^\beta, \mod(\partial r),$$

$$g_{\beta\bar\alpha} = r_{\beta\bar\alpha} - \frac{r_\beta r_{w\bar\alpha}}{r_w} - \frac{r_{\beta\bar w} r_{\bar\alpha}}{r_{\bar w}} + r_\beta r_{\bar\alpha} \frac{r_{w\bar w}}{r_w r_{\bar w}},$$

where the hermitian matrix $g_{\beta\bar\alpha}$ represents the Levi form, the assertion is clear. We denote by $L_x(M) \subset H_x(M)$ the Levi-null space at $x \in M$. It follows that $L_x M$ is preserved as a complex vector space, and $H_x M$ as a real vector space, under contact transformation.

The sign of a non-zero eigenvalue of the Levi form can be reversed by a contact transformation. To see this assume $0 \in M$ and

(2.5) $$r = \frac{i}{2}(w - \bar{w}) + \sum_{\alpha=1}^{n-1} \lambda_\alpha z_\alpha \bar{z}_\alpha + O(3),$$

so that the λ_α are the Levi eigenvalues at 0. If $\lambda_1 \neq 0$ we substitute the inverse of the Ampere transform (1.8) into r and R_α,

$$r = \frac{i}{2}(w^* - \bar{w}^* - p_1^* z_1^* + \bar{p}_1^* \bar{z}_1^*) + \lambda_1 p_1^* \bar{p}_1^* + \sum_{\mu=2}^{n-1} \lambda_\mu z_\mu^* \bar{z}_\mu^* + O(3),$$

$$R_1 = -\lambda_1 \bar{p}_1^* + \frac{i}{2} z_1^* + O(2).$$

Using $R_1 = R_{\bar{1}} = 0$, we rewrite r as

$$r = \frac{i}{2}(w^* - \bar{w}^*) - \frac{1}{2\lambda_1} z_1^* \bar{z}_1^* + \sum_{\mu=2}^{n-1} \lambda_\mu z_\mu^* \bar{z}_\mu^* + O(3).$$

It follows that the projection of the transform of $\tilde{M}$ is a real hypersurface M^* given by r^* of the form (2.5) with

$$\lambda_1^* = -1/2\lambda_1, \ \lambda_\mu^* = \lambda_\mu, \ 2 \leq \mu \leq n-1.$$

Now we consider an arbitrary real submanifold $K \subset P$. For a point $x \in K$ set

$$T_x = T_x K, \ E_x \equiv E(T_x) = T_x + J T_x, \ H_x = T_x \cap J T_x,$$

where $J(J^2 = -I)$ is the complex structure on TP. Suppose that (2.3) holds on K; then we have locally a smooth function λ with

(2.6) $\bar{\theta} = \lambda\theta, \ \lambda\bar{\lambda} = 1.$

Thus, for a suitable (non-holomorphic) function $\mu \neq 0$, $\theta_1 = \mu\theta$ satisfies $\bar{\theta}_1 = \theta_1$, and hence $d\bar{\theta}_1 = d\theta_1$, on K. $d\theta_1$ is a symplectic (2,0)-form on G. Since

$$(\operatorname{Im} d\theta_1)^{2n-2} = (-\mu\bar{\mu}/4)^{n-1}(dz \wedge dp)^{n-1} \wedge (d\bar{z} \wedge d\bar{p})^{n-1} \neq 0,$$

on G, $\operatorname{Im} d\theta_1$ is a real symplectic form on G. Since it vanishes on each $T_x \cap G_x$, which has real codimension 1 in T_x, we have $\dim K \leq 2n - 1$, as asserted above.

The complex tangents to a real-contact Lagrangian have a purely contact-geometric characterization. Since θ_1, as above, is of type (1,0), i.e. $\theta_1(Jv) = i\theta_1(v)$, and real on K, we must have $\theta(v) = 0$ if both v and Jv are in T_x. Thus, $H_x \subset G_x \cap T_x$. Similarly, since $d\theta_1$ is of type (2,0) on G, we must have $d\theta_1(v, w) = 0$ if $v \in H_x$, $w \in G_x \cap T_x$. Thus, H_x is contained in $\perp_{d\theta} (T_x \cap G_x)$, the annihilator of $T_x \cap G_x$ relative to $d\theta$. By complex linearity in w, we even have $H_x \subset \perp_{d\theta} E(T_x \cap G_x)$. Since $\operatorname{Im} d\theta_1$ vanishes on a subspace of T_x if and only if $d\theta_1$ does, and since $T_x \cap G_x$ is a maximal subspace annihilated by $d\theta_1$, we have $\perp_{d\theta} (T_x \cap G_x) \subset \perp_{\operatorname{Im} d\theta_1} (T_x \cap G_x) = T_x \cap G_x$. The first subspace is J-invariant, so must be contained in H_x. Hence,

(2.7) $H_x = \perp_{d\theta} (T_x \cap G_x) = \perp_{d\theta} E(T_x \cap G_x).$

Since $d\theta_1 = \mathrm{Re}\, d\theta_1$ when restricted to K, it follows that K is totally real if and only if $\mathrm{Re}\, \theta|_K$ is a real contact form. In case $K = \tilde{M}$, M a nondegenerate real hypersurface, this corresponds to the usual contact form on M.

Proposition(2.1) *All real analytic, totally real, real-contact Lagrangians are locally equivalent under holomorphic contact transformation.*

To see this let K and K^* be two such submanifolds of P. Then the factors μ, μ^* above are real analytic on K, K^*, so continue to local holomorphic functions on P. By Darboux's theorem there is a local real analytic diffeomorphism, $\varphi : K \to K^*$, and a real analytic factor $\eta \neq 0$, with $\varphi^* \mathrm{Re}\, \theta_1^* - \eta \mathrm{Re}\, \theta_1 \equiv \mathrm{Re}\, (\varphi^* \theta_1^* - \eta \theta_1) = 0$. Both φ and η continue holomorphically to P. Since $\mathrm{Im}\, (\varphi^* \theta_1^* - \eta \theta_1) = \varphi^* \mathrm{Im}\, \theta_1^* - \eta \mathrm{Im}\, \theta_1 = 0$ on K, it follows that the holomorphic $(1,0)$-form $\varphi^* \theta_1^* - \eta \theta_1$ must vanish identically, as it does so on a maximal totally-real submanifold. Thus φ is a holomorphic contact transformation.

If $K \subset P$ is totally real, real analytic and of dimension $2n-1$, then it is locally the fixed-point set of an anti-holomorphic involution, $\rho : P \to P$. Suppose that K is also real isotropic. Then ρ must be an anti-contact transformation,

$$(2.8) \qquad\qquad \rho^* \theta = \lambda \bar{\theta}, \ \ \lambda \neq 0.$$

In fact, since λ is real analytic in (2.6), and $\rho^* \bar{\theta} = \bar{\theta} = \lambda \theta$ on K, continuation to P gives (2.8). With variables as in (1.4), equation (2.8) suggests at least one generating relation among $(z, w, \bar{z}^*, \bar{w}^*)$. In case $K = \tilde{M}$, where $M : r(z, w, \bar{z}, \bar{w}) = 0$ is an analytic non-degenerate real hypersurface, this relation is $r(z, w, \bar{z}^*, \bar{w}^*) = 0$ $(\text{see}[7])$.

3. Generic CR submanifolds

A generic real submanifold $M^{2n-l} \subset \mathbf{C}^n$ (of codimension l) is one for which $T_x + JT_x = T_x \mathbf{C}^n$ $(T_x = T_x M)$ for all $x \in M$. Thus, $H_x = H_x M$ always has complex dimension $n - l$. By analogy with section 1 we might consider all complex hyperplanes $V^{n-1} \subset T_x \mathbf{C}^n$ which contain H_x. However, this set turns out to be too large. Since T_x can't be contained in any V by the generic hypothesis, $V \cap T_x$ has codimension 1 or 2 in T_x. If the codimension is 1, we say, as in the introduction, that V has real contact with M at x. We again denote by $\tilde{M}$ the set of all such V.

If $\varphi = 0$ is a complex linear equation for V, then $\varphi \not\equiv 0$ on T_x, but $\mathrm{Re}\, \varphi$ and $\mathrm{Im}\, \varphi$ are linearly dependent on T_x. It follows that V contains

H_x and is uniquely determined by its image in T_x/H_x. Also, there is a unique real hyperplane containing both T_x and V. Hence, $\pi : \tilde{M} \to M$ is a fibering by real projective $(l-1)$-spaces, and $\dim \tilde{M} = 2n-1$. If $\xi, \eta \in T_{(x,V)}\tilde{M}$, then it follows that

$$\varphi \wedge \bar{\varphi}(\pi_*\xi, \pi_*\eta) = 0, \quad \varphi \circ \pi_* \not\equiv 0,$$

which is just another way of stating (2.3). Thus, $\tilde{M}$ is a real-contact Lagrangian.

Suppose that M is given by the real equations

(3.1) $\qquad r^{n-l+1} = 0, \ldots, r^n = 0; \ \partial r^{n-l+1} \wedge \cdots \wedge \partial r^n \neq 0.$

For any real parameters $t_j, n-l < j \leq n$, not all zero, we set $r^t = \sum t_j r^j$ and $M_t = \{r^t = 0\}$. Then M_t is a real hypersurface containing M, and $V = H_x M_t$, for $x \in M$, gives an element of $\tilde{M}$. By varying the t's and x we get all points of $\tilde{M}$. If $r_w^t \neq 0$, then V has fiber coordinates p, where

(3.2) $\qquad r_z^t + p r_w^t \equiv \sum t_t (r_z^j + p r_w^j) = 0.$

$\tilde{M}$ is the integral submanifold of

$$r^j = 0, \quad \theta = \sum \sigma_j \partial r^j,$$

gotten by eliminating the σ's as in (1.3).

For a given $(x, V) \in \tilde{M}$ we may choose the functions r^j so that $t_j = 0, j < n, t_n = 1$. We may further assume that $x = 0$, that $z^\mu, 1 \leq \mu \leq n - l$, span H_0, and that $(z^j = x^j + iy^j)$

(3.3) $\qquad r^j = -y^j + h^j(z^\mu, x^i), \ n-l < i,j \leq n,$

$$h^j = \sum_{\mu,\nu=1}^{n-l} b_{\mu\bar\nu}^j z^\mu \bar{z}^\nu + O(3).$$

The hermitian matrices b^j represent the vector-valued Levi form of M at 0. V is given by $\partial r^n(0) \equiv (i/2)dz^n = 0$, and we want to find $H_{(0,V)}\tilde{M}$. By (2.7) it suffices to find the null space of $d\theta$.

By (3.2) and (2.4) we have $d\theta = dz^\alpha \wedge dp_\alpha = (-r_w)^{-1} g_{\alpha\bar\beta}^t dz^\alpha \wedge d\bar{z}^\beta - (r_\alpha^t/r_w^t)_{t_j} dz^\alpha \wedge dt_j$. At $(0, V)$ we have $g_{\alpha\bar\beta}^t = g_{\alpha\bar\beta}^n = r_{\alpha\bar\beta}^n(0)$, and $(r_\alpha^t/r_w^t)_{t_j}(0) = \delta_\alpha^j$. Thus,

$$d\theta|_{(0,V)} = 2i \sum_{\mu,\nu=1}^{n-l} b_{\mu\bar\nu}^n dz^\mu \wedge d\bar{z}^\nu - \sum_{j<n} dx^j \wedge dt_j,$$

$$\theta|_{(0,V)} = dx^n.$$

If $\xi \in H_{(0,V)}\tilde{M}$, then $\pi_*(\xi) \in H_0 M$, so we may write

$$\xi = \sum_{\mu=1}^{n-l}(\xi^\mu \partial_\mu + \bar{\xi}^\mu \partial_{\bar\mu}) + \sum_{j<n}\xi_j \partial_{t_j}.$$

If $\eta \in T_{(0,V)} \cap G_{(0,V)}$ then

$$\eta = \sum_{\mu=1}^{n-l}(\eta^\mu \partial_\mu + \bar{\eta}^\mu \partial_{\bar\mu}) + \sum_{j<n}(\eta^j \partial_{x^j} + \eta_j \partial_{t_j}).$$

We have

$$d\theta_{(0,V)}(\xi, \eta) = 2i \sum b^n_{\mu\bar\nu}(\xi^\mu \bar\eta^\nu - \eta^\mu \bar\xi^\nu) + \sum \xi_j \eta^j,$$

which must vanish for all such η. Since we may replace η^μ by $i\eta^\mu$, this is equivalent to $\xi_j = 0$, $\sum b^n_{\mu\bar\nu}\xi^\mu = 0$. Thus, π_* is in the null space of $\partial\bar\partial r^n$ restricted to $H_0 M$, and we have

Proposition(3.1) *If the matrix b^n in (3.3) is non-degenerate, then $\tilde{M}$ is totally real at $(0, V)$.*

In case M is also real analytic, we have the local reflection ρ (2.8) about $\tilde{M}$ near $(0, V)$. It can be used, just as in [7], to extend biholomorphic maps past M under suitable conditions. It should be noted that this method of reflection is different from, and in a sense more natural than the one used in [8]. That one was defined on a proper submanifold of the set of tangent $(n - l)$-planes, and not on P as here. The conditions of applicability for the two are in general different.

4. CR singularities

A CR-singularity on a real submanifold M is a point where $H_x M$ jumps up in dimension. In general, this creates topological singularities in $\tilde{M}$. We shall not attempt a general theory, but rather briefly examine this phenomenon in the simplest case. Thus, we consider a surface $M^2 \subset \mathbf{C}^2$ near a non-degenerate complex tangent [6]. We may represent M by

$$w = F(z, \bar{z}) \equiv q_\gamma + B, \quad B = O(3),$$

(4.1)
$$q_\gamma = z\bar{z} + \gamma(z^2 + \bar{z}^2), \quad 0 \le \gamma < \infty,$$

$$q_\infty = z^2 + \bar{z}^2.$$

γ is Bishop's invariant [2].

For a point $x = (z, w) \in M$ a complex line V has real contact with M at x if and only if $V \cap T_x$ is either a real line or all of T_x. Thus, the fiber $\pi^{-1}(x)$ of $\pi : \tilde{M} \to M$ is a circle for a totally real point x and a point if T_x is a complex line. $\tilde{M}$ is given by

$$R = \bar{R} = 0, \ \ dR \wedge d\bar{R} \wedge \theta \wedge \bar{\theta} = 0,$$

where $R = F - w$ and $\theta = dw - p\,dz$. The latter condition is

(4.2) $$F_z \bar{F}_{\bar{z}} - F_{\bar{z}} \bar{F}_z - p\bar{F}_{\bar{z}} - \bar{p}F_z + p\bar{p} = 0.$$

In the case of a quadric, $B \equiv 0$, (4.1) and (4.2) show $\tilde{M}$ to be a graph over the real quadratic cone $p\bar{p} = 2\mathrm{Re}\,(pq_{\bar{z}})$ in $(z, \bar{z}, p, \bar{p})$-space. Significantly, this cone is non-degenerate except when $\gamma = 1/2$, which is the parabolic case. $\tilde{M}$ is totally real away from its singularity. In the elliptic case, $0 \le \gamma < 1/2$, there are analytic discs bounding on M [2]. Taking tangents lifts these to discs bounding on $\tilde{M}$.

5. Third order invariants

We return to the case of a real hypersurface M in $\mathbf{C}^n$. For biholomorphic point transformations a complete system of local invariants was given by Chern and Moser [3] in the non-degenerate case. We consider the lowest order invariants in the degenerate case.

If M is given as in (2.1) near a point $0 \in M$, we know that its Levi form at 0 is given by

$$l_0(X_0, Y_0) = i\partial r([X, \bar{Y}])_0,$$

where X_0, Y_0 are tangent $(1,0)$-vectors and X, Y are tangent $(1,0)$-fields extending them. l_0 is a well-defined hermitian form on $T_0^{(1,0)}(M) \cong H_0(M)$, which changes by a non-zero real factor when r is changed. As before,

$$L_0 = \{Z_0 \in H_0 | l_0(X_0, Z_0) = 0, \forall X_0 \in H_0\}.$$

The *cubic form* q_0 at $0 \in M$ is given by

(5.1) $$q_0(X_0, Y_0; Z_0) = \partial r([X, [Y, \bar{Z}]])_0,$$

$$q_0 : H_0 \times H_0 \times L_0 \to \mathbf{C},$$

for tangential (1,0)-fields extending the vectors. It is well defined, symmetric complex bilinear in X_0, Y_0, and anti-linear in Z_0. To see this note that $\partial r([\lambda X, [Y, \bar{Z}]]) = \lambda \partial r([X, [Y, \bar{Z}]])$ for any function λ. The symmetry in X_0, Y_0 follows from the Jacobi identity and the Levi-orthogonality of Z_0 and $[X, Y]_0$. Similarly, $\partial r([X, [Y, \bar{\lambda} \bar{Z}]])_0 = \bar{\lambda}(0)\partial r([X, [Y, \bar{Z}]])_0$, so that we have tensoriality at 0 in all three variables. Clearly q_0 changes by the same factor as does l_0, if r is changed.

To compute q_0 we take 0 to be the origin, and as in (2.5) $(w = u + iv)$

$$(5.2) \qquad\qquad r = -v + b + h,$$

$$b = b_{\alpha\bar{\beta}} z^\alpha \bar{z}^\beta, h = O(3) = h_3 + O(4).$$

We split the variables $(z^\alpha) = (z^\mu, z^i)$, where the z^α-space is H_0 and the z^μ-space is L_0, and use the index ranges

$$(5.3) \quad 1 \le \alpha, \beta, \gamma < n - 1; \ 1 \le \mu, \nu \le n - l; \ n - l < i, j \le n - 1,$$

as well as the summation convention for each. We have $b_{\alpha\bar{\mu}} = 0, \det(b_{i\bar{j}}) \neq 0$.

The terms in h_3 which are purely holomorphic in z, multiplied by powers of u, can all be removed as in [3] by a holomorphic change of the w-variable. Thus, we may assume

$$h_3 = 2\mathrm{Re}\,\{a_{\alpha\beta\bar{\gamma}} z^\alpha z^\beta \bar{z}^\gamma + u a_{\alpha\bar{\beta}} z^\alpha \bar{z}^\beta\},$$

where $a_{\alpha\beta\bar{\gamma}}$ is symmetric, $a_{\alpha\bar{\beta}}$ hermitian symmetric in α, β. A change

$$z^\alpha \to z^\alpha + Q^\alpha, \ Q^\alpha = Q^\alpha_{\beta\gamma} z^\beta z^\gamma + Q^\alpha_\beta z^\beta w,$$

results in

$$a_{\alpha\beta\bar{\gamma}} \to a_{\alpha\beta\bar{\gamma}} + Q^\delta_{\alpha\,\beta} b_{\delta\bar{\gamma}},$$

$$a_{\alpha\bar{\beta}} \to a_{\alpha\bar{\beta}} + Q^\gamma_\alpha b_{\gamma\bar{\beta}} + b_{\alpha\bar{\gamma}} Q^{\bar{\gamma}}_{\bar{\beta}}.$$

Since $b_{i\bar{j}}$ is non-singular, we can achieve, noting (5.3),

$$(5.4) \qquad\qquad h_3 = 2\mathrm{Re}\,\{a_{\alpha\beta\bar{\mu}} z^\alpha z^\beta \bar{z}^\mu + u a_{\mu\bar{\nu}} z^\mu \bar{z}^\nu\}.$$

For a basis of tangent (1,0)-fields we take

$$X_\alpha = \partial_\alpha - (r_\alpha/r_w)\partial_w \equiv \partial_\alpha - (r_\alpha/2r_w)\partial_u,$$

so that $X_\alpha(0) = \partial_\alpha$ span H_0 and $X_\mu(0) = \partial_\mu$ span L_0. One readily computes (see (2.4))

$$[X_\alpha, X_{\bar{\beta}}] = \frac{1}{2}g_{\alpha\bar{\beta}}\partial_u,$$

$$[X_\beta, [X_\alpha, X_{\bar\mu}]] = \frac{1}{2}(X_\beta g_{\alpha\bar\mu} + g_{\alpha\bar\mu}r_\beta/2r_w)\partial_u.$$

Since $\partial r(0) = idu/2, g_{\alpha\bar\mu} = r_{\alpha\bar\mu} + O(2), r_\alpha = O(1)$, we get

(5.5) $$q_0(\partial_\alpha, \partial_\beta; \partial_\mu) = \frac{i}{2}a_{\alpha\beta\bar\mu}.$$

Thus, in general, the coefficients $a_{\alpha\beta\bar\mu}$ can't all be removed by a point transformation.

We denote the Levi degeneracy set by

(5.6) $$N = \{\det g_{\alpha\bar\beta} = 0\}.$$

If $d(\det g_{\alpha\bar\beta})_0 \neq 0$, then N is smooth and L_0 is one-dimensional, spanned by a single vector Z_0. In the current set-up $Z_0 = \partial_1$ and one computes $\partial_1(\det g_{\alpha\bar\beta})_0 = 2a_{11\bar1}\det b_{i\bar j}$. If $q_0(Z_0, Z_0; \bar Z_0) \neq 0$, we say that the Levi-form degeneracy is *generic*. L_0 is then transverse to N.

In the case of $M^3 \subset \mathbf{C}^2$, we have

$$r = -v + 2\mathrm{Re}\,\{az^2\bar z + buz\bar z\} + O(4),$$

with $a \neq 0$ being the condition for generic degeneracy. By scaling the z-variable we can make $a = 1$. Then, a change $z \to z + \beta w$ results in $b \to b + 2\beta$. Thus we can make $b = 0$, and then remove the spurious cubic terms by changing w as before. Hence, we may arrange

(5.7) $$r = -v + z^2\bar z + z\bar z^2 + O(4).$$

6. Partial contact normalization

We consider an analytic real hypersurface M with defining function r as in (5.2), and look for a contact transformation Φ which reduces r to a simpler form. A generating function

$$Q = w^* - w + z \cdot z^* + S(z, w, z^*), \quad S = O(3),$$

gives a small perturbation of the Legendre transform L (1.7). Composition with L^{-1} gives a candidate for Φ. Writing $(z^*, w^*, p^*) = \Phi(z, w, p)$, we readily find from (1.6)

$$z^* = z + S_{z^*}(z, w, p^*),$$

(6.1) $$w^* = w + p^* \cdot S_{z^*}(z, w, p^*) - S(z, w, p^*),$$

$$p^* = p - S_z(z, w, p^*) - pS_w(z, w, p^*).$$

The last equation has to be solved for p^* and substituted into the first two to get

(6.2) $\qquad z^* = z + O(2), \ w^* = w + O(3), \ p^* = p + O(2).$

$M^* \equiv \pi(\Phi(\tilde{M}))$ is clearly a small perturbation of M and has a defining function $r^* = -v + b + h^*$. The substitution (6.1) gives

$$r^* \circ \Phi = -\text{Im} \{w + p^* \cdot S_{z\bullet} - S\} + b(z + S_{z\bullet}) + h^*$$

$$= -v + b(z) - \text{Im} \{p^* \cdot S_{z\bullet} - S\} + 2\text{Re} \, b(S_{z\bullet}, z) + b(S_{z\bullet}) + h^*(z^*, u^*).$$

Restricting to $\tilde{M}$ makes this vanish, so we get

(6.3) $\qquad h = h^* + \text{Re} \{i(p^* \cdot S_{z\bullet} - S) + 2b(S_{z\bullet}, z)\} + b(S_{z\bullet}).$

In this equation the *-variables are removed via (6.2), and p, v via (2.2), (5.2), so that a power series equation in $(z, \bar{z}, u)$ results.

We write h as a sum of homogeneous polynomials

(6.4) $$h = \sum_{s=3}^{\infty} h_s, \quad h_s = \sum_{k=0}^{s} h_{s,k} u^k,$$

$$h_{s,k} = \sum h_{A\bar{B}} z^A \bar{z}^B, \quad \bar{h}_{A\bar{B}} = h_{B\bar{A}}.$$

Here, in multi-index notation, $A = \alpha_1 \cdots \alpha_\mu$ and $B = \beta_1 \cdots \beta_\nu$, $\mu + \nu = s - k$, and $h_{A\bar{B}}$ is symmetric in A and in B. Assuming h_t already normalized for $3 \le t < s$, we choose $S = S_s$ a homogeneous polynomial of degree s,

(6.5) $$S = \sum_{k=0}^{s} S_k w^k, \quad S_k = \sum_{|A|+|B|=s-k} S_A^B z^A p_B^*.$$

This gives $\Phi = I + O(s - 1)$, and (6.3) shows that the previous normalizations are not disturbed. Since $p = 2ib\bar{z} + O(2)$ and $w = u + O(2)$, (6.3) gives

(6.6) $\qquad h_s(z, u) = h_s^*(z, u) + \text{Re} \{i[2ib\bar{z} \cdot S_{z\bullet} - S] + 2b(S_{z\bullet}, z)\}$

$$= h_s^*(z, u) + \text{Re} \, iS(z, u, 2ib\bar{z}).$$

With the further notation

$$b_{B\bar{C}} = (2ib_{\beta_1\bar{\gamma}_1}) \cdots (2ib_{\beta_\mu\bar{\gamma}_\mu}), \quad S_{A\bar{C}} = iS_A^B b_{B\bar{C}},$$

we may write (6.6) in the form

$$(6.7) \qquad h_{A\bar{B}} = h^*_{A\bar{B}} - \frac{1}{2}(S_{A\bar{B}} + \overline{S_{A\bar{B}}}).$$

Now we assume that the Levi form has nullity $n - l$, and that the coordinates have been chosen as just above (5.3). From (6.7) it follows that, by a proper choice of S, we can make $h^*_{A\bar{B}} = 0$ if either A or B contains an index $j > n - l$. S will be unique if we take it independent of p^*_μ, $\mu \le n - l$. Thus, every term in h^*_s will have a factor $z^\mu \bar{z}^\nu$, $\mu, \nu \le n - l$. As in [9] we choose a sequence of contact transformations $\Phi_s = I + O(s - 1)$, with polynomial generating functions, achieving this condition up to each successive degree. The composition of all these maps yields a *formal* transformation, and we have the following.

Proposition(6.1) *Suppose M is an analytic real hypersurface in $\mathbb{C}^n$ having Levi rank l at a point z_0. Then there is a formal contact transformation taking z_0 to 0, and M into the surface $r = 0$, where*

$$(6.8) \qquad r = -v + \sum_{j=n-l}^{n-1} |z^j|^2 + \sum_{\mu,\nu=1}^{n-l-1} h_{\mu\bar{\nu}} z^\mu \bar{z}^\nu.$$

Here, the $\bar{h}_{\mu\bar{\nu}} = h_{\nu\bar{\mu}} = O(1)$ are formal power series in all the variables $(z, \bar{z}, u)$.

In particular, the third order terms are as in (5.4) with $a_{\alpha\beta\bar{\mu}} = 0$, unless $\alpha, \beta < n - l$.

7. Generic Levi degeneracies

If M is generically degenerate at the point 0, then the normalization in (6.8) can be carried much further. For simplicity we restrict to the case of two complex variables, and assume that r has the form (5.7), or alternately $(z = x + iy)$,

$$(7.1) \qquad r = -v + \frac{8}{3}x^3 + h(z, \bar{z}, u), \quad h = O(4).$$

Theorem(7.1) *There is a formal holomorphic contact transformation Φ taking M into the real hypersurface*

$$(7.2) \qquad v = \frac{1}{3}x^3.$$

For the proof we again assume Φ given by (6.1). In place of (6.3) we get the functional relation

(7.3)
$$h(z, u) = h^*(z^*, u^*) + \text{Im}\,\{S - p^* S_{z^\bullet}\} +$$

$$8x^2 \text{Re}\,S_{z^\bullet} + 8x(\text{Re}\,S_{z^\bullet})^2 + \frac{8}{3}(\text{Re}\,S_{z^\bullet})^3.$$

From (2.2), (7.1) we get $p = 8ix^2 + O(3)$ on $\tilde{M}$, and from (6.2) $p^* = O(2)$. Hence, we assign the *weights*,

$$wt\,z = wt\,w = 1,\quad wt\,z^* = 2.$$

We assume $h = O(s) = h_s + O(s+1)$ and take for $S(z, w, z^*)$ a polynomial of weight s. Since $p^* S_{z^\bullet} = 8ix^2 S_{z^\bullet} \cdot (z, u, 8ix^2) + O(s+1)$, after all substutions, we get

(7.4)
$$h_s(z, u) = h_s^*(z, u) + \text{Im}\,S(z, u, 8ix^2).$$

The decompositions (6.4),(6.5) give

(7.5)
$$h_{s,k}(z) - h_{s,k}^*(z) = \text{Im}\,S_k(z, 8ix^2),$$

where $deg\,h_{s,k} = s - k$, $wt\,S_k = s - k$.

Setting $s - k = l$, we see that the right hand side of (7.5) defines a real linear operator L from the space of holomorphic polynomials,

$$S(z, z^*) = \sum_{j=0}^{[l/2]} a_j z^{l-2j} z^{*j},$$

of weight l to the space of real polynomials,

$$h(z, \bar{z}) = \sum_{j=0}^{l} b_j z^{l-j} \bar{z}^j,\quad b_{l-j} = \bar{b}^j,$$

homogeneous of degree l in $(z, \bar{z})$. For l even the space of S's has real dimension $l + 2$, and the space of h's has dimension $l + 1$. For l odd both have dimension $l + 1$.

To show that L is surjective, we analyse its null space. If $LS = 0$, then $S(z, z^*)$ is real when $z^* = x^* + iy^*$ satisfies $x^* = 0, y^* = 8x^2$. In other words the function $z \to S(z, iy^*)$ is real when $x = \pm\sqrt{(y^*/8)} \equiv \epsilon$, for any $y^* > 0$. Assume that $a_j = 0$ for $j < k \le l/2$. Then

$$S(\epsilon + iy, iy^*) = (iy^*)^k a_k (\epsilon + iy)^{l-2k} + \cdots$$

$$= (iy^*)^k a_k [(iy)^{l-2k} + \epsilon(l-2k)(iy)^{l-2k-1}] + \cdots,$$

where the dots indicate terms of degree $\leq l - 2k - 2$ in y. Since the coefficients of y^{l-2k} and y^{l-2k-1} must be real, it follows that $a_k = 0$ unless l is even and $2k = l$. Hence, L is injective for L odd and has one-dimensional null space for l even. Therefore, l is surjective in both cases.

It follows that we can choose S_k in (7.5) to make $h^*_{s,k} = 0$. Thus we get $h^* = O(s+1)$. As in the last section, we may compose an ∞-sequence of transformations $\Phi_s = I + O((s-1)/2)$ to get a formal contact transformation yielding theorem(7.1). It is most probable that, for a generically chosen real analytic M, any holomorphic contact transformation taking M into the normal form (7.12) must diverge. Results of this kind are given by Mr. Xianghong Gong in his thesis (University of Chicago) for the analogous normal form given in [9].

References

[1] V. I. Arnold, *Mathematical Methods of Classical Mechanics*, Springer-Verlag, New York, 1980.

[2] E. Bishop, Differentiable manifolds in complex Euclidean space, Duke Math. J. 32(1965)1-22.

[3] S. S. Chern and J. K. Moser, Real hypersurfaces in complex manifolds, Acta Math. 133 (1974)219-271.

[4] E. Goursat, *Lecons sur l'integration des equations aux derivee partielle du premier ordre*, Paris (1921).

[5] Sophus Lie, *Theorie des Transformationsgruppen II*, Chelsea, New York (1970).

[6] J. K. Moser and S. M. Webster, Normal forms for real surfaces in $\mathbf{C}^2$ near complex tangents and hyperbolic surface transformations, Acta Math. 150 (1983)255-296.

[7] S. M. Webster, On the reflection principle in several complex variables, Proc. AMS 72 (1978)26-28.

[8] ____ , Holomorphic mappings of domains with generic corners, Proc. AMS 86 (1982)236-240.

[9] ____ , Holomorphic symplectic normalization of a real function, Ann. Sc. Norm. Sup. Pisa, series 4, vol. 19 (1992) 69-86.

University of Chicago